T0405764

Building Systems in Interior Design

Building Systems in Interior Design takes an entirely new approach to teaching this essential topic for architects, designers and building engineers. Written to prepare students for the real world and packed with practical examples, the book will foster an understanding of specific issues that are critical to those features of technical systems that most directly affect design. The book stresses the ever-present nature of these systems: they are everywhere, all the time.

Taking a design-oriented view, it outlines what can and cannot be done, and provides the student with the know-how and confidence to defend and promote their design intent when working with other industry professionals.

Covering lighting, HVAC, plumbing and much more, the book is packed with key features to aid learning including:

- numerous illustrations, plans and photographs;
- key terms defined in an extensive glossary;
- chapter introductions that identify key concepts and chapter summaries to revisit those key concepts;
- professional design tips; and
- a detailed bibliography and web links.

This book is not only a core text for interior design, building systems engineering, and architecture students but will become an essential working reference throughout their careers.

Samuel L. Hurt, PE, RA, RID, LC, LEED® AP, HFDP is a registered architect, interior designer and engineer. He has taught on related topics as an adjunct or part-time instructor for more than 15 years at universities including Purdue, Ball State, and Indiana State.

Building Systems in Interior Design

Samuel L. Hurt

LONDON AND NEW YORK

First published 2018
by Routledge
2 Park Square, Milton Park, Abingdon, Oxon OX14 4RN

and by Routledge
711 Third Avenue, New York, NY 10017

Routledge is an imprint of the Taylor & Francis Group, an informa business

British Library Cataloguing-in-Publication Data
A catalogue record for this book is available from the British Library

Library of Congress Cataloging-in-Publication Data
Names: Hurt, Samuel L., author.
Title: Building systems in interior design / Sam Hurt.
Description: New York : Routledge, 2018. | Includes bibliographical references and index.
Identifiers: LCCN 2017017997| ISBN 9781138723368 (hardback : alk. paper) | ISBN 9781315193052 (ebook : alk. paper)
Subjects: LCSH: Buildings—Environmental engineering. | Buildings—Mechanical equipment—Design and construction. | Buildings—Electric equipment—Design and construction.
Classification: LCC TH6021. H825 2018 | DDC 729—dc23
LC record available at https://lccn.loc.gov/2017017997

ISBN: 978-1-138-72336-8 (hbk)
ISBN: 978-1-315-19305-2 (ebk)

Typeset in Univers
by Keystroke, Neville Lodge, Tettenhall, Wolverhampton

Thanks to my ever patient wife Carmen who allows me to pursue book writing in addition to my already busy business life; to innumerable students at IUPUI, The Art Institute of Indianapolis, Ball State University, Purdue University, and Indiana State University, and especially to Sherin Christopher, a former student who prepared many of the illustrations.

Contents

Preface

Design is the subject of this book, in the sense that building systems are every bit as much a part of interior design as are overall design concept, space planning, and the specific decisions that support the concept and the plan: design of built-in millwork, development of the color palette, selection of finishes, etc. Building systems are often thought of as the boring part of design; we put in plumbing and HVAC because we have to, not because we want to or because we believe that plumbing and HVAC support design (the term "necessary evil" comes to mind). But plumbing and HVAC do indeed support design (they can certainly ruin a good design), as do all other building systems. So building systems are merely the topics that will be used to assist readers in their quest to become more complete designers. The greatest design idea can be ruined if the space is too hot or too cold; too bright or too dark; too loud or too quiet; if the office door will not stay closed or will not close at all; or if the toilet will not flush, or a door handle falls off, or the drawers in the expensive audio-visual cabinet in the board room stick, or even if the wallcovering or carpet looks likes it is 20 years old when it is only five years old. Quite literally, mountains of catalogs and brochures and an unending flood of on-line information are available to cover these systems, but it is very easy to get overwhelmed and to be confused. My goal is to take away that confusion and to boil these subjects down to their essence; mostly to talk about principles, using details only to clarify and to familiarize.

Talking about design first in a context like this might seem to be a little unexpected, but I believe completely that this emphasis will make the subject matter more relevant and more easily understood. I am a designer first and foremost and both my B.S. and M.Arch. degrees are in Architectural Design. But I have practiced for nearly 10 years as a registered architect, about seven years in commercial (mostly) interior design (and as a registered interior designer in Indiana since October 2009), and more than 18 years as a professional engineer (in mechanical, electrical, and plumbing systems), giving me a uniquely broad and deep point-of-view about interior design, architecture, and architectural engineering. I have also taught in interior design as an adjunct or part-time instructor for more than 15 years. Over more than 34 years of practice and thousands of individual projects, I have seen the full gamut of project types and sizes in both new construction and renovation: correctional, educational (day-care, pre-school, K-12, and higher education—public and private), governmental (Federal, State of Indiana, and numerous local entities), historical (both adaptive re-use and restoration), healthcare (dental, medical offices, outpatient surgery, endoscopy, imaging (X-ray, mammography, CT, cyberknife, linear accelerators, and MRI), etc.), hospitality (motels and hotels, restaurants, convention centers), industrial (warehousing and light manufacturing), offices, recreational (YMCAs, swimming pools and spray parks, parks facilities, etc.), religious (Christian, Jewish, Muslim, and even Buddhist—and for both worship and educational facilities), residential (single-family and multi-family), and retail (small shops to department stores and even including tenant coordination of a major mall), and across the range of sizes from single rooms through new high-rise multi-use buildings.

Since I began my career in 1981, I have found that many designers (architects, interior designers, and engineers alike) are sometimes intimidated by the vast amount of information involved in these systems, and sometimes by their specialized practitioners (like me—except for the specialized part). But this really is not correct. At the most fundamental level, design is complex problem solving of the highest order: the synthesis of a multitude of diverse factors into a cohesive whole that meets a complicated (and not always consistent) set of objectives. To design a project that is beautiful but which fails to heat, cool, and ventilate properly is to have failed. To design a project that has technically functional lighting but is ill-suited to the owner's needs and desires is also to have failed. Design success, at least in my view, is defined by the completeness of the solution, across the full spectrum of design issues: meaning, beauty, functionality, cost, and yes, all of the building systems.

I make no apologies for my practitioner's point-of-view, despite my extensive experience in teaching. Experience is priceless and wins out over all else in the long run. Of course, this is apparent only *after* amassing long experience.

So this book will take an unusual point-of-view, emphasizing the true design aspects of building systems and the integration of such systems into day-to-day design work. My goal is to boil these systems down to their basics—what really matters. Why do we use walls and doors? Are door frames actually necessary? When to use troffers and recessed downlights. And then to show how those principles apply to various design challenges. Along the way, numerous illustrations will be used to show what this or that looks like (we designers are visual people, after all), but it is important to remember that all air-handlers might not look like the ones in the pictures, all door frames are not alike, and that not all acoustical treatments have to be fabric-wrapped rigid foam.

If you see something that bothers you, look it up somewhere else. If you still disagree, send me a note. I am learning too, and I certainly want to avoid confusion to the greatest degree possible.

Samuel L. Hurt, RID, PE, RA
April 1, 2017

Introduction

Much of the time, the term "building systems" is used by interior designers to describe the "technical" systems in a given space, which usually means mechanical, electrical, and plumbing systems; however, this narrow definition is limiting and this book will take a much broader point-of-view. In addition to these basic (and sometimes highly technical) systems that are usually designed by consulting engineers or contractors, other systems include the following:

- legal (codes, regulations, and standards);
- structural (mostly from the point-of-view of when structures can and cannot be modified and when they must be modified);
- architectural (walls, windows, doors, millwork, stairs, etc.);
- acoustics.

Each of these will be summarized briefly in Chapter 1. After that, each one will be covered in detail in Chapters 3 through 11. Electrical systems will be divided into two chapters: Chapter 8 for lighting (plus daylighting) and Chapter 9 for power.

In today's world of deep concern about water resources and planet-wide climate change (global warming, the greenhouse effect, ozone depletion, water shortages, and more), it is vital to keep the concepts of sustainable design at the forefront of any discussion of building design. Ironically, this apparently new phenomenon is not really due to the recent interest in climate change; instead, this is due to our having lost our previously well-understood sense of how to build buildings using minimal resources in a manner that is compatible with the local climate. Indigenous architecture (localized architecture developed over a long period of time without the input of design professionals) has always been responsive to climate and it has tended to use renewable resources—a pretty good definition of sustainable design. Here are few simple examples.

In the deserts of the world, indigenous buildings are relatively tall (to encourage stack-effect **ventilation**); thick (to exclude heat gain from the sun); heavy, and made of stone, sand and mud (all highly sustainable materials); and without windows on the sunny side(s) (also to exclude heat gain from the sun). People have lived and worked

Figure I.A
Indigenous desert dwelling

in such buildings for thousands of years without air-conditioning in some of the hottest and most inhospitable climates in the world. Is it "cool" (meaning 72 °F) in these buildings on hot days? Of course not. Is it possible for humans to adapt to living in such buildings in such a climate? Or course it is.

At the opposite extreme, in the Arctic, igloos are built using thick snow and ice (also highly sustainable materials) to help internal heat sources (bodies and small fires) heat the buildings; as domes (for structural strength and to enclose the maximum volume of space with the minimum amount of materials); without windows (to keep internal heat in and external cold out), and with a door protected by a tunnel on the lee-ward (away from the prevailing wind) side. Again, this almost perfectly defines a sustainable building for the Arctic climate. Is it "warm" (meaning 70 °F) in such buildings on cold days? Of course not. Is it possible for humans to adapt to living in such buildings in such a climate? Of course it is.

But in the middle—temperate zones that cover most of the populated areas of the planet—this is also true. Even in the Midwest in the United States, this can be seen. Farmhouses that are more than 100 years old or more often have a row of evergreen (usually pine) trees planted very close together and setback from the west side (usually,

Figure I.B
Indigenous Arctic dwelling

Figure I.C
Indigenous temperate dwelling

but the side for prevailing winter winds in all cases); these trees protect the house from the wind, especially in the winter, but allow in the sun. Such houses will also usually have few windows on the north side and limited windows on the east and west sides. The front of the house will face south in most cases. Is it cool on hot days and warm on cold days in such a building in such a location? Of course not. Is it feasible for humans to adapt to living in such buildings in such a climate? Or course it is.

In the deep south in the United States, the climate is described as "hot and humid" in the summertime (and merely warm in the winter). Adaptive indigenous buildings (developed over a few hundred years and not over thousands of years) in this region are usually raised above the ground and they have large openings on all sides (and at the top of the roof) to allow for maximum natural ventilation; for days having little wind, large ceiling fans are often used to mimic the effect of natural ventilation; shutters (real working shutters—not fake plastic shutters as are used all around the country) are used over the openings to block sun while allowing the wind through and for storm (mostly hurricane) protection. The thick and heavy construction used in the desert is not used here because the high relative humidity would cause the buildings to be damp and moldy—entirely unhealthy for human habitation.

So what happened? What don't we build buildings like this today? The answer is very simple: cheap energy, or the perception of cheap energy beginning in the mid-to-late 19th century (the Industrial Revolution), accelerating rapidly following the end of World War II. During the economic boom times in the 1940s through 1960s, it was widely believed (wrongly we now know) that oil production would never end and that, even if it did, nuclear power would produce power "too cheap to meter" (to borrow a common, but unfortunate phrase from the period). So the intention was to use more energy, not less. To build bigger buildings, not smaller ones. To ignore solar orientation. To use new high-tech materials, and not time-tested old ones. To build as though energy really did not matter at all.

Figure I.D
Indigenous tropical dwelling

So today, virtually every building is air-conditioned, and most of you who are reading this have probably never lived or worked in a non-air-conditioned building. Yet air-conditioning, even in the United States, is a relatively new development. Air-conditioning was very rare in single-family houses in the 1950s, rare in the 1960s, uncommon in the 1970s, common in the 1980s, and virtually everywhere by the 1990s for new construction. Many older houses do not have "central air-conditioning" even now. For commercial buildings, air-conditioning started in the 1930s (mostly in theaters and other large public spaces), gradually moving into office buildings and department stores in the 1940s and 1950s, and becoming commonplace (expected really) by the 1970s. But keep in mind that the 1970s were only 50 years ago—only two generations or so, and a blip on the geological time scale.

Before the dominance of air-conditioning, buildings were different. Schools had natural ventilation systems (using operable windows and vertical ventilation shafts, à la indigenous buildings in the Middle East) and were largely illuminated by the sun; office buildings were thin, primarily for daylighting, and had operable windows (and **transoms** over corridor doors) for natural ventilation. Even factory buildings were usually daylit. And operating hours were aligned with sunlight hours: no night shifts anywhere for any reason. Where it was necessary to work at night—for emergency medical procedures and the like—the artificial lighting was quite poor, as it was generated by candles or gas lights.

These issues vary by region too. Many buildings in central coastal California (the San Francisco Bay area, plus or minus a few hundred miles north and south) do not have air-conditioning today because the climate is so mild that it is needed on only rare occasions. Similarly, many buildings in very cold climates have no air-conditioning, because it gets very hot only for very short periods of time. Also, many buildings in tropical (near the equator plus or minus a few hundred miles) areas have no heating at all because it very rarely gets cold enough to need it. But across the temperate zone in the

United States—from the middle of Florida (or even a little farther south) to upper Maine, North Dakota, northern Idaho, and Washington state—heating is required in most cases and cooling is highly desirable.

So the goal of the next generation is to go forward to the past; to revisit the well-understood strategies of sustainable design from 100, or even 1,000, years ago. To build buildings using fewer (and hopefully longer lasting) materials, recycled materials, and recyclable materials, in a manner that respects the local climate: solar orientation and protection from prevailing winds (especially in the winter). And to use free natural daylight whenever feasible. In this book, these concepts will not be covered by a separate chapter or in special sections; instead, they will be woven throughout the text, into each chapter, each topic, and each sub-topic. After all, if we cannot make sustainable design simply an integral part of design, the next generation will still be fighting the same uphill battle. We simply must go back to that place where there is only one fundamental way to build in any given location. This is not to be taken literally; no one is proposing that new buildings in Arizona should be built using adobe and without air-conditioning. But there are basic approaches that make sense in each climate zone and those approaches need to form the basis of design for every project.

Overall, the primary purpose of this book is to provide a handy reference guide to the full range of these systems for interior design students and interior design practitioners. The illustrations have been planned to avoid projecting the idea that "all" water closets look like "the one in the photo" or all cabinets like the ones in the illustration. In fact, there are wide variations across virtually all of these systems. The specifics of a particular piece of equipment are not what matters; instead, what matters are the principles of the systems—what types of components are used in what types of locations taking up roughly how much space.

So we are going back to basics here, and each topic (and sub-topic) will begin with a very simple question: why? Why do we use walls? Why are they transparent, translucent, or solid? Why are they made the way they are? Why are tall spaces noisier than low spaces? Why are most fluorescent light fixtures large and why is it that this no longer matters? Why do we use urinals in restrooms for men (but rarely for women)? Why are water-source heat pumps noisy?

The system for the illustrations was developed specifically for this purpose. The level of detail is deliberately low so as to focus on the principles mentioned above. Additional documentation will be provided to illustrate other facets, but it is key to remain focused on principles. In addition to illustrations (photographs and drawings) and "Design Tips", web-links will be provided throughout the book, mostly to guide the reader to specific product (or more generally manufacturer) web-sites. In today's world, the most accurate and up-to-date product information is usually found on web-sites, and this device will assist the reader in finding detailed information whenever such information is called for.

To cover everything in this book will require a major effort, but everything that is included is important to the practice of interior design so it is worth a major effort to get through all of this. In truth, there really is much, much more needed to cover all of these topics in detail, but one has to start somewhere.

Chapter 1

What are building systems?

Objectives

1.1 To understand what each specific building system includes.

1.2 To understand the areas of expertise of each design team member: **interior designer**, **Registered Architect**, **Landscape Architect**, **Professional Engineer** (civil, structural, mechanical, electrical, plumbing), **code consultant**, **lighting designer**, **IT consultant**, **acoustical consultant**, **construction manager**, and **contractors**.

1.3 To understand team dynamics and professional courtesy in interior design practice.

1.1 Building systems definitions

1.1.1 Legal building systems

Codes, regulations, and standards are a subject in themselves (refer to *Codes, Regulations, and Standards in Interior Design*, also by Samuel L. Hurt), but a brief summary is appropriate for introductory purposes. For now, suffice to say that this complicated subject requires career-long involvement for any design professional.

In all states, Registered Architects (RA, usually) and Professional Engineers (PE, usually) are licensed professionals, who sign and stamp (certify) construction documents for "code compliance." In most cases, PEs are not licensed in a specific engineering discipline (civil, fire protection, **geotechnical**, **environmental**, structural, mechanical, electrical, or plumbing) and each individual's practice is a matter of professional ethics; if the individual PE believes that he or she is competent to practice electrical engineering, there is nothing to prevent them, even if the individual's educational background is in another discipline. (In the case of the author, he took the licensing exam as a civil engineer, with a specialization in Foundations and Geotechnical, neither of which he practices. He practices mechanical, electrical, and plumbing engineering, due to many years of experience and a solid understanding of the subject matter.) In some states, interior designers are also licensed and can sign and stamp (certify) construction documents. This responsibility should not be taken lightly at all, because a substantial code violation (at the time of construction or discovered in the future) can threaten the license of the individual who certified the drawings. That a particular rule is not enforced in a particular jurisdiction (which is quite common) does not mean that it is OK to violate that rule; it is not—period.

Codes and regulations (as contrasted with standards) are laws and there can be real legal consequences for violating them.

1.1.2 Structural building systems

Certainly, an interior designer will never design a structural component or modifications to a structural component. Nevertheless, structural issues do arise on interior design projects on a regular basis.

- Is it possible to put an opening through a **bearing** (structural) **wall**? (Almost certainly yes, but potentially at very high cost.)
- Is it dangerous to put a vault or large safe on the 27th floor of an existing high-rise office building somewhere? (Probably not, but it is usually necessary to add structural reinforcement, which is sometimes very costly.)
- Does a large file room, or simple storage room, require structural reinforcement? (Quite often, yes, and usually at substantial cost.)
- What should be done if demolition uncovers an existing diagonal steel brace where it was planned to put in a new doorway? (Move the door or modify the structural bracing.)

Structural engineers (PEs in most states and structural engineers, SE, in some states) perform most structural engineering, but RAs are also licensed to do this work in most states. That said, few RAs undertake full structural design these days. In states where interior designers are licensed, such licensure explicitly excepts structural design, which is never part of interior design.

1.1.3 Architectural building systems—walls, windows, doors, millwork, stairs, etc.

This area covers a lot of territory, but it is necessary for interior designers to understand how to design and specify all elements in a project, including the walls (solid, hollow, **fire-rated**, non-fire-rated, partial height, full height, transparent, translucent, etc.), windows, window frames, window hardware, doors, door frames, door hardware, millwork, millwork hardware, stairs, and whatever else might come along. And, again, more questions come to mind:

- When is it acceptable to design a full custom door? (When it adds substantially to the design, the client can afford it, and they do not mind paying for it.)
- When are **full-height walls** justifiable? (Only when necessary, mostly due to high costs.)
- What to do if something heavy is going to hang on a wall? (Reinforce the wall with **blocking** of some sort—wood or metal, vertical or horizontal, or both.)
- Is oil-rubbed bronze finished door hardware more costly? (Yes.)
- Should one specify exterior windows for interior uses? (Not required but maybe.)
- Should millwork be **shop-finished** or **field-finished**? (Shop-finished if at all possible.)
- How can a full custom wood molding profile be justified for a project? (When it adds substantially to the design, the client can afford it, and does not mind paying for it.)

1.1.4 Architectural acoustics

Everyone has experienced poor architectural acoustics, usually in the form of hearing a neighbor (next to, above, or below) when one would prefer not to, not being able to understand speech (maybe in a presentation in a gymnasium), or maybe just due to excessive loudness (in an office or other workspace). There are good techniques to minimize problems in all of these categories, but the costs can be very high and ideal solutions are therefore often not affordable. Many beautiful projects have been ruined by poor acoustics.

More questions:

- Do some spaces require special treatment for **sound isolation** from other adjoining spaces? (Certainly for noise-sensitive spaces—some conference rooms, human resources offices, medical offices, etc.)
- Do some spaces require **noise reduction** treatment? (Yes, sometimes to decrease **speech intelligibility** in open office environments or to increase speech intelligibility in assembly spaces.)
- Should one locate a small toilet room directly adjacent to a large corporate boardroom? (Only if the goal is embarrassment.)
- What to do in a home theater? (Various things, including ...)

Who practices as an acoustical consultant? Sometimes PEs, sometimes RAs, and sometimes individuals with no specific license or educational credentials. There are no laws (of which the author is aware) that restrict the practice of "acoustics," so many different people promote such services. This is complicated by the fact that there are two paths to education in acoustical design: physics and architecture. Dr. Amar G. Bose (of the Bose Corporation, famous for the Bose WaveRadio and many other consumer products) was a Professor of Physics at MIT for many years, and he practiced what is best described as "physical acoustics," or acoustics from the perspective of physics. Robert B. Newman and Leo Beranek were two of the founders of Bolt, Beranek, and Newman, which has been one of the leading architectural acoustics consulting firms in the world from the 1960s until the present. Bob Newman was an RA and Leo Beranek was a scientist/engineer (who came from the perspective of acoustical physics). Leo Beranek did important research in the field (and wrote a seminal textbook on the subject) and designed acoustics for real projects, while Bob Newman taught acoustics (at both Harvard University and at MIT—where the author was his student) and designed acoustics for real architectural projects.

The author has performed acoustical design services for several projects, some as an independent consultant. But this was limited to relatively basic situations and he would never take on the design of a new concert hall or other highly sophisticated (and potentially risky) acoustical design.

In many programs, the study of architectural acoustics comprises its own separate subject.

1.1.5 Mechanical building systems

Like codes, mechanical systems—really Heating, Ventilating, and Air-Conditioning (HVAC) systems—could easily be a topic in themselves. Air-conditioning includes air-filtration and moisture control (usually called de-humidification) and cooling; that is why it is not called "cooling." These days, numerous options are available for just about any type of project, and energy codes in most jurisdictions have a strong impact on the selection and design of such systems (and of the building envelope that contains such systems). These systems can be very complicated, and there are many different ways to do them.

We can use old-style **direct expansion** (DX) **cooling**; **air-conditioning** coupled with natural gas-fired heating (the most common residential system in many parts of the country) for small and large projects alike; gas-electric rooftop systems (**centralized** or **de-centralized**, **constant air volume** or **variable air volume** (VAV)); or centralized all air systems (constant or variable air volume, with various sources for heating and cooling), or centralized air-and-water systems (4-pipe **chilled and hot water** systems, 2-pipe chilled and hot water changeover systems, closed and open loop water-source **heat pump** systems), or semi-centralized **refrigerant**-and-air systems (Variable Refrigerant Flow—VRF). If a DX system is run backwards, it produces heat instead of producing cooling and it is called an "air-to-air heat pump."

We can add **energy recovery ventilators**. We can use energy recovery chillers. And we can use high-efficiency fans, motors, pumps, **compressors**, and on and on.

On a day-to-day basis, interior designers will not design these systems, but it is necessary to understand the basic types of systems and how it is possible to modify such systems for renovation projects. It is also necessary to understand the space (both horizontal and vertical) impacts of various systems so that those systems can "fit" into their projects.

Who does design HVAC systems? Primarily consulting engineers (PEs and their staffs, as noted above) but RAs are also authorized to certify documents for these systems. In fact, while some RAs design small and simple systems, few of them take on large or complicated systems, leaving that work to consulting engineers. Interior designers rarely venture into this area, with the likely exception of relocating existing air supply diffusers or return air grilles here and there in very simple situations.

1.1.6 Lighting building systems

Since the 1880s when electric lighting first became feasible, electric lighting has been the fastest changing of all of the building systems. Since Edison perfected the carbon filament lamp, a number of different technologies have come along—tungsten filament lamps (shortly after carbon filament lamps); halogen gas inside the bulb of tungsten filament lamps; neon; fluorescent lamps (long, round, U-shaped, long compact, short compact, and self-ballasted compact); cold-cathode; high-intensity discharge (mercury vapor, metal halide, low pressure sodium, high pressure sodium, and ceramic metal halide); induction; plasma; and then solid state (usually called LEDs). Fluorescent was invented in the 1930s and LEDs were invented in the 1960s, so the pace of change was very fast. LEDs did not become practical for architectural lighting until after 2000 and the pace of change has accelerated rapidly since then. Even now, LEDs are changing almost monthly, and it is difficult to keep up.

Once electric lighting became feasible, especially after the advent of "efficient" fluorescent lighting, the necessity, and desire, for daylighting in buildings fell away more and more as the years went by. Eventually, it got to the point where many buildings had no windows at all (especially industrial buildings) and even buildings with lots of glass (such as high-rise office buildings) were not designed for daylighting. After the energy crisis of the early 1970s, the sustainable design community did begin to talk about daylighting again, and today it has become an important part of design for many projects. But to drive the point home, the author designed his first "daylight compensating automatic fluorescent dimming system" in 1987 (in what then passed as a high-performance library building) and his second in 2011. Such systems are called "daylight harvesting" systems today, and they have become far simpler and less costly than they once were (this is partly due to the use of LEDs).

Interior designers can, and should, be involved in lighting design, which is also a complete subject unto itself:

- Should **uplighting** or **downlighting** be used in a space? Or both?
- Should a space be very **bright**? Or very **dark**? Or neither?

- Is **glare** a critical concern?
- What is the budget?
- Does the owner want simple controls? Or do they enjoy high-tech gadgets (such as turning on the lights at home or pre-heating the oven from a smart phone while travelling)?

Who designs lighting systems? Primarily consulting engineers (PEs and their staffs, as noted above) and lighting designers, but RAs are allowed to certify these systems too. Just as in HVAC though, few RAs will take on large or complex lighting systems, leaving that work to their consulting engineers or lighting designers instead. Interior designers are highly unlikely to take on wiring design or even layout of large or complex systems, but some basic activities are done to direct the "look" of the project: locating **convenience receptacles**, light switches, **occupancy sensors** (maybe), selecting **luminaire** types, locating luminaires, etc.

1.1.7 Power building systems

Power systems include power to the building (called the power service) and power distribution within the building, using various types of hardware (usually **panelboards** with or without **transformers**), to all loads—equipment, HVAC, water heating (maybe), and artificial lighting. There are several different wiring systems that might be applicable to a given situation.

Just as in mechanical systems, interior designers will not be working out power **circuit** layout, large equipment selection, or wire sizing. But some aspects of power system design need to be understood:

- How much space is needed to add a branch circuit panelboard? (It all depends ...)
- What happens if a large dry-type transformer is needed? (Put it in a 1-hour rated room with appropriate clearances.)
- Can existing power equipment be moved? (Certainly, but often at very high cost.)

Who designs power systems? Primarily consulting engineers (PEs and their staffs, as noted above) but RAs are allowed to certify these systems too. Just as in HVAC though, few RAs will take on large or complex power systems, leaving that work to their consulting engineers instead. Interior designers are highly unlikely to take on wiring design or even layout of large or complex systems, but some basic activities are done: locating convenience receptacles, light switches, occupancy sensors (maybe), selecting luminaire types, locating luminaires, etc.

1.1.8 Plumbing building systems

As in the other cases, it is not expected that an interior designer will be sizing **domestic water** piping or even a **water heater** for that matter. But interior designers do select plumbing fixtures on a regular basis, so it is important to understand the differences between types, materials, grades, and operating characteristics.

And, as with HVAC systems, it is important to understand the space impacts of different types of water heaters and other plumbing equipment (domestic water pressure booster pumps, **mixing valves**, meters, reduced pressure zone backflow preventers (**RPZBP**), **recirculation pumps**, **fire pumps**, **fire risers**, **fire standpipes**, etc.).

Who designs plumbing? Similarly to HVAC and electrical, mostly consulting engineers, but RAs are allowed to certify these systems too. Interior designers are likely to select fixtures (sinks, lavatories, showers, bathtubs, **water closets**, **urinals**, etc.) but are unlikely to design piping or to calculate flow rates.

1.1.9 Fire protection building systems

Fire protection building systems overlap between electrical and plumbing because fire alarm systems are electrical but fire protection sprinkler systems are plumbing. But they are directly related to one another—fire alarm systems monitor fire sprinkler systems—and it makes sense to cover them separately.

Who designs fire protection systems? Most fire alarm systems are designed by consulting engineers or architects (or design-build specialty vendors or contractors) and most sprinkler systems are designed by specialty design-build contractors because only fire protection PEs are allowed to certify sprinkler drawings in most jurisdictions. Most commonly, such engineers work for the contractors but there are independent consulting engineers who offer these services too.

1.2 Professional roles

As far as professional responsibilities go, there can be a lot of overlap between interior design and architecture. Architecture is defined, by law in most states, as the general practice of building design, engineering, and construction, and RAs are therefore defined as general practitioners (which is why they are licensed to perform structural engineering). This general practice also includes interior design in most cases. Interior design is defined as the design of interior non-structural components of buildings by most states, CIDA (Conference on Interior Design Accreditation), IIDA (International Interior Design Association), and ASID (American Society of Interior Design).

Even though only RAs and PEs are licensed to certify documents for building envelope elements (exterior walls, windows, doors, roofs, etc.) and structural systems, that does not mean that only RAs and PEs can design such elements. Most registration (licensure) laws state that RAs and PEs (and Registered Interior Designers (RIDs) where applicable) can certify only documents produced directly by them or under their direct supervision. The latter describes most of the work that gets done, because the RAs and/or PEs (and RIDs) who certify documents very often do not actually draw anything; instead, they are in supervisory roles (as is the author, who has been in management since 1989, supervising various different groups of interior designers, architects, and/or engineers). Practically speaking, this means that interior designers can really design just about anything, as long as an RA or a PE is willing to certify the resulting work.

As noted above, each design team member can have numerous roles:

- *Interior designer*: design of non-structural interior building elements, including walls, doors, interior windows, millwork, finishes, plumbing fixtures, luminaires, etc.
- *Registered Architect*: the broadest role of all, encompassing all of building design and construction.
- *Landscape architect*: a specialist in outdoor plant materials and ecosystems. Some states offer title registration for landscape architects (Registered Landscape Architect, RLA, is the regulated title in the State of Indiana, for example), and others do not. It is very common to see landscape architects acting as informal civil engineers; in other words, they design site grading and drainage under the supervision of a PE.
- *Professional Engineer*: another very broad category encompassing the disciplines of civil (which itself include structural, environmental, and geotechnical sub-disciplines), fire protection, mechanical, electrical, and plumbing. As noted above, a PE can practice any discipline, or sub-discipline, for which she or he considers themself qualified—in many jurisdictions.
- *Code consultant*: Code consultants might be RAs or PEs or not. There are no regulatory (legal) requirements of which the author is aware. Most code consultants are individuals who understand the details of enforcement in a given jurisdiction, or jurisdictions. Much of their efforts go to writing code analysis reports for upcoming projects, completing **modification** applications, and working directly with enforcement officials.
- *Lighting designer*: there are no legal requirements for lighting designers (so far), but there is a well-established national credential—Lighting Certified (LC), which is administered by the National Council on Qualifications in the Lighting Professions and a newer credential—Credentialed Lighting Designer (CLD), which is experience-based. There are also two major lighting organizations: IES (Illuminating Engineering Society) and IALD (International Association of Lighting Designers), but neither organization has legal standing. (Similarly, architects have the American Institute of Architects (AIA) and the Association of Licensed Architects (ALA); landscape architects have the Association of Landscape Architects (ALA also, unfortunately); and PEs have the National Society of Professional Engineers (NSPE) and the American Society of Civil Engineers (ASCE).) For all intents and purposes, anyone can be a lighting designer, and it is common to see RAs, PEs, RIDs, interior designers, and others fulfilling that role.
- *IT consultant*: this is an individual who designs computer networks and/or telecommunications systems (i.e. phones) or a combination of both. Most of these folks have an engineering oriented background, but there is no requirement for them to be PEs or to have any professional standing whatsoever.
- *Acoustical consultant*: as noted above, acoustical consultants can be RAs, PEs, or neither, with no particular qualifications being legally required. Acoustical consultants

also commonly design audio-visual systems, but AV systems can be designed by AV specialists, RAs, or PEs as well.

- *Contractor*: a person, or organization, who provides a "trade" service under contract, including everything from concrete forming and placement, to wood framing, to steel framing, to painting, plumbing, HVAC, wiring, and on and on.
- *General contractor (GC)*: a person, or organization, who packages a complete project for an owner. All trade contractors work for the GC in this arrangement, and the owner works directly only with the GC. In theory, GCs make projects run more smoothly by having direct control over all work on the job-site. But many GCs do not really control all of the trade contractors who work for them, and it is common to see disputes and **change orders** on projects like this. The response to this was the construction manager.

    Construction managers (CMs) take several forms:

- *CM, general*: individuals (or firms) who assist owners from the outset of a project—selecting and acquiring property, writing a program, hiring a design team, and hiring trade contractors. Individuals doing CM can be RAs or PEs, but there are no legal requirements.
- *CM, coordinating*: individuals (or firms) who assist owners in packaging and bidding their projects. This form derived from unhappiness with GCs, and the selling point was that hiring the CM would reduce the overall cost and streamline the project by eliminating **mark-ups**; unfortunately, this frequently failed and this approach to projects is seen less and less often these days. A CM in this role has no financial risk, because each individual contractor works directly for the owner.
- *CM, at risk*: this was the result of unhappiness with coordinating CMs. In this case, the CM does have financial risk and the individual contractors work for the CM. In practice, this looks very similar to GC, with the main difference being that a CM at risk will show the owner all of the bids, and sub-bids, so that all information is known. A CM at risk most commonly charges their direct costs (supervision, cleaning, job facilities, etc.) to the project and applies a "cost plus" fee mark-up to cover profit. These costs are identified at the start of the project.

GCs came about in the late 19th century when some architects decided that they did not want to be CM (general or coordinating) anymore, which had been the model for hundreds of years previously, although not with the "CM" label prior to this time; it was simply an "architect" who handled both design and construction of a project. GCs manage the whole construction process (identifying, hiring, and managing **sub-contractors**; supervising the work on the site, etc.) but they do not operate in an open environment. In other words, when a GC bids a project, no one knows the amounts that were bid by the sub-contractors. Much of that information comes out in a "Schedule of Values" during construction (on some projects) but there is no way to confirm that the Schedule of Values figures match the bid day figures, and there is good reason to believe that they often do not match. (If a GC can "buy out" the contract for numbers that are lower than they were on bid day, the GC gets to keep the savings, however much it might be.) Individuals working as GCs can be RAs or PEs, but there are no legal requirements.

1.3 Team dynamics and professional courtesy

It is vital for all design team players—RA, SE, PE, RID, interior designer, LA, RLA, code consultant, lighting designer, acoustical designer, IT consultant, etc.—to understand and respect each role and its reasonable boundaries. This includes more general job categories as well: principal, associate principal, associate, project director, project manager, project engineer, project architect, project interior designer, senior interior designer, senior space planner, interior designer, space planner, junior interior designer, junior space planner, drafter, BIM (Building Information Management) specialist, etc.

Primarily, this takes the form of respecting other individuals' areas of expertise and authority. For example, if an interior designer has a strong opinion about the requirements of the Building Code for fire alarm annunciation but the consulting engineer doing the electrical design has a differing opinion, the interior designer should defer to the consulting engineer because the fire alarm system falls under the PE's certification. Similarly, if the consulting engineer designing the plumbing believes that the interior designer's space plan does not comply fully with the accessibility requirements of the Building Code, and the interior designer disagrees, the consulting engineer should defer to the interior designer (keeping in mind that the interior designer would have to convince the RA, RID, or PE who is certifying the project that her/his opinion is correct).

This is nothing more than common decency and courtesy, but it is not all that uncommon to see problems in this area. None of us is right about everything and we should not force our opinion onto others who have either more responsibility than we do (i.e. certification), equal or better qualifications, or both.

Once a project moves into construction (or earlier for CM projects, and even some GC projects), the team expands to include additional project managers, on-site superintendents, on-site foremen, and individual workmen, and everyone on the expanded team also needs to be treated with professional respect.

Summary

The main idea for this book is to cover a lot of territory but in a way that is direct, appropriate, and understandable. There will be very little required calculation, but examples will be provided for how to calculate any number of different things: stair **riser** height and stair length; basic heat transfer and heating, cooling, and ventilation loads; reverberation, noise reduction, and sound transmission; power loads; and lighting (ambient and accent).

Outcomes

1.1 Understanding what each specific building system includes, as noted above, for legal, structural, architectural, acoustical, mechanical, lighting, power, plumbing, and fire protection systems.

1.2 Understanding the areas of expertise of each design team member: interior designer, RID, RA, LA, RLA, PE, code consultant, lighting designer, IT consultant, acoustical consultant, CM, and GC as noted above.

1.3 Understanding team dynamics and professional courtesy in interior design practice as general common decency and courtesy.

Chapter 2

Why do building systems matter?

Objectives

2.1 To understand that building systems have an unavoidable impact on design and built projects, whether we like it or not.

2.2 To understand that occupants care about temperature, humidity, light, running water, noise, safety, and even beauty, and that building systems are the means to make all of the occupants happy, healthy, and satisfied. Well, maybe not all of them, which is impossible. But as many as possible for each and every project.

2.1 Comfort and sustainability

Comfort and sustainability matter. And comfort alone covers a lot of territory—temperature, relative humidity, "draftiness," "stuffiness," odors, heat from direct sun, glare from direct sun, glare from electric lights, excessive lighting, inadequate lighting, the height of a worksurface or keyboard, noise (from outside, co-workers, mechanical equipment, lighting), vibrations, bad plumbing fixtures, leaky pipes, wobbly walls, sticking doors, truth, beauty, and charm, etc. Hundreds of years ago, buildings were easy—no artificial lighting, no air-conditioning, maybe even no heating. But today's buildings must meet many requirements simultaneously, and for multiple occupants who all have different opinions about what is "right."

As noted in the Introduction, it is vital that we move sustainable design beyond its status as a "fringe element" in design (where clients pick and choose how "green" they want to be, within the bounds imposed by the applicable Energy Code, or LEED® or some other similar standard) into the mainstream. We simply need to stop building large areas of west-facing glass in Indianapolis (due to the brutal summertime western sun exposure); we need to tolerate wider temperature ranges in indoor spaces (maybe 60°F in winter to 85°F in summer instead of 72°F year-round); we need to stop using low-cost more-or-less disposable materials (which only add to landfill problems); we need to stop growing turf lawns in deserts (à la Las Vegas) and wasting large amounts of water on irrigation everywhere else (we should grow crops where it rains, not where there is milder weather but no rain—and we will just have to get used to not having fresh produce all year in cold climates); and we should build the amount of space that we actually need, as opposed to building more space just because we want to. None of this is meant to be taken literally; there are exceptions where sophisticated building technology can be leveraged to use more glass; there are situations where tight temperature control makes sense, etc. But the broader point that we should break relatively recent habits and go back to older habits does make sense. One of the most meaningful and recognizable mantras of the sustainable design movement is "Reduce, Re-use, Re-cycle" which is entirely consistent with this larger point—go back to basics and keep it simple.

Back to specific requirements—take temperature and relative humidity for example. The author is an over-weight middle-aged guy who likes cool conditions, cool meaning 70°F or so, unless the relative humidity is unusually low. The ratio of his surface-to-volume is smaller than for a very thin person. So a thin person often prefers warmer conditions—but not always. And we vary from time to time. We might like it warmer, or cooler, in the mornings, which could change from season-to-season, if we live in a part of the country where there are seasons. Our preferences depend upon the time of year, our location, events of the moment, and even our mood. No wonder HVAC designers are wary of complaints.

Relative humidity is a term that we should not even use. Relative humidity is simply the ratio of the amount of water vapor in air at a given temperature as compared to the maximum amount of water vapor that the air can hold. When it is raining, relatively humidity is 100% and the water literally falls out of the air. When people from moderate to highly humid climates visit a hot desert (think Las Vegas), they often have sinus problems due to excessive dryness. When people from very dry climates visit moderate to highly humid climates, they can find the conditions almost insufferable.

When discussing this subject, it would be better to use absolute—not relative—water content (usually measured in grams of water per pound of dry air—now that is an awkward unit) because those numbers do not change if the temperature changes. If the air contains 40g/lb and the temperature goes *up*, the relative humidity goes *down*, because warmer air can hold more moisture. If the temperature of that same air goes down, the relatively humidity goes UP for the same reason. This can cause any number of problems, especially when regulations are involved.

For example, in the hospital world, most operating rooms are required to be maintained at the following conditions: 68–73 °F at 30–60% relative humidity (RH), but many doctors believe that the "maximum" RH allowed in an operating room is 60%—which is not true. The maximum RH in the operating room is 60% when the dry-bulb temperature is between 68 and 73 °F. If the room is warmer than 73 °F, the maximum RH actually would be *less* than 60%, and the maximum RH is MORE than 60% if the temperature drops below 68 °F. (It is very common for operating rooms to be set at 65 °F, or even less, so this is a real issue.) What these regulations actually say is that the absolute moisture content of the air in an operating room cannot be less than 30 g/# (the condition of 30% RH at 68 °F) and not more than 74 g/# (the condition of 60% at 73 °F). This is much clearer than talking about RH, so, of course, everyone talks about the far more confusing RH.

"Draftiness" is a vague term that people use when they feel like cool, or cold, air is blowing on them when it shouldn't be (as in winter when one wants to feel "warm"). In truth, this is about simple temperature variations. The cores of our bodies should be somewhere pretty close to 98.6 °F most of the time (with a degree or two of variation throughout the day), and our surface temperature (on our skin) is probably somewhere close to 95 °F. If air that is 30 °F warmer than our skin blows on us, we perceive that to be "heat," but if the same air is only 95 °F or colder, we perceive that to be a "draft." (The ultimate draft is cold air blowing in through an old leaky window in the winter.) Can we warm a room that is 70 °F up to 75 °F by using air that is 95 °F? Certainly, that would work just fine. But if you are sitting in that 95 °F air-stream, you might not feel like the room is actually getting warmer. (This is a common complaint for people who have air-to-air heat pumps at home, which deliver air in the 95 °F range, or sometimes even less.) The answer to this issue is to design the air delivery system to minimize direct air contact, where that is feasible.

"Stuffiness" occurs when spaces are under-ventilated, and what it really means is that odors, gases, even particulates have built up in the air to such a degree that breathing is impaired. Spaces that are merely "stuffy" are generally not life-threatening (due to suffocation), but stuffiness, left unresolved, can lead to very unhealthy conditions. As we breathe, we take in air (which is mostly nitrogen but with a very important amount of oxygen); our lungs use the oxygen in the air, and we exhale a mixture of gases that is heavy in carbon dioxide (CO_2). As CO_2 builds up due to our mere presence in a space, breathing becomes more and more difficult and it is possible to suffocate under high concentrations. In addition, we perspire (sweat) and respire (breathe), releasing water vapor into the air; we give off various odors due to perspiration, perfumes, after shave, cologne, eating, burping, flatulence, etc.; and we give off particulates—mostly small flakes of skin, but also hair and other substances. It is not hard to imagine what the atmosphere would be like if all of this material stayed in the air. So we use ventilation to change the air in a given space, by one of two basic methods: natural and mechanical. Natural ventilation means opening a window or door and letting the wind do the work (but if there are no counter-facing windows to allow for "flow-through" wind movement or high openings to take advantage of vertical stack-effect, this might not be very effective). Mechanical ventilation means that we use mechanical equipment (fans, ducts, grilles, louvers, etc.) to force air into and out of a space or building. Natural ventilation is allowed (by Code) for most projects if some stringent rules can be met for size of openings, etc.

As noted above, we produce a number of different odors ourselves, but we also produce odors through the use of chemicals in the production and use of a multitude of different products: food, beverages, soaps, detergents, conditioners, hair gels, hair spray, air fresheners, toothpaste, mouthwash, drain cleaner, plants, pets, off-gassing furniture and/or carpets and rugs (off-gassing is the process whereby chemicals are released from products for a long period after manufacturing), paint, printer toner, printer ink, mold, mildew, and on and on. If these chemicals are not ventilated out of a space, they will remain there for a long time. (The author once evaluated a very smelly, and stuffy, 40-year-old commercial office building that had no operable windows and no mechanical ventilation; every odor or chemical that had ever entered the building over 40 years was essentially still there! One can imagine how noxious that would be. The recommendation was very simple: mechanically ventilate the building as heavily as possible for as long as possible before letting new occupants move in.)

Heat from direct sun is nearly impossible to control in buildings. Even on a very cold winter day, if one stands facing the sun for a while, the front of one's body will warm up (not the back though—radiation is one-directional); even sophisticated window glass has a very hard time dealing with that much energy. Ideally, buildings should be planned so that occupants are not exposed to direct sun for extended periods of time. (When supermarkets first started putting in skylights to reduce the need for electric lighting, some stores used clear glass in their skylights and had problems with melting frozen food in enclosed cases with glass doors due to sun exposure!) Furniture placement should take this into account to avoid over-heating occupants.

In addition to producing a massive amount of heat, the sun also produces a massive amount of light. Day-time light levels on a sunny day can reach 10,000 **foot-candles** (fc) (for reference, most offices are illuminated to 30 fc or so these days), and daylight penetration into highly glazed buildings can be disabling for occupants. This must also be taken into account when planning furniture placement.

Even though electric lights cannot produce (well, not given practical amounts of power) quantities of light that are comparable to the sun, it is still possible to create glare problems. Glare is simply excessive unneeded light, and it is not all that difficult to put light in the wrong place for the wrong reason. And some people see glare where other people see appreciated brightness. Some people like it darkish; other people think that more is never enough. But we still have rules and regulations for this broad and unclear subject.

Do you like to have a desk at +30″ above the floor, the "standard"? Or would you prefer 28″ or 31″ or some other height? How high is your chair? How high should it be? How wide is your chair? Is it too narrow, just right, or too wide? Is it comfortable at all? Do you work standing up? Or sitting down? In your kitchen, does a 36″-high counter really work for you? Do you like "standard" height toilets or tall ones (supposedly for the disabled)? The subject of ergonomics has been relevant to design for a long time, but it has become even more important in recent years, at least in part because we have a wider variety of body types and sizes in the United States now than anywhere else or at any previous time. We also have a strong commitment to making public facilities accessible to the disabled (the Americans with Disabilities Act and *ICC/ANSI A117.1-2003 Accessible and Useable Facilities and Buildings* to cite the two most commonly used documents), and there is a movement out there for universal design—everything accessible to everyone all the time. Universal design requires having desks at 26″, 35″, 31″, and 32″ (and probably more), and chairs of various sizes, shapes, and heights, among other things.

2.2 Noise

Are you bothered by a family member, fellow student, or a co-worker who talks loudly, on the phone or otherwise? Do background noises (traffic outside the building, the mechanical system, music, etc.) bother you? Do you worry about being overheard when you are talking to your doctor? Do you hear the 60 Hz (hertz—cycles per second) hum from electrical equipment? Does it bother you? Do you know why there are dividers between the drop-off counters at your pharmacy? (It is required by the federal government's HIPPA healthcare privacy law—and it does not work at all, by the way.) Have you been to a wedding where you could not hear anything due to the poor acoustics and/or poor sound system in the church? All of these are design issues.

Does your desk vibrate when a large truck drives by? Is there a large machine that vibrates the floor under you? Can you feel the floor vibrating when someone walks nearby? These vibration issues are probably related to structural or mechanical systems, but they are real and should also be accounted for in your design.

2.3 Plumbing

Do you have a 1.6 gallon per flush (as required by the federal government's Energy Policy Act) toilet that does not work, so you have a plunger nearby? Do you like low-flow shower heads? Do you know when to use accessible **faucet** handles (virtually everywhere, except in residences and sometimes there too)? Should you or should you not specify urinals for men's restrooms? If so, what kind? Should you consider waterless urinals? How about urinals in women's restrooms? When do you have to have floor drains? What is a trap primer? How many options are there for lavatory bowls for bathrooms? (Too many to list.) How many ways are there to build a kitchen sink? (Again, too many to list here.) How much space do you need between back-to-back restrooms? (Usually just a few inches will do; there is no need for a chase unless it is actually big enough to work in—say 4′ 0″ wide.)

What happens to your pristine white and smooth gypsum board ceiling if a pipe springs a leak directly above it? (It will discolor, stain, and eventually collapse—unless the leak is fixed quickly.)

Can you run a sanitary waste pipe over the ceiling of an operating room in a hospital? (No, unless it runs above a protective tray or inside a protective sleeve.)

What is PEX piping? (Cross-linked polyethylene—flexible plastic tubing.)

Is copper piping more or less costly than the alternatives? (More, usually.)

Should PVC (polyvinylchloride) piping be used for sanitary waste lines in noise-sensitive locations? (No, because it transfers noise very effectively. If someone flushes an upstairs water closet (toilet) into a vertical PVC pipe in the wall of your office, you might think that you have a waterfall. Cast iron is 17 times more dense, and greatly reduces sound transmission.)

2.4 Appearance

Appearance matters. We are all designers and we want to design beautiful spaces, as subjective as that statement is. What should you do if a mechanical engineer wants to put a large and ugly supply air diffuser in the middle of your shallow plaster dome over the middle of the church sanctuary? (Say no—please. There is no reason to cooperate with such a crude approach. Mechanical engineers have many different options available to them to deliver air to a space and they should take your design into consideration when they select products and locate equipment. See Chapter 7.)

Under what circumstances is it reasonable to design full custom doors and/or door frames? What gauge of metal studs should be used? What size of metal studs should be used? Is gypsum board the right answer to every wall question? Or any wall question? What are the alternatives to gypsum board? What is appropriate backing for ceramic tile on walls? Should your floor tile be set with thin set adhesive, medium set adhesive, or on a mud bed? How much blocking is required to support wall hung shelving and cabinets? What about crown mold and other moldings?

How can a group of light switches or wall box dimmers be organized? Is it reasonable to use screw-less (i.e. snap-on) electrical device cover plates? Do you really have to have that power panelboard out in the open?

How can you assure that the lighting design is appropriate for the overall design? Do you want 8″ aperture recessed downlights? Or 2″? Or some other size? Do you like 2 × 4 recessed troffers? What is a 2 × 4 recessed troffer? Should the space have an average level of 10 fc or 30 fc? Is task lighting needed or desired? How can you reduce maintenance of a lighting system? Are automated, or semi-automated, **lighting controls** justifiable in retail stores, restaurants, or anywhere else? Should you use LEDs everywhere? Or nowhere? Compact fluorescents? Ceramic metal halide? Induction? Plasma? And what about power-over-ethernet (POE)?

All of these questions (and many, many more) relate to various building systems, and they are all crucial to design. The ultimate goal is a beautiful design that is well executed technically. But many beautiful designs get executed poorly, completely ruining the intended effect. On the other hand, an ugly design that is well executed is still ugly. So is it more critical to have a strong design concept or good technical execution? The answer varies from time to time, place to place, and project to project, but poor execution is always a problem, good design or otherwise. One of the hallmarks of high-quality design is high-quality execution, and projects cannot be executed with high quality without careful attention to all of the various building systems. It does not take very many poorly located HVAC diffusers or badly selected and placed luminaires to ruin a lot of good design work. The goal of this book is to provide resources for interior designers to avoid such problems.

2.5 Lighting

Architectural lighting makes or breaks every design project. This simple truth might be unpleasant, even a little scary maybe, but it is a truth. So good designers really need to understand the basic lingo of lighting design: lumen, footcandle, centerbeam candela, luminaire, uplighting, down lighting, wall washing, lens, reflector, refractor, lamp, diode, circuit board, LED, optics, brightness ratio, zonal cavity, iso-footcandle plot, troffer, downlight, pendant, track, and so on.

And designers need to understand the differences between uplighting, downlighting, and side-lighting; the basic "rules" for accent lighting; how to manage light color and quality; what is too bright and what is too dark. And, perhaps most importantly of all, what is affordable.

There are wide variations in costs across most elements of construction, but none is wider than the potential range for architectural lighting. For comparison purposes, the installed cost for commercial carpet is probably between $15.00/square yard and $45.00/square yard—a 3:1 ratio. But 3:1 in lighting is a SMALL difference and 10:1 is not a particularly high difference. So one has to be very careful when selecting lighting hardware, lest the owner go crazy about the cost. No one likes beautiful lighting hardware more than the author, but busted budgets never help anyone. Never. Never. Never.

Summary

This budget issue is key. Projects fail because they do not meet the owner's needs or expectations or because they do not meet the budget. If a project goes over budget, the most common result is re-design (even if that is after hard bidding by contractors). This process is often called "value engineering" but what it means is that the contractors delete work or reduce quality without giving the owner the full value back as credits; this is neither a process of engineering nor does it improve value. In truth, value engineering is a design function, and it is a requirement for designers to hit their budgets. But it is very hard to do that, so it is common for it not to work out.

If a client approaches a designer (interior designer, architect, or engineer) with a project that is underfunded/under-budgeted, the designer should point that out to the owner. If the owner insists upon going forward anyway and the project goes over budget, that does give the designer a fighting chance to argue for additional compensation for re-design. But it is far from certain that such compensation will be forthcoming.

One avoids such problems by designing to budget. Period.

And it does not matter which part of the project is involved because this affects everything with no exceptions. It is sometimes possible to shift resources around within a budget—a little less here to pay for a little more there—but that is not always feasible either.

Specific to building systems that are not designed by the interior designer, this means that the consulting engineers (or whoever else is involved) need to commit to budgets early in the process and stick to them. Allowing the electrical budget to creep up because better lighting was added late in the process will probably cause a budget bust. This is why it is so important to establish budgets and expectations at the beginning of the project.

Outcomes

2.1 Understanding that building systems have an unavoidable impact on design and built projects, whether we like it or not.

2.2 Understanding that occupants care about temperature, humidity, light, running water, noise, safety, and even beauty, and that building systems are the means to make all of the occupants happy, healthy, and satisfied.

Chapter 3

What are legal building systems?

Objectives

3.1 Introduction
   3.1.1 To understand "standards."
   3.1.2 To understand "regulations."
   3.1.3 To understand "codes."
3.2 To understand basic code history.
3.3 To understand the code adoption process.
3.4 To understand code enforcement.
3.5 To understand code appeals.
3.6 To understand the role(s) of code consultants.
3.7 To understand code documents.
3.8 To understand code research.
3.9 To understand key concepts, including use and occupancy classification; special detailed requirements based on use and occupancy; types of construction; fire-resistance-rated construction; interior finishes; means of egress, accessibility, and other requirements.
3.10 To understand other codes, including the 2015 International Residential Code and the 2015 International Energy Conservation Code.
3.11 To understand other regulations (planning and zoning, etc.).

3.1 Introduction

Legal "systems" in interior design specifically include codes, regulations, and standards. For the purposes of this book, the following definitions should be used:

- *Code*: rules for the design and construction of buildings and building elements.
- *Regulation*: a law that applies only to a select group for a specific purpose.
- *Standard*: a set of practices defined by a trade group or research body.

These will be addressed in reverse order in the following sub-sections.

3.1.1 Standards

Standards are written by many different groups, including, but certainly not limited to, the following:

- American National Standards Institute (ANSI), best known for accessibility standards for disabled persons and elevator standards
- American Society of Heating, Refrigeration, and Air-Conditioning Engineers (ASHRAE), best known for standards for HVAC (and refrigeration)
- Architectural Woodwork Institute (AWI), best known for woodworking standards for doors, frames, standing and running trim, paneling, and cabinetry
- Association for Contract Textiles (ACT), best known for commercial textiles
- ASTM International, best known for testing protocols (for textiles, among many other things)
- Building Hardware Manufacturers' Association (BHMA), best known for door hardware
- Factory Mutual (FM), best known for fire protection
- Gypsum Association (GA), best known for gypsum board assemblies
- International Code Council (ICC), best known as the primary source for model codes
- National Fire Protection Association (NFPA), best known for extensive fire protection standards
- NSF International, best known for food service
- Underwriters' Laboratories (UL), best known for testing protocols (for many products and systems)
- Window and Door Manufacturers' Association (WDMA), best known for windows and doors

Primarily, these documents are written to provide guidance to members of the organization (in most cases). In many cases, the requirements are binding on members of the organization, but not on anyone else. For example, if a millworker is a member of AWI, it is necessary for that millworker to follow the AWI standards as though they were law; if the millworker is not a member of AWI, the standards are not binding at all—unless they are made binding by special contract provision, in the technical specifications, or through some other means.

It is possible for standards to be adopted directly into law by a rule-making body of some kind, and it is quite common to see various NFPA documents adopted as such by state or local rule-making entities. The standards published by the ICC are intended to be adopted as codes, as are some of the documents published by NFPA (*NFPA 70 National Electrical Code*, is the most common of these) and ASHRAE (*ASHRAE 90.1 Energy Standard for Buildings Except Low-Rise Residential Buildings*, is the most common of these).

If a standard has not been adopted into law by an enacting authority, it remains merely recommended practice and it is not legally binding on anyone.

3.1.2 Regulations

Regulations are local, state, or federal laws that are targeted at specific activities (smog control, spill of outdoor lighting onto a neighbor's property, workplace safety, etc.) or specific groups of people (industries who contribute to smog, property owners who spill light onto their neighbor's property, employers with dangerous workplaces, etc.). So the federal Occupational Safety and Health Administration (OSHA) regulations (and their similar state regulations) are applicable only to some employers (most employers actually, but most of the regulations affect workplaces with unusual hazards—climbing, excavating, high heat, etc.) and not to the population as a whole.

To take another example, the federal Department of Energy has enacted limitations on the efficacy (the proper technical term for efficiency in light sources) of various "light bulbs," which directly affect only those companies who manufacture light sources. Everyone else is affected indirectly because some sources leave the marketplace due to inadequate efficacy. (Despite much coverage in the news media late in 2010 regarding the banning of the "incandescent light bulb," this process actually began in the mid-1990s and numerous different lamps have already been eliminated from the market, including some fluorescent and high-intensity discharge lamps in addition to a number of conventional incandescent lamps. This process is neither new nor radical, nor has it ever "banned" any specific product. Instead, the process has simply set minimum efficacy standards—which have been increased several times over the years—and the result has been that non-complying products simply are not manufactured any more.)

The federal regulation that most directly affects building design and construction is the "Americans with Disabilities Act Accessibility Guidelines." The Americans with Disabilities Act (ADA) itself is not a regulation at all; in fact, it is a federal civil rights law, which is the highest (and broadest) level of federal law. The Accessibility Guidelines are regulations that have been attached to the ADA to assist designers (and others) in understanding how to make facilities accessible under the ADA. Despite the fact that the ADA is not a code per se (see 3.1.3 Codes), it is applicable to *everyone* in the United States because it is a federal civil rights law.

The key issue here is that regulations are not universal, covering everyone, everywhere, all the time. But it is vital to understand when regulations are applicable, and which regulations are applicable, to avoid any number of different potential legal problems in the future.

3.1.3 Codes

Codes—building rules—are adopted only by local and state governments in the United States. There is no federal authority to regulate building design and construction on a national basis, so there are no national codes. There are federal regulations that affect the design and construction of buildings (most notably the ADA), but such regulations only come into play through other mechanisms, civil rights law in the case of the ADA, or simple contract provisions. For example, all projects for the federal government require compliance with NFPA 101.

All codes are regulations, but all regulations certainly are not codes. Codes are regulations that are specifically targeted toward building construction, so anyone who is not directly involved in the design and construction of buildings is not directly affected by codes. Of course, all of us are indirectly affected by codes as we work in, use, and live in buildings, so the primary reason for having codes is to protect public health and safety. This protection is achieved primarily by limiting the spread of fire and preventing structural collapse, but there are many nuanced aspects to both requirements.

3.2 Building code history

The first building codes were written thousands of years ago, mostly to prevent structural collapse, but modern building codes did not really appear until early in the 20th century. Building technology changed very little from thousands of years ago until the middle of the 19th century, when it became feasible to build much larger and taller buildings, mostly through the use of cast iron (and then steel). When buildings started to get much larger and taller, the risks of fire and structural collapse increased dramatically. But the management of those risks was still not well understood.

After the great fire in Chicago, Illinois in 1871, the re-building of the city was begun with essentially no changes in building regulations. But after the Coconut Grove night club fire disaster in Boston, Massachusetts in 1942 (in which 492 people died), rule improvements were greatly accelerated.

Who can write such rules? Who has the knowledge and experience to develop, write, and publish detailed rules for building construction that can cover all potential situations? Very few governmental units have such expertise. So three "code writing" entities came into being: the Building Officials and Code Administrators International, Inc. (BOCA), the International Conference of Building Officials (ICBO), and the Southern Building Code Congress International (SBCCI). Over a period of several decades, each of these organizations developed and published a complete Building Code and several other more specialized documents—mechanical code, plumbing code, etc.—with the intention of having state and local governments adopt one or more of the documents as state or local codes. (Usually, adopting entities issue "amendments" with adopted model codes, which alter—delete, modify, or add—the model code text according to state or local requirements.) But by the 1990s, it had become clear that none of these three competing organizations would ever really dominate code publishing, so they decided to merge. The merger happened in 1997 and the new entity is called the International Code Council (ICC), which publishes the most extensive and authoritative series of model codes that is available. The only dominant model code that is not published by the ICC is *NFPA 70 National Electrical Code*; the latter is the most dominant of all model code books throughout the US.

Today, every jurisdiction (state or locality) has adopted building rules, usually based on ICC model codes. (The city of Chicago, Illinois has its own unique set of rules that are not based on a model code.)

In parallel with the development of the codes themselves, there has also been a process of defining credentials for licensed design professionals—usually architects and engineers. In many jurisdictions, this process actually occurred before code development. The thought process has been that some design professionals (those in charge of projects, mostly) should be licensed via a strict process so as to protect the health and welfare of the public. Each state initially developed its own approach to this, but national organizations have emerged (both for licensed architects and engineers) to make the process more uniform across the country. For architects, this consists (usually) of minimum educational requirements (usually a "professional" degree—a Master of Architecture (usually abbreviated M.Arch.) mostly but sometimes a Bachelor of Architecture (usually abbreviated B.Arch.)); minimum working requirements (usually three years under the direct supervision of a licensed architect), and passing a very long and broad-ranging examination. For engineers, this consists (usually) of minimum educational requirements (highly variable from state to state and time to time, but most commonly requiring a four-year on-campus ABET-accredited degree in engineering); minimum working requirements (also highly variable but most commonly four years), and passing a short but challenging examination in a very specific discipline. (The PE exam is actually in two parts: general and specific. Everyone takes the same general examination—called the Fundamental of Engineering (FE) exam, but then each candidate selects a specific examination, based on discipline.) An engineer who has passed the FE and who is working toward the PE exam is usually given the title of Engineer in Training (EIT). In jurisdictions that have Registered (and/or Licensed) interior designers, the process is similar: minimum educational and work requirements and passing an examination (usually the NCIDQ exam, but sometimes the Architect's Registration Exam).

The fact that licensing laws usually say that only licensed design professionals have the education, knowledge, and experience to interpret and apply codes does tend to compete with the arrangement of the enforcement community (which is not generally based on licensed professionals). This came about at least in part because licensed design professionals were not necessarily all that interested in learning about enforcing code requirements themselves; the resulting vacuum (rules with no enforcement) has been filled by expanding enforcement bureaucracies, which should come as no surprise to anyone. There is no clear way to resolve this tension, so all practitioners simply must make do.

3.3 Code adoption

In any given jurisdiction, the government (state or local) adopts building rules—usually building, fire, mechanical, plumbing, fuel gas, electrical, and energy codes.

Every state falls into one of three categories for organization of building rules:

1. Centralized (where the state has complete control over the entire process): In a centralized state, the state sets all rules and enforcement procedures, although enforcement is usually done at the local level. This means that localities (counties, parishes, townships, cities, towns, villages, etc.) have no say in rule adoption or interpretation, beyond basic enforcement.
2. Local (where the state has no control over the entire process): In a local state, localities (counties, parishes, townships, cities, towns, villages, etc.) set all rules and enforcement procedures.
3. Mixed (where the state has some control over the entire process): In a mixed state, there are both state and local rules and enforcement practices, sometimes in the same location for the same project.

The adoption process varies considerably from one location to another, but the basic process goes something like this most of the time:

1 Identify the need for new regulations.
2 Write proposed regulations (probably simply amendments for a pre-selected model code).
3 Conduct public hearings about the proposed regulations.
4 Modify the proposed regulations based on public commentary (or not).
5 Adopt the proposed regulations via whatever process is required in the jurisdiction.
6 Set up regulatory administration (usually required only for new regulations; for updates to existing regulations, such procedures are usually already in place).
7 Begin enforcement of new regulations.

In most cases, this process will take a considerable amount of time, and it is quite common to see these procedures taking well over a year from step 1 to step 7.

3.4 Code enforcement

Code enforcement can be a sensitive topic. The number and extent of the rules (thousands of pages of sometimes highly technical and difficult-to-read material) mean that virtually no one knows all of them. In fact, few individuals know the details of more than one of the various books. On the other hand, a number of individuals (including enforcement officials) believe themselves to be "experts" in the requirements of one specific code (most commonly the electrical code, followed by the plumbing code); sometimes this is true and sometimes it is not true.

The generalized enforcement process is usually done in two steps:

1 Pre-construction plan review.
2 On-site inspection during construction.

During plan review, state or local officials, or sometimes both, review submitted construction documents for code compliance. If non-compliant (or claimed non-compliant) items are found, some kind of correction document is usually issued (sometimes a simple e-mail these days), and some form of written response is required. Such responses might also require changes to the construction documents, which might be required to be re-submitted. When the plan reviewer decides that the project is acceptable, some kind of approval document is issued.

This process is highly imperfect in most cases. Many, if not most, plan reviewers are not highly qualified design professionals (RAs, PEs, or RIDs) and they usually have limited time for their reviews. This means that reviews tend to be limited to a "check-list" of "key items," which does mean that many other issues can easily be overlooked. It is also very common to see erroneous comments from plan reviewers, due to inaccurate reading of the construction documents and/or errors in code interpretation.

In general, code enforcement is good, but inaccurate code enforcement is bad. But many officials really do not understand why the codes are the way that they are, and misinterpretations are quite common.

In many cases, on-site inspection by local officials pre-dated state or local plan review, which is part of the reason that local officials often attempt to exceed their authority (often by trying to enforce rules that do not exist, misinterpreting rules that do exist, or ignoring rules that exist). On-site inspection is almost always a local function, even in a centralized state, because the state simply does not have the resources to handle inspection at dozens, hundreds, or even thousands of job-sites simultaneously. Some local authorities have very limited resources in this area too, so very detailed inspection is not necessarily the norm.

Some localities issue a "certificate of occupancy" or a similar document at the completion of construction, and the owner is not allowed to use the facility until such permission is granted. Other localities do not do this at all. In areas where such documents are issued, it is not uncommon to see officials asking for work to be done that is not required and then refusing to issue the certificate if that work is not completed. These situations are unfortunate and compliance is often the only viable alternative (at the owner's cost, of course).

Unfortunately, multiple officials, most commonly multiple local officials such as the Fire Marshall and the Building Inspector, might even disagree about code provisions, which can make compliance all the more challenging. When such situations occur, it is best to look for a higher authority to encourage the two local authorities to agree. But that might not be possible in a local or mixed state.

That an owner might have to pay for additional work for code compliance (whether at the last minute or not) is the reason that officials should be resisted when they ask for work that is not required. Owners hire design professionals, at least in part, because such professionals are required for code compliance reasons (only RAs, PEs and some RIDs can certify that projects are code-compliant in most jurisdictions), and owners often ask design professionals to pay for items that are added by code officials. After all, in the owner's mind, the designer told them that the project is code-compliant. If an official adds one exit sign that is worth $100.00, that is one thing (although there is no reason that a designer should pay for it if it is not required), but if the official asks for something that costs $10,000.00, that is entirely different. Few designers can afford to pay for such things, especially if they are not required. If they are required and the designer simply made a mistake, it might be necessary for the designer to pay for the item, or some portion of the item, no matter what it costs.

3.5 Code appeals

When an official verbally "cites" a project for non-compliance, the designer should ask for documentation. This means that the official should provide a detailed written citation that explains which section of code applies and why the official sees a violation. (In the author's experience, merely asking for documentation for a verbal request will frequently cause the official to drop the request; unfortunately, it is very common for officials to ask for items that are not required, especially if they are doing so on-site and verbally.) The official is required to provide this documentation, so they should not resist providing it. If the official does resist providing a citation, one should be suspicious about the validity of the claim.

If the official provides the written citation (which they usually do), the designer should read the citation and verify the accuracy and applicability of the code language in question. If the official is correct, the designer simply must make whatever changes are necessary, even if that results in an expense for the designer. If the official is incorrect, then it is necessary to verify whatever appeals process exists in that particular jurisdiction.

Appeals of code official opinions are not necessarily easy to do. Sometimes, there is a higher authority, and sometimes there is not. If there is, it is worth getting an interpretation from the higher authority. If the higher authority agrees that the local official is incorrect, then the local official should (the key word being "should") accept the ruling and the project should proceed. If the local official refuses to accept a higher authority's opinion (which is not as rare as it probably should be), then the only options are (1) an official appeal or (2) do the unnecessary work. If the higher authority agrees with the lower authority, the designer simply must make whatever changes are necessary, even if that results in an expense for the designer.

In most jurisdictions, there is a means of legal appeal, probably an administrative law court (or similar entity). But such procedures tend to be long (think in terms of several months or longer) and few owners can tolerate such long delays. If an official simply digs in and refuses to cooperate, doing the unnecessary work is often the only viable answer.

In most jurisdictions, there is some mechanism for code "modifications," which is a process that is used to gain approval for something that is not strictly code-compliant. Modifications often require offsetting work (doing something that is *not* required to offset not doing something that *is* required), and it can be very difficult to know when a modification is possible and when it is not, which is where code consultants often come in.

3.6 Code consultants

Code consultants perform two primary functions in most jurisdictions:

1 Pre-design code analysis (which sometimes continues on into design).
2 Modifications.

The codes have become so extensive and complicated over the past several decades that many practicing design professionals simply do not know all of the rules any more. So they tend to rely on outside experts (independent code consultants) to determine what the code requirements are in a given situation. So code consultants often do code research and write summary reports, especially for very large or unusual projects. (This is especially common in historic rehabilitation/restoration work, for obvious reasons.) Even in these cases, these reports are usually limited to the Building Code, and maybe portions of the Fire Code. (Consulting engineers usually do their own code research, with or without code consultants, for structural, mechanical, electrical, fire protection (fire alarm and/or sprinklers), plumbing, and fuel gas.) Energy codes affect nearly the whole design team, so they might, or might not, be included in the code consultant's report.

When modifications are necessary (or sometimes desirable), code consultants often have the best knowledge about the likelihood of success for a given request. Code consultants are usually familiar with previously granted modifications (in their main jurisdiction of practice) and they are usually the most familiar with enforcement staff and approval processes. So code consultants are usually involved in modification requests.

In many jurisdictions, including Indiana, there are no formal requirements for credentials to be a "code consultant" and many of those who practice are former state or local plan reviewers. Typically, plan review positions require only high school graduation and a short class (perhaps a day or two); nevertheless, former plan reviewers often have a good understanding of whether or not a modification might be accepted so they can be useful code consultants. However, if the issues at hand are highly technical or simply unusually complex, that might not be the case. This is controversial in some jurisdictions and proposals have been made to limit the arguing of modifications to licensed design professionals only. That was once attempted in Indiana but the requirement was not adopted.

3.7 Code documents

The most current ICC model code documents are:

- *2015 International Building Code* (this document, or a previous edition—2000, 2003, 2006, 2009, or 2012—is used in most jurisdictions in the US.)

- *2015 International Fire Code*
- *2015 International Energy Conservation Code*
- *2015 International Electrical Code*
- *2015 International Fuel Gas Code*
- *2015 International Mechanical Code*
- *2015 International Plumbing Code*
- *2015 International Residential Code* (used primarily for one- and two-family dwellings)
- *2015 International Existing Buildings Code*
- *2015 International Performance Code*
- *2015 International Private Sewage Code*
- *2015 International Property Maintenance Code*
- *2015 International Wildland-Urban Interface Code*
- *2015 International Zoning Code*

(Remember: none of these documents holds the force of law unless it has been adopted by a governmental entity. They become true codes only when they are adopted; they are not codes just because the ICC says so.)

Many of these documents have older editions that are still available, possibly including 2000, 2003, 2006, 2009, and/or 2012 so even if a jurisdiction is using the "IBC," it is necessary to know which version of the IBC is in use. This is true for the other codes as well.

Commonly used NFPA documents include:

- *NFPA 13-2016 Installation of Sprinkler Systems* (many older editions exist too)
- *NFPA 13R-2016 Standard for the Installation of Sprinkler Systems in Residential Occupancies Up to and Including Four Stories in Height* (many older editions here too)
- *NFPA 13D-2016 Standard for the Installation of Sprinkler Systems in One- and Two-Family Dwellings and Manufactured Homes*
- *NFPA 70-2017 National Electrical Code*
- *NFPA 72-2016 National Fire Alarm Code* (referenced by the IBC, so this is usually indirectly adopted, if not directly adopted)
- *NFPA 99-2015 Healthcare Facilities* (the electrical portions of this document are included in NFPA 70-2010 and this is also often adopted, indirectly or directly)
- *NFPA 101-2015 Life Safety Code* (this document is virtually never adopted as code per se, but it is applicable to many projects due to its inclusion within any number of different federal and state regulations or by owner requirement)
- *NFPA 5000 Building Construction and Safety Code* (this is the NFPA's version of the International Building Code (IBC), but it has largely failed to catch on and it is rarely used)

One ANSI document is very widely used:

- *ICC/ANSI A117.1-2003 Accessible and Useable Buildings and Facilities*

It should be noted that this document is co-published by ANSI and the ICC, so it should come as no surprise that it is adopted by reference within the IBC (since 2006).

Other accessibility regulations include:

- 2010 Americans with Disabilities Act Accessibility Guidelines
- 2006 ABA (Architectural Barriers Act) Accessibility Standards for Federal Facilities

For federally funded housing projects:

- 1991 FHA (Fair Housing Act) Guidelines
- 2001 CRHA (Code Requirements for Housing Accessibility)

For licensed (most jurisdictions have some form of licensing for hospitals and various other healthcare facilities) healthcare facilities, the most commonly used regulations are the *Guidelines for Design and Construction of Hospital and Health Care Facilities*, published jointly by The American Institute of Architects Academy of Architecture for Health, The Facilities Guidelines Institute, and the US Department of Health and Human Services (2010 is the most current edition, but older editions are still in use in various jurisdictions).

In addition to codes, in most jurisdictions, there are specialty regulations for various special building types: healthcare, food service (restaurants and commercial and institutional kitchens), schools, and day care facilities. These must be verified one by one in each jurisdiction.

3.8 Code research

Code research is usually a 2-step process:

1. Determine what are the applicable codes in the jurisdiction where the project will be located.
2. Determine applicable requirements of the applicable codes to the project.

For Step 1, it is necessary to contact officials to verify what the applicable documents might be. These days, it is often feasible to do this on the internet, by visiting the web-sites for the state and/or local jurisdictions. But if that will not work for some reason, the next avenue is to call the local (not state) officials and simply to ask them what the applicable codes are in their jurisdiction. If they are using state codes, they should simply say so. If they are using local codes, they should say exactly what they are, including amendments. Whatever you do, do not forget to find and cross-check amendments.

For Step 2, it is necessary to read the applicable documents to find provisions that are applicable to the project at hand. This will usually not require reading the entire Building Code (the 2015 IBC is 700 pages), but it could require reading a large portion of it. When reading codes, it is vital to follow the "threads," or strings of references, as they occur. It is very common for code provisions to refer to other provisions (oftentimes in other chapters) and it is very important to follow all of the references until they end. It is also critical to read all exceptions and footnotes to tables to make sure that important items are not missed.

While proceeding with code research, it is advisable to take notes, so as to build a report as one goes. Trying to re-construct a report after reading through dozens of pages is too difficult and is likely to lead to error.

There is no need to remember code language and to cite it from memory. These are legal documents and it is important to understand that misquoting code is worse than not quoting it at all. (Among other things, this is a good reason why officials should not quote code language either.) When it is necessary to quote code language, it must be looked up, verified, and checked carefully to make sure that it is completely accurate—down to the punctuation.

3.9 Key concepts

Of the 35 chapters and 12 appendices of the 2015 IBC, several chapters are of key importance in the day-to-day practice of interior design. These are the chapters that cover:

- Use and Occupancy Classification (Chapter 3);
- Special Detailed Requirements Based on Use and Occupancy (Chapter 4);
- Types of Construction (Chapter 6);
- Fire and Smoke Protection Features (Chapter 7);
- Interior Finishes (Chapter 8);
- Means of Egress (Chapter 10); and
- Accessibility (Chapter 11).

Chapter 5 "General Building Heights and Areas" and Chapter 9 "Fire Protection Systems" are of lesser importance.

3.9.1 Use and occupancy

Everything in the code world revolves around use and occupancy. In other words, what we use a facility for determines how we need to build it, protect it from fire, and set up exiting. There are a number of occupancy groups in Chapter 3, as follows:

A Assembly

A-1 "Assembly uses, usually with fixed seating, intended for the production and viewing of the performing arts or motion pictures ..."

A-2 "Assembly uses intended for food and/or drink consumption ..."

A-3 "Assembly uses intended for worship, recreation or amusement and other assembly uses not classified elsewhere in Group A ..."

A-4 "Assembly uses intended for viewing of indoor sporting events and activities with spectator seating ..."

A-5 "Assembly uses intended for participation in or viewing outdoor activities ..."[1]

B Business[2]

E Educational[3] (K-12 only)

F Factory

F-1 Moderate-Hazard

F-2 Low-Hazard[4]

H Hazardous[5]

I Institutional

I-1 "Institutional Group I-1 occupancy shall include buildings, structures or parts thereof housing more than 16 persons, excluding staff, who reside on a 24-hour basis, in a supervised residential environment and receive custodial care ..."

I-2 "Institutional Group I-2 occupancy shall include buildings and structures used for medical care on a 24-hour basis for more than five persons who are incapable of self-preservation ..."

I-3 "Institutional Group I-3 occupancy shall include buildings and structures that are inhabited by more than five persons who are under restraint or security. A group I-3 facility is occupied by persons who are generally incapable of self-preservation due to security measures not under the occupants' control ..."

I-4 "Institutional Group I-4 occupancy shall include buildings and structures occupied by more than 5 persons of any age who receive custodial care for fewer than 24 hours per day by persons other than parents or guardians, relatives by blood, marriage or adoption, and in a place other than the home of the person cared for ..." "308.61 Classification as Group E. A child day care facility that provides care for more than five but not more than 100 children $2\frac{1}{2}$ years or less of age, where the rooms in which the children are cared for are located on a level of exit discharge serving such rooms and each of these child care rooms has an exit door directly to the exterior, shall be classified as Group E."[6]

M Mercantile[7]

R Residential

R-1 "Residential Group R-1 occupancies containing sleeping units where the occupants are primarily transient in nature ..."

R-2 "Residential Group R-2 occupancies containing sleeping units or more than two dwelling units where the occupants are primarily permanent in nature ..."

R-3 "Residential Group R-3 occupancies where the occupants are primarily permanent in nature and not classified as R-1, R-2, R-4, or I ..."

R-4 "Residential Group R-4 occupancies shall include buildings, structures, or portions thereof for more than five but not more than 16 occupants, excluding staff, who reside on a 24-hour basis in a supervised residential environment and receive custodial care."[8]

S Storage

S-1 Moderate-hazard

S-2 Low-hazard[9]

U Utility and Miscellaneous[10]

Within these groups and sub-groups, there are more fine distinctions and exceptions, and a full explanation is far beyond the scope of this chapter. For now, suffice to say that higher hazards (more people, sleeping people, incarcerated people, dangerous industrial activities, storage of dangerous products, etc.) result in higher requirements, which only makes sense.

3.9.1.1 Mixed/multiple occupancies

In reality, while many facilities fall into the single-use category (churches, K-12 schools, some office buildings, some mercantile buildings, etc.), many buildings contain more than one occupancy group. One of the most common arrangements is to have assembly occupancies in the same building with other uses, such as:

- A-1 and B—a theater in a university performing arts classroom building
- A-2 and R-1—a restaurant in a hotel
- A-3 and B—a conference center in an office

But it is also possible to have retail and offices together (especially in multi-story buildings); parking garages and residential, or retail, or offices, or all three; or many other combinations. In mixed use facilities, each use is treated separately, and fire-separation is required under some conditions, as determined by Table 3.1.

3.9.2 Special detailed requirements based on use and occupancy

In Chapter 4, many different special requirements are spelled out for various different special occupancies:

- Covered mall buildings (risk: large numbers of people)
- High-rise buildings (risk: large numbers of people far above the ground)
- Atriums (risk: rapid spread of fire, especially vertically)
- Underground buildings (risk: difficult exiting)
- Motor-vehicle related occupancies (risk: fuel)
- Group I-2 (risk: incapacitated or incarcerated persons)
- Group I-3 (risk: incapacitated or incarcerated persons)
- Motion picture projection rooms (risk: flammable film)
- Stages, platforms, and technical production areas (risk: spread of fire)
- Special amusement buildings (risk: large numbers of people)
- Aircraft related occupancies (risk: fuel)
- Combustible storage (risk: spread of fire)
- Hazardous materials (risk: spread of fire)
- Application of flammable finishes (risk: spread of fire)
- Dying rooms (risk: spread of fire)

Table 3.1

Required separation of occupancies (hours) (2015 IBC, Table 508.4)

OCCUPANCY	A,E		I-1[a], I-3, I-4		I-2		R[a]		F-2, S-2[b], U		B[e], F-1, M, S-1		H-1		H-2		H-3, H-4		H-5	
	S	**NS**	**S**	**NS**	**S**	**NS**	**S**	**NS**	**S**	**NS**	**S**	**NS**	**S**	**NS**	**S**	**NS**	**S**	**NS**	**S**	**NS**
A, E	N	N	1	2	2	NP	1	2	N	1	1	2	NP	NP	3	4	2	3	2	NP
I-1[a], I-3, I-4	—	—	N	N	2	NP	1	NP	1	2	1	2	NP	NP	3	NP	2	NP	2	NP
I-2	—	—	—	—	N	N	2	NP	2	NP	2	NP	NP	NP	3	NP	2	NP	2	NP
R[a]	—	—	—	—	—	—	N	N	1[c]	2[c]	1	2	NP	NP	3	NP	2	NP	2	NP
F-2, S-2[b], U	—	—	—	—	—	—	—	—	N	N	1	2	NP	NP	3	4	2	3	2	NP
B[e], F-1, M, S-1	—	—	—	—	—	—	—	—	—	—	N	N	NP	NP	2	3	1	2	1	NP
H-1	—	—	—	—	—	—	—	—	—	—	—	—	N	NP	NP	NP	NP	NP	NP	NP
H-2	—	—	—	—	—	—	—	—	—	—	—	—	—	—	N	NP	1	NP	1	NP
H-3, H-4	—	—	—	—	—	—	—	—	—	—	—	—	—	—	—	—	1[d]	NP	1	NP
H-5	—	—	—	—	—	—	—	—	—	—	—	—	—	—	—	—	—	—	N	NP

S = Buildings equipped throughout with an automatic sprinkler system installed in accordance with Section 903.3.1.1.

NS = Buildings not equipped throughout with an automatic sprinkler system installed in accordance with Section 903.3.1.1.

N = No separation requirement.

NP = Not permitted.

a. See Section 420.

b. The required separation from areas used only for private or pleasure vehicles shall be reduced by 1 hour but not to less than 1 hour.

c. See Section 406.3.4.

d. Separation is not required between occupancies of the same classification.

e. See Section 422.2 for ambulatory care facilities.

- Organic coatings (risk: spread of fire)
- Live/work units (risk: combined business and residential)
- Groups I-2, R-2, R-3, and R-4 (risk: sleeping persons)
- Hydrogen fuel gas rooms (risk: spread of fire/explosion)
- Ambulatory healthcare facilities (risk: limited self-preservation in some cases)
- Storm shelters
- Children's play structures
- Hyperbaric facilities
- Combustible dusts/grain processing and storage

The responses to these special risks vary from category to category. Sometimes, more fire alarm protection is required; sometimes, more fire protection (fire-rated construction and/or fire sprinklers) is required; or more exits, or smaller buildings, or larger exits, etc. If any project includes any of these special categories, it is very important to include Chapter 4 in the code research.

3.9.3 Types of construction

In Chapter 6 of the 2015 IBC, five basic construction types are identified:

I Non-combustible, fire-protected, sub-classes I-A and I-B
II Non-combustible, fire-protected sub-class II-A and non-protected sub-class II-B
III Combination non-combustible/combustible, partial fire-protection sub-class III-A, less partial fire-protection sub-class III-B
IV Combustible heavy timber
V Combustible, fire-protected sub-class V-A, non-protected sub-class V-B.

Without question, the costliest classification is I-A, which is used for large high-rise buildings and some very large special buildings (with large numbers of occupants). Type V-B is without question the least-cost classification; after all, it is unprotected light wood framing (just like most houses). All of the other classifications fall somewhere in between on the cost continuum. The height and number of stories that are allowed for a particular building are determined by Tables 3.2 and 3.3.

The allowable area of a building is determined by Table 3.4.

Table 3.4 provides basic areas only, and there are various techniques that are used to increase these areas. A full explanation is beyond the scope of this chapter, but the short version is that it is generally pretty simple to double or even triple basic areas. For some occupancy groups, area can be unlimited if type I-A or type I-B construction is used.

3.9.4 Fire-resistance-rated construction

Fire-resistance-rated construction is required to various degrees for various reasons. The degree of fire-resistance protection is measured in hours of protection, ranging from

Table 3.2
Allowable building height in feet above grade plane (2015 IBC, Table 504.3)

OCCUPANCY CLASSIFICATION	SEE FOOTNOTES	TYPE OF CONSTRUCTION								
		TYPE I		TYPE II		TYPE III		TYPE IV	TYPE V	
		A	B	A	B	A	B	HT	A	B
A, B, E, F, M, S, U	**NS**[b]	UL	160	65	55	65	55	65	50	40
	S	UL	180	85	75	85	75	85	70	60
H-1, H-2, H-3, H-5	NS[c, d]	UL	160	65	55	65	55	65	50	40
	S									
H-4	NS[c, d]	UL	160	65	55	65	55	65	50	40
	S	UL	180	85	75	85	75	85	70	60
I-1 Condition 1, I-3	NS[d, e]	UL	160	65	55	65	55	65	50	40
	S	UL	180	85	75	85	75	85	70	60
I-1 Condition 2, I-2	NS[d, f, e]	UL	160	65	55	65	55	65	50	40
	S	UL	180	85						
I-4	NS[d, g]	UL	160	65	55	65	55	65	50	40
	S	UL	180	85	75	85	75	85	70	60
R	NS[d, h]	UL	160	65	55	65	55	65	50	40
	S13R	60	60	60	60	60	60	60	60	60
	S	UL	180	85	75	85	75	85	70	60

For SI: 1 foot = 304.8 mm.

Note: UL = Unlimited; NS = Buildings not equipped throughout with an automatic sprinkler system; S = Buildings equipped throughout with an automatic sprinkler system installed in accordance with Section 903.3.1.1; S13R = Buildings equipped throughout with an automatic sprinkler system installed in accordance with Section 903.3.1.2.

a. See Chapters 4 and 5 for specific exceptions to the allowable height in this chapter.

b. See Section 903.2 for the minimum thresholds for protection by an automatic sprinkler system for specific occupancies.

c. New Group H occupancies are required to be protected by an automatic sprinkler system in accordance with Section 903.2.5.

d. The NS value is only for use in evaluation of existing building height in accordance with the *International Existing Building Code.*

e. New Group I-1 and I-3 occupancies are required to be protected by an automatic sprinkler system in accordance with Section 903.2.6. For new Group I-1 occupancies Condition 1, see Exception 1 of Section 903.2.6.

f. New and existing Group I-2 occupancies are required to be protected by an automatic sprinkler system in accordance with Section 903.2.6 and Section 1103.5 of the *International Fire Code.*

g. For new Group I-4 occupancies, see Exceptions 2 and 3 of Section 903.2.6.

h. New Group R occupancies are required to be protected by an automatic sprinkler system in accordance with Section 903.2.8.

Table 3.3

Allowable number of stories above grade plane (2015 IBC, Table 504.4)

OCCUPANCY CLASSIFICATION	SEE FOOTNOTES	TYPE OF CONSTRUCTION								
		TYPE I		TYPE II		TYPE III		TYPE IV	TYPE V	
		A	B	A	B	A	B	HT	A	B
A-1	NS	UL	5	3	2	3	2	3	2	1
	S	UL	6	4	3	4	3	4	3	2
A-2	NS	UL	11	3	2	3	2	3	2	1
	S	UL	12	4	3	4	3	4	3	2
A-3	NS	UL	11	3	2	3	2	3	2	1
	S	UL	12	4	3	4	3	4	3	2
A-4	NS	UL	11	3	2	3	2	3	2	1
	S	UL	12	4	3	4	3	4	3	2
A-5	NS	UL	UL	UL	UL	UL	UL	UL	UL	UL
	S	UL	UL	UL	UL	UL	UL	UL	UL	UL
B	NS	UL	11	5	3	5	3	5	3	2
	S	UL	12	6	4	6	4	6	4	3
E	NS	UL	5	3	2	3	2	3	1	1
	S	UL	6	4	3	4	3	4	2	2
F-1	NS	UL	11	4	2	3	2	4	2	1
	S	UL	12	5	3	4	3	5	3	2
F-2	NS	UL	11	5	3	4	3	5	3	2
	S	UL	12	6	4	5	4	6	4	3
H-1	NS[c, d]	1	1	1	1	1	1	1	1	NP
	S									
H-2	NS[c, d]	UL	3	2	1	2	1	2	1	1
	S									
H-3	NS[c, d]	UL	6	4	2	4	2	4	2	1
	S									
H-4	NS[c, d]	UL	7	5	3	5	3	5	3	2
	S	UL	8	6	4	6	4	6	4	3
H-5	NS[c, d]	4	4	3	3	3	3	3	3	2
	S									
I-1 Condition 1	NS[d, e]	UL	9	4	3	4	3	4	3	2
	S	UL	10	5	4	5	4	5	4	3
I-1 Condition 2	NS[d, e]	UL	9	4	3	4	3	4	3	2
	S	UL	10	5						
I-2	NS[d, f]	UL	4	2	1	1	NP	1	1	NP
	S	UL	5	3						
I-3	NS[d, e]	UL	4	2	1	2	1	2	2	1
	S	UL	5	3	2	3	2	3	3	2

(continued)

Table 3.3

Allowable number of stories above grade plane (2015 IBC, Table 504.4) ***(continued)***

OCCUPANCY CLASSIFICATION	TYPE OF CONSTRUCTION									
	SEE FOOTNOTES	TYPE I		TYPE II		TYPE III		TYPE IV	TYPE V	
		A	B	A	B	A	B	HT	A	B
I-4	NS[d, g]	UL	5	3	2	3	2	3	1	1
	S	UL	6	4	3	4	3	4	2	2
M	NS	UL	11	4	2	4	2	4	3	1
	S	UL	12	5	3	5	3	5	4	2
R-1	NS[d, h]	UL	11	4	4	4	4	4	3	2
	S13R	4	4						4	3
	S	UL	12	5	5	5	5	5	4	3
R-2	NS[d, h]	UL	11	4	4	4	4	4	3	2
	S13R	4	4						4	3
	S	UL	12	5	5	5	5	5	4	3
R-3	NS[d, h]	UL	11	4	4	4	4	4	3	3
	S13R	4	4						4	4
	S	UL	12	5	5	5	5	5	4	4
R-4	NS[d, h]	UL	11	4	4	4	4	4	3	2
	S13R	4	4						4	3
	S	UL	12	5	5	5	5	5	4	3
S-1	NS	UL	11	4	2	3	2	4	3	1
	S	UL	12	5	3	4	3	5	4	2
S-2	NS	UL	11	5	3	4	3	4	4	2
	S	UL	12	6	4	5	4	5	5	3
U	NS	UL	5	4	2	3	2	4	2	1
	S	UL	6	5	3	4	3	5	3	2

Note: UL = Unlimited; NP = Not Permitted; NS = Buildings not equipped throughout with an automatic sprinkler system; S = Buildings equipped throughout with an automatic sprinkler system installed in accordance with Section 903.3.1.1; S13R = Buildings equipped throughout with an automatic sprinkler system installed in accordance with Section 903.3.1.2.

a. See Chapters 4 and 5 for specific exceptions to the allowable height in this chapter.
b. See Section 903.2 for the minimum thresholds for protection by an automatic sprinkler system for specific occupancies.
c. New Group H occupancies are required to be protected by an automatic sprinkler system in accordance with Section 903.2.5.
d. The NS value is only for use in evaluation of existing building height in accordance with the *International Existing Building Code.*
e. New Group I-1 and I-3 occupancies are required to be protected by an automatic sprinkler system in accordance with Section 903.2.6. For new Group I-1 occupancies, Condition 1, see Exception 1 of Section 903.2.6.
f. New and existing Group I-2 occupancies are required to be protected by an automatic sprinkler system in accordance with Section 903.2.6 and Section 1103.5 of the *International Fire Code.*
g. For new Group I-4 occupancies, see Exceptions 2 and 3 of Section 903.2.6.
h. New Group R occupancies are required to be protected by an automatic sprinkler system in accordance with Section 903.2.8.

Table 3.4
Allowable area factor (2015 IBC, Table 506.2)

OCCUPANCY CLASSIFICATION	TYPE OF CONSTRUCTION									
	SEE FOOTNOTES	TYPE I		TYPE II		TYPE III		TYPE IV	TYPE V	
		A	B	A	B	A	B	HT	A	B
A-1	NS	UL	UL	15,500	8,500	14,000	8,500	15,000	11,500	5,500
	S1	UL	UL	62,000	34,000	56,000	34,000	60,000	46,000	22,000
	SM	UL	UL	46,500	25,500	42,000	25,500	45,000	34,500	16,500
A-2	NS	UL	UL	15,500	9,500	14,000	9,500	15,000	11,500	6,000
	S1	UL	UL	62,000	38,000	56,000	38,000	60,000	46,000	24,000
	SM	UL	UL	46,500	28,500	42,000	28,500	45,000	34,500	18,000
A-3	NS	UL	UL	15,500	9,500	14,000	9,500	15,000	11,500	6,000
	S1	UL	UL	62,000	38,000	56,000	38,000	60,000	46,000	24,000
	SM	UL	UL	46,500	28,500	42,000	28,500	45,000	34,500	18,000
A-4	NS	UL	UL	15,500	9,500	14,000	9,500	15,000	11,500	6,000
	S1	UL	UL	62,000	38,000	56,000	38,000	60,000	46,000	24,000
	SM	UL	UL	46,500	28,500	42,000	28,500	45,000	34,500	18,000
A-5	NS	UL	UL	UL	UL	UL	UL	UL	UL	UL
	S1									
	SM									
B	NS	UL	UL	37,500	23,000	28,500	19,000	36,000	18,000	9,000
	S1	UL	UL	150,000	92,000	114,000	76,000	144,000	72,000	36,000
	SM	UL	UL	112,500	69,000	85,500	57,000	108,000	54,000	27,000
E	NS	UL	UL	26,500	14,500	23,500	14,500	25,500	18,500	9,500
	S1	UL	UL	106,000	58,000	94,000	58,000	102,000	74,000	38,000
	SM	UL	UL	79,500	43,500	70,500	43,500	76,500	55,500	28,500

(continued)

Table 3.4
Allowable area factor (2015 IBC, Table 506.2) ***(continued)***

OCCUPANCY CLASSIFICATION	TYPE OF CONSTRUCTION									
	SEE FOOTNOTES	TYPE I		TYPE II		TYPE III		TYPE IV	TYPE V	
		A	B	A	B	A	B	HT	A	B
F-1	NS	UL	UL	25,000	15,500	19,000	12,000	33,500	14,000	8,500
	S1	UL	UL	100,000	62,000	76,000	48,000	134,000	56,000	34,000
	SM	UL	UL	75,000	46,500	57,000	36,000	100,500	42,000	25,500
F-2	NS	UL	UL	37,500	23,000	28,500	18,000	50,500	21,000	13,000
	S1	UL	UL	150,000	92,000	114,000	72,000	202,000	84,000	52,000
	SM	UL	UL	112,500	69,000	85,500	54,000	151,500	63,000	39,000
H-1	NSc S1	21,000	16,500	11,000	7,000	9,500	7,000	10,500	7,500	NP
H-2	NSc S1 SM	21,000	16,500	11,000	7,000	9,500	7,000	10,500	7,500	3,000
H-3	NSc S1 SM	UL	60,000	26,500	14,000	17,500	13,000	25,500	10,000	5,000
H-4	NSc,d	UL	UL	37,500	17,500	28,500	17,500	36,000	18,000	6,500
	S1	UL	UL	150,000	70,000	114,000	70,000	144,000	72,000	26,000
	SM	UL	UL	112,500	52,500	85,500	52,500	108,000	54,000	19,500
H-5	NSc,d	UL	UL	37,500	23,000	28,500	19,000	36,000	18,000	9,000
	S1	UL	UL	150,000	92,000	114,000	76,000	144,000	72,000	36,000
	SM	UL	UL	112,500	69,000	85,500	57,000	108,000	54,000	27,000

I-1	NSd,e	UL	55,000	19,000	10,000	16,500	10,000	18,000	10,500	4,500
	S1	UL	220,000	76,000	40,000	66,000	40,000	72,000	42,000	18,000
	SM	UL	165,000	57,000	30,000	49,500	30,000	54,000	31,500	13,500
I-2	NSd,f	UL	UL	15,000	11,000	12,000	NP	12,000	9,500	NP
	S1	UL	UL	60,000	44,000	48,000	NP	48,000	38,000	NP
	SM	UL	UL	45,000	33,000	36,000	NP	36,000	28,500	NP
I-3	NSd,e	UL	UL	15,000	10,000	10,500	7,500	12,000	7,500	5,000
	S1	UL	UL	45,000	40,000	42,000	30,000	48,000	30,000	20,000
	SM	UL	UL	45,000	30,000	31,500	22,500	36,000	22,500	15,000
I-4	NS[d, g]	UL	60,500	26,500	13,000	23,500	13,000	25,500	18,500	9,000
	S1	UL	121,000	106,000	52,000	94,000	52,000	102,000	74,000	36,000
	SM	UL	181,000	79,500	39,000	70,500	39,000	76,500	55,500	27,000
M	NS	UL	UL	21,500	12,500	18,500	12,500	20,500	14,000	9,000
	S1	UL	UL	86,000	50,000	74,000	50,000	82,000	56,000	36,000
	SM	UL	UL	64,500	37,500	55,500	37,500	61,500	42,000	27,000
R-1	NS[d, h]	UL	UL	24,000	16,000	24,000	16,000	20,500	12,000	7,000
	S13R									
	S1	UL	UL	96,000	64,000	96,000	64,000	82,000	48,000	28,000
	SM	UL	UL	72,000	48,000	72,000	48,000	61,500	36,000	21,000
R-2	NS[d, h]	UL	UL	24,000	16,000	24,000	16,000	20,500	12,000	7,000
	S13R									
	S1	UL	UL	96,000	64,000	96,000	64,000	82,000	48,000	28,000
	SM	UL	UL	72,000	48,000	72,000	48,000	61,500	36,000	21,000
R-3	NS[d, h]	UL	UL	UL	UL	UL	UL	UL	UL	UL
	S13R									
	S1									
	SM									

(continued)

Table 3.4
Allowable area factor (2015 IBC, Table 506.2) ***(continued)***

OCCUPANCY CLASSIFICATION	TYPE OF CONSTRUCTION									
	SEE	TYPE I		TYPE II		TYPE III		TYPE IV	TYPE V	
	FOOTNOTES	A	B	A	B	A	B	HT	A	B
R-4	NS[d, h]	UL	UL	24,000	16,000	24,000	16,000	20,500	12,000	7,000
	S13R									
	S1	UL	UL	96,000	64,000	96,000	64,000	82,000	48,000	28,000
	SM	UL	UL	72,000	48,000	72,000	48,000	61,500	36,000	21,000
S-1	NS	UL	48,000	26,000	17,500	26,000	17,500	25,500	14,000	9,000
	S1	UL	192,000	104,000	70,000	104,000	70,000	102,000	56,000	36,000
	SM	UL	144,000	78,000	52,500	78,000	52,500	76,500	42,000	27,000
S-2	NS	UL	79,000	39,000	26,000	39,000	26,000	38,500	21,000	13,500
	S1	UL	316,000	156,000	104,000	156,000	104,000	154,000	84,000	54,000
	SM	UL	237,000	117,000	78,000	117,000	78,000	115,500	63,000	40,500
U	NS	UL	35,500	19,000	8,500	14,000	8,500	18,000	9,000	5,500
	S1	UL	142,000	76,000	34,000	56,000	34,000	72,000	36,000	22,000
	SM	UL	106,500	57,000	25,500	42,000	25,500	54,000	27,000	16,500

Note: UL = Unlimited; NP = Not permitted;

For SI: 1 square foot = 0.0929 m2.

NS = Buildings not equipped throughout with an automatic sprinkler system; S1 = Buildings a maximum of one story above grade plane equipped throughout with an automatic sprinkler system installed in accordance with Section 903.3.1.1; SM = Buildings two or more stories above grade plane equipped throughout with an automatic sprinkler system installed in accordance with Section 903.3.1.1; S13R = Buildings equipped throughout with an automatic sprinkler system installed in accordance with Section 903.3.1.2.

a. See Chapters 4 and 5 for specific exceptions to the allowable height in this chapter.
b. See Section 903.2 for the minimum thresholds for protection by an automatic sprinkler system for specific occupancies.
c. New Group H occupancies are required to be protected by an automatic sprinkler system in accordance with Section 903.2.5.
d. The NS value is only for use in evaluation of existing building area in accordance with the *International Existing Building Code*.
e. New Group I-1 and I-3 occupancies are required to be protected by an automatic sprinkler system in accordance with Section 903.2.6. For new Group I-1 occupancies, Condition 1, see Exception 1 of Section 903.2.6.
f. New and existing Group I-2 occupancies are required to be protected by an automatic sprinkler system in accordance with Section 903.2.6 and Section 1103.5 of the *International Fire Code*.
g. New Group I-4 occupancies see Exceptions 2 and 3 of Section 903.2.6.
h. New Group R occupancies are required to be protected by an automatic sprinkler system in accordance with Section 903.2.8.

$\frac{1}{3}$ hour (20 minutes) up to 4 hours, with $\frac{3}{4}$ hour (45 minutes), 1 hour, $1\frac{1}{2}$ hours, 2 hours, and 3 hours in between. The requirements for fire-resistance for major building components are provided in Table 3.5.

Note the fourth line down on the table: *Nonbearing walls and partitions, Interior*. As you can see, such walls are never required to be fire-rated, according to this table; however, footnote d says "Not less than the fire-resistance rating required by other sections of this code." So it is possible that interior partitions might have to be fire-rated in some circumstances. This simple example demonstrates why it is so critical to read footnotes, exceptions, etc.

Table 3.5
Fire-resistance rating requirements for building elements (hours) (2015 IBC, Table 601)

BUILDING ELEMENT	TYPE I		TYPE II		TYPE III		TYPE IV	TYPE V	
	A	**B**	**A**	**B**	**A**	**B**	**HT**	**A**	**B**
Primary structural framef (see Section 202)	3[a]	2[a]	1	0	1	0	HT	1	0
Bearing walls Exterior[e,f]	3	2	1	0	2	2	2	1	0
Interior	3[a]	2[a]	1	0	1	0	1/HT	1	0
Nonbearing walls and partitions Exterior	See Table 602								
Nonbearing walls and partitions Interiord	0	0	0	0	0	0	See Section 602.4.6	0	0
Floor construction and associated secondary members (see Section 202)	2	2	1	0	1	0	HT	1	0
Roof construction and associated secondary members (see Section 202)	$1\frac{1}{2}$[b]	1[b,c]	1[b,c]	0[c]	1[b,c]	0	HT	1[b,c]	0

For SI: 1 foot = 304.8 mm.

a. Roof supports: Fire-resistance ratings of primary structural frame and bearing walls are permitted to be reduced by 1 hour where supporting a roof only.
b. Except in Group F-1, H, M and S-1 occupancies, fire protection of structural members shall not be required, including protection of roof framing and decking where every part of the roof construction is 20 feet or more above any floor immediately below. Fire-retardant-treated wood members shall be allowed to be used for such unprotected members.
c. In all occupancies, heavy timber shall be allowed where a 1-hour or less fire-resistance rating is required.
d. Not less than the fire-resistance rating required by other sections of this code.
e. Not less than the fire-resistance rating based on fire separation distance (see Table 602).
f. Not less than the fire-resistance rating as referenced in Section 704.10.

There are two basic methods to determine fire-resistance rating for a given type of construction. First, there are tables (and other information) in Chapter 7 of the 2015 IBC to gauge the performance of generic materials (e.g. concrete, plaster, brick, etc.) The most significant is Table 721.1(1) Minimum Protection of Structural Parts Based on Time Periods for Various Non-combustible Insulating Materials. Second, most ratable wall (and roof and floor) systems have been tested by Underwriters' Laboratories (UL) and they have "UL numbers." These numbers are unique for each test and can be used to verify fire-resistance performance. The method of using the UL numbers is far better and far less likely to cause problems with enforcement officials.

When openings are made in fire-resistance-rated construction, it is generally necessary to fire-protect the openings too. The general rule is that opening protection is 1 level down from the rating of the wall (or ceiling or floor), which means that openings in 2-hour rated walls are usually rated for 1.5 hours. (There are exceptions to everything, of course, so this should not be assumed for every condition.) Fire-rating of openings is done by UL number too, and there is no effective alternative. This can apply to doors and door frames, windows and window frames, duct openings, etc. It is by far the most challenging for windows.

A generation ago, the only viable alternative for fire-rated glass was "wired glass," which is two thin panes of glass laminated together with a 1″ × 1″ diagonal wire grid in between (this used to be widely used in schools, hospitals, and other institutional buildings). It is also possible to build fire-rated openings using traditional hollow glass block (this requires special head, jamb, and sill details), and, more recently, many fire-ratable "glass" products have come on the market. These products are not really glass; instead, they are transparent ceramic materials, and they can be used in relatively large areas. (In very special circumstances, sprinklers can be used to create curtains of water to protect glass areas too. For examples see: www.safti.com (Safti*First*, a Division of O'Keeffe's); www.pilkington.com/fire ("Pyrostop" products by Pilkington); www.us.schott.com ("Pyran" products by Schott).)

In order to meet the requirements for fire-resistance ratings, it is first necessary to determine what has to be rated and to what degree; then to design the appropriate assembly, and then to make sure that the appropriate assembly is used where it is needed in the design. This will probably require special detailing and special notations on the construction drawings, especially for protection of openings.

3.9.5 Interior finishes

Clearly, the selection and specification of interior finish materials is of primary importance to interior designers. But the code ramifications are really very simple and straightforward. Basic requirements are shown in Table 3.6.

As you can see from Table 3.6, these requirements are *not* related to construction type, which means that they are the same for all construction types. Whether or not the building is sprinklered is a major issue, as is occupancy group. If the facility is a mixed occupancy, then multiple requirements could apply. To use this table, simply identify the occupancy group along the left column, and follow the line to the right

Table 3.6
Interior wall and ceiling finish requirements by occupancy (2015 IBC, Table 803.11)

GROUP	SPRINKLERED[l]			NONSPRINKLERED		
	Interior exit stairways and ramps and exit passageways[a, b]	Corridors and enclosure for exit access stairways and ramps	Rooms and enclosed spaces[c]	Interior exit stairways and ramps and exit passageways[a, b]	Corridors and enclosure for exit access stairways and ramps	Rooms and enclosed spaces[c]
A-1 & A-2	B	B	C	A	A[d]	B[e]
A-3[f], A-4, A-5	B	B	C	A	A[d]	C
B, E, M, R-1	B	C	C	A	B	C
R-4	B	C	C	A	B	B
F	C	C	C	B	C	C
H	B	B	C[g]	A	A	B
I-1	B	C	C	A	B	B
I-2	B	B	B[h, i]	A	A	B
I-3	A	A[j]	C	A	A	B
I-4	B	B	B[h, f]	A	A	B
R-2	C	C	C	B	B	C
R-3	C	C	C	C	C	C
S	C	C	C	B	B	C
U	No restrictions			No restrictions		

For SI: 1 inch = 25.4 mm, 1 square foot = $0.0929m^2$.

a. Class C interior finish materials shall be permitted for wainscotting or paneling of not more than 1,000 square feet of applied surface area in the grade lobby where applied directly to a noncombustible base or over furring strips applied to a noncombustible base and fireblocked as required by Section 803.13.1.
b. In other than Group I-3 occupancies in buildings less than three stories above grade plane, Class B interior finish for nonsprinklered buildings and Class C interior finish for sprinklered buildings shall be permitted in interior exit stairways and ramps.
c. Requirements for rooms and enclosed spaces shall be based upon spaces enclosed by partitions. Where a fire-resistance rating is required for structural elements, the enclosing partitions shall extend from the floor to the ceiling. Partitions that do not comply with this shall be considered enclosing spaces and the rooms or spaces on both sides shall be considered one. In determining the applicable requirements for rooms and enclosed spaces, the specific occupancy thereof shall be the governing factor regardless of the group classification of the building or structure.
d. Lobby areas in Group A-1, A-2 and A-3 occupancies shall not be less than Class B materials.
e. Class C interior finish materials shall be permitted in places of assembly with an occupant load of 300 persons or less.
f. For places of religious worship, wood used for ornamental purposes, trusses, paneling or chancel furnishing shall be permitted.
g. Class B material is required where the building exceeds two stories.
h. Class C interior finish materials shall be permitted in administrative spaces.
i. Class C interior finish materials shall be permitted in rooms with a capacity of four persons or less.
j. Class B materials shall be permitted as wainscotting extending not more than 48 inches above the finished floor in corridors and exit access stairways and ramps.
k. Finish materials as provided for in other sections of this code.
l. Applies when protected by an automatic sprinkler system installed in accordance with Section 903.3.1.1 or 903.3.1.2.

to the appropriate vertical column. To determine the requirements for corridors in a non-sprinklered A-3 occupancy, go to the second line and move across to the second column from the right side—Class A finishes are required for corridors in this situation. If the building were sprinklered, this would change to Class B.

Would it make sense just to use only Class A finishes? It certainly could not hurt anything. However, someone (a vendor, even the owner) might point out that less costly Class B or even Class C finishes might be allowable in a certain situation, and one cannot argue against that.

These classifications are determined by performance under various ASTM, NFPA, and/or UL testing protocols; a thorough discussion of those issues is beyond the scope of this chapter.

Ultimately, it simply boils down to checking Table 803.11 and following its requirements; it really is that simple.

3.9.6 Means of egress

The two most important concepts in codes (for interior designers) are Occupancy Group and Means of Egress, primarily because these two areas determine exit quantities and placement, and exit way widths (including corridors and stairs), all of which have a key impact on space planning.

The general requirements are pretty simple. The number of exits is provided in Tables 3.7, 3.8, 3.9, and 3.10.

As you can see, in most typical occupancies (excluding H, I-1, I-3, I-4, R, and S), it is acceptable to have one exit for up to 49 occupants, unless the space is on the second floor or higher. What does that mean? How much space does that relate to? Also, three exits are required above occupant load 500, and four exits are required above occupant load 1,000.

Answering that question requires Table 3.10, which is arguably the most important table in the code for interior designers.

Table 3.11 shows how to calculate occupant load (better put as occupant count, but occupant load is the term that is used throughout the codes), by using actual occupancy—not occupancy group. This can be confusing. Note that the table includes "Classroom area" under Educational; this is not Group E Educational, because many educational occupancies are Group B (all higher education, for example). This classroom designation applies within any occupancy Group where a space is used as a classroom, and that could happen in a B, E, F, H, I, or M occupancy group. Also, it must be noted that some obvious space types are not on the list: break room? restroom? When a space is not on the list, one is supposed to use the closest space to it that is on the list. In the case of a break room, that is probably "Assembly without fixed seats, Unconcentrated (tables and chairs)", or 15 net square feet (sf) per occupant.

The assembly categories on this table bring up one of the most controversial areas of code interpretation. Many code officials are fond of applying "Assembly without fixed seats, Concentrated (chairs only—not fixed)" to almost any large open space that they run across—up to and including a full-sized gymnasium. A typical full-sized gymnasium

Table 3.7
Spaces with one exit or exit access doorway (2015 IBC, Table 1006.2.1)

OCCUPANCY	MAXIMUM OCCUPANT LOAD OF SPACE	MAXIMUM COMMON PATH OF EGRESS TRAVEL DISTANCE (feet)		
		Without Sprinkler System (feet)		With Sprinkler System (feet)
		Occupant Load		
		OL ≤ 30	OL > 30	
A[c], E, M	49	75	75	75[a]
B	49	100	75	100[a]
F	49	75	75	100[a]
H-1, H-2, H-3	3	NP	NP	25[b]
H-4, H-5	10	NP	NP	75[b]
I-1, I-2[d], I-4	10	NP	NP	75[b]
I-3	10	NP	NP	100[a]
R-1	10	NP	NP	75[a]
R-2	10	NP	NP	125[a]
R-3[e]	10	NP	NP	125[a]
R-4[e]	10	75	75	125[a]
S[f]	29	100	75	100[a]
U	49	100	75	75[a]

For SI: 1 foot = 304.8 mm.
NP = Not Permitted.

a. Buildings equipped throughout with an *automatic sprinkler system* in accordance with Section 903.3.1.1 or 903.3.1.2. See Section 903 for occupancies where *automatic sprinkler systems* are permitted in accordance with Section 903.3.1.2.
b. Group H occupancies equipped throughout with an *automatic sprinkler system* in accordance with Section 903.2.5.
c. For a room or space used for assembly purposes having *fixed seating*, see Section 1029.8.
d. For the travel distance limitations in Group I-2, see Section 407.4.
e. The length of *common path of egress travel* distance in a Group R-3 occupancy located in a mixed occupancy building or within a Group R-3 or R-4 *congregate living facility*.
f. The length of *common path of egress travel* distance in a Group S-2 *open parking garage* shall be not more than 100 feet.

Table 3.8
Minimum number of exits or access to exits per story (2015 IBC, Table 1006.3.1)

OCCUPANT LOAD PER STORY	MINIMUM NUMBER OF EXITS OR ACCESS TO EXITS FROM STORY
1-500	2
501-1,000	3
More than 1,000	4

Table 3.9
Stories with one exit or access to one exit for R-2 occupancies (2015 IBC, Table 1006.3.2(1))

STORY	OCCUPANCY	MAXIMUM NUMBER OF DWELLING UNITS	MAXIMUM COMMON PATH OF EGRESS TRAVEL DISTANCE
Basement, first, second or third story above grade plane	R-2[a, b]	4 dwelling units	125 feet
Fourth story above grade plane and higher	NP	NA	NA

For SI: 1 foot = 3048 mm.
NP = Not Permitted.
NA = Not Applicable.

a. Buildings classified as Group R-2 equipped throughout with an automatic sprinkler system in accordance with Section 903.3.1.1 or 903.3.1.2 and provided with emergency escape and rescue openings in accordance with Section 1030.
b. This table is used for R-2 occupancies consisting of dwelling units. For R-2 occupancies consisting of sleeping units, use Table 1006.3.2(2).

Table 3.10
Stories with one exit or access to one exit for other occupancies (2015 IBC, Table 1006.3.2(2))

STORY	OCCUPANCY	MAXIMUM OCCUPANT LOAD PER STORY	MAXIMUM COMMON PATH OF EGRESS TRAVEL DISTANCE (feet)
First story above or below grade plane	A, B[b], E F[b], M, U	49	75
	H-2, H-3	3	25
	H-4, H-5, I, R-1, R-2[a, c], R-4	10	75
	S[b, d]	29	75
Second story above grade plane	B, F, M, S[d]	29	75
Third story above grade plane and higher	NP	NA	NA

For SI: 1 foot = 304.8 mm.
NP = Not Permitted.
NA = Not Applicable.

a. Buildings classified as Group R-2 equipped throughout with an automatic sprinkler system in accordance with Section 903.3.1.1 or 903.3.1.2 and provided with emergency escape and rescue openings in accordance with Section 1030.
b. Group B, F and S occupancies in buildings equipped throughout with an automatic sprinkler system in accordance with Section 903.3.1.1 shall have a maximum exit access travel distance of 100 feet.
c. This table is used for R-2 occupancies consisting of sleeping units. For R-2 occupancies consisting of dwelling units, use Table 1006.3.2(1).
d. The length of exit access travel distance in a Group S-2 open parking garage shall be not more than 100 feet.

Table 3.11
Maximum floor area allowances per occupant (2015 IBC, Table 1004.1.2)

FUNCTION OF SPACE	OCCUPANT LOAD FACTOR[a]
Accessory storage areas, mechanical equipment room	300 gross
Agricultural building	300 gross
Aircraft hangars	500 gross
Airport terminal	
Baggage claim	20 gross
Baggage handling	300 gross
Concourse	100 gross
Waiting areas	15 gross
Assembly	
Gaming floors (keno, slots, etc.)	11 gross
Exhibit gallery and museum	30 net
Assembly with fixed seats	See Section 1004.4
Assembly without fixed seats	
Concentrated (chairs only—not fixed)	7 net
Standing space	5 net
Unconcentrated (tables and chairs)	15 net
Bowling centers, allow 5 persons for each lane including 15 feet of runway, and for additional areas	7 net
Business areas	100 gross
Courtrooms—other than fixed seating areas	40 net
Day care	35 net
Dormitories	50 gross
Educational	
Classroom area	20 net
Shops and other vocational room areas	50 net
Exercise rooms	50 gross
Group H-5 Fabrication and manufacturing areas	200 gross
Industrial areas	100 gross
Institutional areas	
Inpatient treatment areas	240 gross
Outpatient areas	100 gross
Sleeping areas	120 gross
Kitchens, commercial	200 gross
Library	
Reading rooms	50 net
Stack area	100 gross

(continued)

Table 3.11
Maximum floor area allowances per occupant (2015 IBC, Table 1004.1.2) ***(continued)***

FUNCTION OF SPACE	OCCUPANT LOAD FACTOR[a]
Locker rooms	50 gross
Mall buildings—covered and open	See Section 402.8.2
Mercantile	60 gross
Storage, stock, shipping areas	300 gross
Parking garages	200 gross
Residential	200 gross
Skating rinks, swimming pools	
Rink and pool	50 gross
Decks	15 gross
Stages and platforms	15 net
Warehouses	500 gross

For SI: 1 square foot = 0.0929 m^2, 1 foot = 304.8 mm.
a. Floor area in square feet per occupant.

(without spectator seating) is probably at least 6,000 sf; at 7 sf/occupant, that is an occupant load of 857 for that room! While it is indeed possible to put 857 people into such a space, it is highly unlikely that such an occupant load would ever occur (except in a high school, possibly), and it is ludicrous to design exiting for so many occupants. But this particular misinterpretation is commonly seen. This matters for simple conference rooms too. If you take the 49 occupant limit for one exit and apply the 7 sf/occupant factor, that means that you can have only 343 sf in a conference room with one exit in a typical office space. Again, is it possible to put 49 people into a 343sf space? Yes, but not if there is a conference table in the room. In fact, it would be extremely difficult to get that many chairs in the room without a table (try doing a layout to see how many chairs you can fit in with 36″-wide aisles along one side and in the front).

The point is that concentrated assembly is intended to be used for conference center (or convention center) spaces where the occupant density can be quite high; it is not meant to apply to every open space in someone's office, or school, or hospital, or anywhere else for that matter. But that does not mean that it is necessarily easy to convince enforcement officials, so it is only prudent to plan for the worst if it is reasonably practical to do so. (In the case of the gymnasium, using 15 sf/occupant yields an occupant load of 400, which would require two exits. An occupant load of more than 500 would require three exits, and an occupant load of more than 1,000 would require four exits. In such a case, it would only be prudent to design for three, or even four, exits to avoid potential problems in the future.)

To calculate a single occupancy occupant load, the safest thing to do is to take the gross area of the entire space and divide by the appropriate load factor; in a 20,000-gsf (gross square feet) office floor, that would mean

$$20{,}000\,\text{gsf} \div 100\,\text{gsf/occupant ("business")} = 200\ \text{occupants}.$$

To calculate a multiple occupancy occupant load, one should take each area and divide by the appropriate load factor. However, "accessory occupancies" and "incidental accessory occupancies" must be taken into account first.

An accessory occupancy is a secondary occupancy (say A-3 for some, or all, of the conference rooms within the 20,000-sf office space) that is less than 10% of the primary occupancy. So if we have 1,500 sf of conference rooms within the 20,000-sf office space, those conference rooms, in aggregate, would qualify as an accessory occupancy. For occupant load purposes, they are still counted separately though, it is just that there are no requirements for separation. If an accessory occupancy is less than 750 sf, it is counted as part of the primary occupancy for occupant load purposes. Incidental accessory occupancies are defined by Table 3.12.

As you can see, Table 3.12 identifies specific occupancies that represent special risks and special treatments. The most common spaces that involve this table are furnace rooms (or boiler rooms) and laundry rooms.

Returning to calculating a multiple occupancy occupant load, here is an example:

20,000 gsf overall, sub-divided as follows:
15,000 gsf offices
3,000 gsf circulation
1,500 gsf conference room (assume one large room for now)
500 gsf restrooms

Calculate occupant load this way:
15,000 gsf ÷ 100 gsf/occ. = 150 occupants
3,000 gsf ÷ 100 gsf/occ. = 30 occupants (circulation is not on the table; use business—or just leave it out; after all, circulation spaces are not separately occupied)
1,500 sf ÷ 15 sf/occ. = 100 occupants
Restrooms = 10 occupants (assume that there are five stalls in each of two restrooms)
Total occupant load = 290 occupants

Clearly, the multiple occupancy method resulted in more occupants than did the single occupancy method. In this particular case, the differences (corridor width, exit width, stair width, etc.) between 200 and 290 occupants are insignificant, but that is not always true.

If the 1,500 sf of conference space is divided into three smaller 500-sf rooms, the calculation is different because each room is less than 750 sf (so that the 1,500 sf becomes "business" instead of assembly):

16,500 gsf ÷ 100 gsf/occ. = 165 occupants
3,000 gsf ÷ 100 gsf/occ. = 30 occupants
500 gsf = 10 occupants
Total occupant load = 205 occupants

In truth, all of these methods exaggerate the actual likely number of occupants. Most typical office spaces (except call centers) use far more than 100 sf per

Table 3.12
Incidental uses (2015 IBC, Table 509)

ROOM OR AREA	SEPARATION AND/OR PROTECTION
Furnace room where any piece of equipment is over 400,000 Btu per hour input	1 hour or provide automatic sprinkler system
Rooms with boilers where the largest piece of equipment is over 15 psi and 10 horsepower	1 hour or provide automatic sprinkler system
Refrigerant machinery room	1 hour or provide automatic sprinkler system
Hydrogen fuel gas rooms, not classified as Group H	1 hour in Group B, F, M, S and U occupancies; 2 hours in Group A, E, I and R occupancies.
Incinerator rooms	2 hours and provide automatic sprinkler system
Paint shops, not classified as Group H, located in occupancies other than Group F	2 hours; or 1 hour and provide automatic sprinkler system
In Group E occupancies, laboratories and vocational shops not classified as Group H	1 hour or provide automatic sprinkler system
In Group I-2 occupancies, laboratories not classified as Group H	1 hour and provide automatic sprinkler system
In ambulatory care facilities, laboratories not classified as Group H	1 hour or provide automatic sprinkler system
Laundry rooms over 100 square feet	1 hour or provide automatic sprinkler system
In Group I-2, laundry rooms over 100 square feet	1 hour
Group I-3 cells and Group I-2 patient rooms equipped with padded surfaces	1 hour
In Group I-2, physical plant maintenance shops	1 hour
In ambulatory care facilities or Group I-2 occupancies, waste and linen collection rooms with containers that have an aggregate volume of 10 cubic feet or greater	1 hour
In other than ambulatory care facilities and Group I-2 occupancies, waste and linen collection rooms over 100 square feet	1 hour or provide automatic sprinkler system
In ambulatory care facilities or Group I-2 occupancies, storage rooms greater than 100 square feet	1 hour
Stationary storage battery systems having a liquid electrolyte capacity of more than 50 gallons for flooded lead-acid, nickel cadmium or VRLA, or more than 1,000 pounds for lithium-ion and lithium metal polymer used for facility standby power, emergency power or uninterruptable power supplies	1 hour in Group B, F, M, S and U occupancies; 2 hours in Group A, E, I and R occupancies.

For SI: 1 square foot = 0.0929 m^2, 1 pound per square inch (psi) = 6.9 kPa, 1 British thermal unit (Btu) per hour = 0.293 watts, 1 horsepower = 746 watts, 1 gallon = 3.785 L, 1 cubic foot = 0.0283 m^3.

occupant; figures like 225 sf/occupant to 250 sf/occupant are more typical. The code is conservative like this in the interest of making buildings safer, rather than less safe, and there is no reason to get into an unwinnable argument with an enforcement official over the difference between 205 and 290 occupants.

Once an occupant load has been calculated, egress requirements can be evaluated. When two exits are required, in non-sprinklered buildings, they must be separated by at least $\frac{1}{2}$ of the largest diagonal measurement of the area served; in sprinklered buildings, the minimum separation is $\frac{1}{3}$ of the greatest diagonal measurement of the area served.[11] This applies to a single conference room or classroom having two exits or to an entire building alike.

If two or more exits are required, it is necessary to have exit signs and emergency egress lighting.[12] (Egress lighting is required for all occupied spaces whenever they are occupied.) Emergency egress lighting covers all areas required to have two or more exits (but not spaces with one exit), up to and including the landings outside grade-level exit discharge doors. (If NFPA 101 is applicable for some reason, emergency egress lighting is required to the public way, which could be much farther than the door landing in some cases.)

It is allowable to exit one space through another space, as long as the intervening space is not a kitchen, storage room, closet, or space used for a similar purpose. It is not allowable to exit into a corridor, through a room (that is not a kitchen, etc.), and back into a corridor; once one enters into a primary exit path (such as a corridor), one must remain within that path until out of the building. (There are some exceptions for first-floor lobbies.)

The minimum width for stairways is 0.3 inches for each occupant, and the minimum width for other exit components (doors and corridors mostly) is 0.2 inches for each occupant,[13] but no less than 32 inches for doors and 44 inches for corridors and stairs (unless occupant load is less than 50). (To go back to the example: 200 occupants × 0.3 = 60 inches for two stairs, or 30 inches each, so the minimum stair width is 44 inches. For 290 occupants × 0.3 = 87 inches for two stairs, or 43.5 inches each, so the minimum stair width is 44 inches. This is why there is no real difference between 200 and 290 occupants in the example. The calculations for doors and corridors are similar.)

Corridors are required to be fire-resistance-rated according to Table 3.13. As you can see, for sprinklered buildings, corridors are not required to be rated for A, B, E, F, I-2, I-4, M, S, and U occupancies.

3.9.7 Accessibility

There are two completely separate categories of disability: physical and mental. Historically, accessibility regulations for buildings have been limited to physical disabilities, which is still the case under the 2015 IBC. The 2015 IBC includes, by reference, ICC/ANSI A117.1-2003 (noted previously), and the ICC/ANSI document refers only to physical disabilities. But the Americans with Disabilities Act covers physical *and* mental disabilities, which can

Table 3.13
Corridor fire-resistance rating (2015 IBC, Table 1020.1)

OCCUPANCY	OCCUPANT LOAD SERVED BY CORRIDOR	REQUIRED FIRE-RESISTANCE RATING (hours)	
		Without sprinkler system	With sprinkler system[c]
H-1, H-2, H-3	All	Not Permitted	1
H-4, H-5	Greater than 30	Not Permitted	1
A, B, E, F, M, S, U	Greater than 30	1	0
R	Greater than 10	Not Permitted	0.5
I-2[a], I-4	All	Not Permitted	0
I-1, I-3	All	Not Permitted	1[b]

a. For requirements for occupancies in Group I-2, see Sections 407.2 and 407.3.
b. For a reduction in the *fire-resistance rating* for occupancies in Group I-3, see Section 408.8.
c. Buildings equipped throughout with an *automatic sprinkler system* in accordance with Section 903.3.1.1 or 903.3.1.2 where allowed.

change things significantly. The primary effects of mental disabilities on facility design relate to fire alarm visual devices ("strobes," as they are usually called) because such devices can trigger seizures in some people. Other effects are far more subtle and hard to identify.

The main purpose of physical accessibility regulations is to make the world more accessible to people who are not fully able-bodied, for whatever reason. There have been accessibility regulations (on a state and local basis) for a long time (certainly back to the 1970s or earlier) and ANSI has been the leader in publishing accessibility standards. Such standards spell out, in detail, how to make facilities accessible—how large doors should be, how to arrange them for movement (especially for wheelchairs), how high devices should be mounted on walls, how high toilets should be, where grab bars are needed (and in what dimensions), etc. When the ADA was enacted in 1990 (going into effect at the beginning of 1992), this process was accelerated because the federal government stepped in and raised the bar, so to speak, taking away some previous protections in the process. Under ADA, no one was forced to renovate anything to make it more accessible; but if a renovation was undertaken, it was understood (because it says so in the law itself) that it would be considered reasonable to spend up to 20% of the overall project cost on accessibility improvements. This is why so many elevators have been improved and so many ramps have been added to existing facilities since 1992. And it has worked and we see many more disabled people travelling around and working than we used to in the past.

The biggest effect of these regulations on interior design practice is in space planning, because more space is required in toilet rooms and in circulation spaces. Doors have to be at least 32 inches wide (clear opening) and they have to have approach clearances to make them useable. Toilet rooms have to have wheelchair turning spaces (the famous 60-inch (5-foot) diameter circle), knee and toe clearances, grab bars, wrist blades (or lever handles) on faucets, etc. There can be no more door knobs; only lever handles are useable without grasping, which is difficult for some disabled people. If ramps are needed (and they are, even in restaurant dining rooms if multiple floor levels are present), they take up large amounts of space.

At the core of it, this is no different from everything else in the codes: determine what the applicable provisions are for the project at hand, and incorporate them into the design—period.

3.9.8 Other requirements

Chapter 5 "General Building Heights and Areas" and Chapter 9 "Fire Protection Systems" include significant provisions that might not be used for every project.

General building height and area requirements apply mostly to planning new buildings, but an understanding of the basic concepts can help an interior designer understand the context of a given project. Is it feasible to use wood framing for partition walls? Can untreated wood blocking be used inside walls?

Fire protection systems come into play if sprinklers (wet pipe or dry pipe), standpipes (for taller buildings), or fire alarm systems are needed. Detailed coverage of these subjects is far beyond the scope of this chapter, but it can be said that sprinklers are placed at 15 feet on center for most typical environments (or 10′ 6″ for slightly riskier spaces, such as mechanical rooms, storage rooms, etc.). Concealed sprinklers (with the flat coverplates) are the most costly (although affordable for limited use on many projects) and pendant sprinklers (the largest and ugliest option) are the least costly. Sprinklers are available in various finishes too. Fire alarm systems come in manual (no automatic sensing of fire via smoke or heat detectors) and automatic (with automatic heat and/or smoke sensing in common use areas and risky areas); automatic systems are generally not required, except in larger assembly occupancies or in high-rise buildings (but always check the specific requirements).

By definition, most interior design projects are done in "existing buildings," even if the building has not actually been built yet. So existing buildings requirements are very important. They are listed here simply because there are just a few simple rules, in most jurisdictions:

1 New work must meet all current rules.
2 Altered work must meet all current rules.
3 Additions must meet all current rules.
4 Existing non-altered work can remain as is (in many, if not most, jurisdictions). But this also must be carefully verified.

For historic buildings, there is a procedure in section 3412 of the 2009 IBC that allows for an alternative system for code compliance, but this chapter has been deleted from the 2015 IBC. This system consists of awarding various points for features—plus points for optional items that increase fire safety (adding a sprinkler system for example) or minus points for existing items that decrease fire safety (unenclosed exit stairs provide a common example). As long as the points add up properly, the project can proceed without "full compliance" with all rules. This is something that also needs to be carefully verified in each jurisdiction.

3.10 Other codes

In addition to the few basic concepts mentioned previously, there are specific and detailed codes for electrical, fuel gas, mechanical, and plumbing systems. Each of these is far too detailed for treatment here.

And then there is the 2015 International Residential Code (IRC), which covers one- and two-family dwellings in most jurisdictions. This code is completely different because it is written from the perspective of *not* having design professionals involved in a project. As a result, it is very highly prescriptive, meaning that it is very specific about exactly what to do and how to do it for each element of the building. The 2015 IBC says very little about minimum gypsum board thickness (except for rated construction), but the 2015 IRC says a great deal about that particular subject, because it can only be assumed that those who are using the code do not know—on their own—whether to use $\frac{1}{2}$-inch or $\frac{5}{8}$-inch gypsum board (or something else entirely). Again, a detailed explanation is far beyond the scope of this chapter.

Finally, there is the 2015 International Energy Code, which affects virtually everyone on the design team—the structural engineer, architect, mechanical engineer, electrical engineer, plumbing engineer, and the interior designer. The greatest impact of this on interior design is in lighting, because the code dictates maximum power usage and requirements for automatic (or semi-automatic) controls. But a complete treatment is beyond the scope of this chapter.

3.11 Other regulations

For new building projects (residential or commercial, in virtually any jurisdiction in the US), many other regulations come into play, including, but not limited to, the following:

- Local planning and zoning regulations (review and approval processes can be onerous and very time consuming)
- Local development standards (review and approval processes can be onerous and time consuming)
- Local lighting regulations
- State and/or local rules for storm water treatment (usually called "Best Management Practice" or "BMP" and dictated by federal Environmental Protection Agency)
- State and/or local rules for sanitary sewer provisions
- Utility regulations for domestic water, fire service (sprinkler system and/or standpipe water), natural gas (or propane), power, and communications (telephone, data, high-speed data, cable television, satellite communications, etc.)

Most of these regulations do not come into play for most interior design projects, but interior designers could encounter some, or all, of them if they are working on new building projects.

Summary

Ultimately, the incorporation of codes into a given design is a simple process (on paper anyway):

1. Determine what the applicable rules are.
2. Determine what the applicable specific provisions of the rules are.
3. Incorporate the applicable provisions into the design.

Of course, in reality, it is not so simple because it can be difficult to determine what all of the applicable rules and specific provisions are, and because navigating enforcement can be bewildering. But there are a few basic principles that should always be kept in mind:

1. Never make a given situation worse, or more dangerous. No matter what you do, your project should either keep conditions as they already are or improve upon those conditions.
2. Improve a given situation whenever feasible to do so. If an owner is willing and able to make improvements (from a codes point-of-view), such improvements should be made.
3. Do not ignore dangerous existing situations. If a dangerous situation is observed, the owner should be notified immediately. If the codes require correcting the situation (which might or might not be the case, depending on the specifics in the particular jurisdiction), then it must be corrected. If the codes do not require correction, but the interior designer believes that the situation should be corrected anyway, then the interior designer should pursue the necessary changes with the owner. If the owner refuses to make the changes, the decision should be clearly documented, in writing, to assure that the interior designer does not get blamed for the problem in the future.

Finally, the incorporation of code requirements into interior design projects is a career-long learning process. No one can or should "know" all about codes right away; nor should anyone ignore codes after a while in practice because they are no longer interesting. Codes change over time. The ICC publishes all new versions of all of its codes every three years; the NFPA publishes new documents frequently, although not necessarily on a regular schedule. Adopting governmental entities can adopt new codes at any time, so practitioners need to be vigilant at all times, looking for new rules as they change, and incorporating such changes as needed.

Outcomes

3.1.1 Understanding that "standards" might or might not be regulations.

3.1.2 Understanding that "regulations" are targeted laws.

3.1.3 Understanding that "codes" are regulations targeted toward building design and construction.

3.2 Understanding basic code history, via the development of model code and professional licensing.

3.3 Understanding that the code adoption process varies from jurisdiction to jurisdiction, even though there are some common principles.

3.4 Understanding that code enforcement can run the gamut from non-existent to overbearing and that unexpected interpretations occur on a regular basis.

3.5 Understanding that there probably is a code appeals process, although the process might vary widely from jurisdiction to jurisdiction.

3.6 Understanding that code consultants are mostly involved in research and modifications.

3.7 Understanding what basic code documents exist.

3.8 Understanding that code research is a two-step process.

3.9 Understanding the key concepts of use and occupancy classification; special detailed requirements based on use and occupancy; types of construction; fire-resistance-rated construction; interior finishes; means of egress, accessibility, and other requirements.

3.10 Understanding that other codes, including the 2009 International Residential Code and the 2009 International Energy Conservation Code, might or might not be applicable.

3.11 Understanding other regulations (planning and zoning, etc.) might or might not be applicable.

Notes

1 2015 IBC, pages 41–42
2 Ibid., page 42
3 Ibid.
4 Ibid., pages 42–43
5 Ibid., pages 45–48
6 Ibid., pages 48–49
7 Ibid., page 49
8 Ibid., pages 49–50
9 Ibid., pages 50–51
10 Ibid., page 52
11 Ibid., page 256
12 Ibid., page 257
13 Ibid., pages 252–253

Chapter 4

What are structural building systems?

Objectives

4.2 To understand when structural reinforcement might be required.
4.3 To understand when structural modifications might be required or appropriate.
4.4 To understand various construction types based on structural systems.

4.1 Introduction

Even though interior designers will never design structural components of buildings, it is important to cover a few vital topics related to structural engineering. This is true mostly because interior design projects sometimes do affect structural systems, and it is necessary for interior designers to know what to do when that happens.

The issues fall into two categories:

1 Structural reinforcement (usually of floors) due to heavy loads.
2 Structural modifications, including modifying or removing bearing walls, modifying structural bracing, and even adding openings for stairways, etc.

In addition, it is useful for interior designers to understand the characteristics of various structural systems, so that they can more easily understand the construction type of existing (or even new) buildings that they are working in. These three sub-topics will be covered in individual sub-sections.

4.2 Structural reinforcement

How heavy does something have to be before it is necessary to reinforce a floor system? That all depends. First, what is the floor system?

If the floor is a slab-on-grade (as is common on the first floor of many buildings these days), only very heavy loads (for large machines usually) would require reinforcement, so an interior designer should not expect to see problems with slabs-on-grade. If it is not a slab-on-grade, then the interior designer should bring in a structural engineer (or maybe an RA for a "quick check") if something that is compact weighs 1,000 pounds or more. What is compact? There is no real definition, but a good rule-of-thumb would be something of 12 sf or so (four 4-drawer vertical letter-sized file cabinets would have a footprint of roughly 12.5 sf, so that would hit the threshold and would require at least a quick check). A large safe could easily weigh more than 1,000 pounds and it could be considerably smaller than 12 sf, so that would require a check too. The safe could also require checking a path from an elevator to the permanent site because it could be necessary to reinforce the path as well as the final location.

High-density file systems (where the vertical shelves or cabinets move horizontally to open and close aisles as needed) nearly always require structural reinforcement, if not on a slab-on-grade. Large concentrations of large fixed shelving sometimes require such reinforcement. The bottom line? If in doubt, ask an architect or a structural engineer.

www.ancom-filing.com (Ancom Business Products, manual and electric options)
www.spacesaver.com (Spacesaver Corporation)

Structural reinforcement is usually easily done in steel-framed buildings (although it might be costly) but it is very difficult (and very costly) to reinforce concrete-framed buildings, so reinforcement should be avoided in concrete buildings if at all possible. Many, if not most, relatively new buildings have areas of their floors that were designed initially for unusually heavy loads (often in core areas, but one never knows), and it is prudent to take advantage of such arrangements in space planning, if that is feasible. In other words, if there is an area on the floor designed for file rooms that is where the heaviest loads should be located.

Reinforcing steel-framed buildings usually requires adding steel to one or more beams to add load capacity. This is often done by welding a continuous plate along the bottom of the beam (under the bottom flange) or sometimes by welding rods on both sides of the bottom of the web (the vertical middle part) of the beam. The structural engineer must know which floor of the building is involved to make sure that the pieces (plates or rods) that are specified can be moved into the space in a reasonable way. (The author once had a structural engineer design reinforcement using rods that were to be 30′ 0″ long—for beams under the 38th floor of a high-rise office building; needless to say, there was no practical way to get such long rods into the space. It was re-designed with a splice detail so that shorter rods could be used.)

It must be kept in mind that reinforcing a floor system requires working in the ceiling space of the floor below, so access issues (even security issues) can come into play. It is not uncommon at all for work like this to be done after hours to minimize disruptions to the space below.

4.3 Structural modifications

There are two broad categories of modifications:

1 Bearing walls
2 Bracing

4.3.1 Bearing walls

Bearing walls are structural walls that support a floor(s) or a roof above. Is it possible to remove a bearing wall? Actually, it is, but it is usually very costly. The cost does depend upon how much of the bearing wall needs to be removed, but removing more than a few feet (as in a door opening) is challenging no matter what. Bearing walls are usually removed by adding some form of beam work with columns, usually in steel (although it could be done in wood in a wood-framed building). The columns usually require foundations or a supporting structure below if not at the lowest level where it is practical to build footings.

The difficulty of removing the bearing wall depends largely on how the wall is constructed and on the load the wall is carrying. If the wall is a 2 × 4 wood stud wall in a small commercial (or residential) building that supports only a roof, removing it should not be particularly difficult. If the wall is a 25-inch-thick solid masonry (brick and plaster) wall and it supports a floor above (and a roof above that), its removal would be far more challenging. But virtually anything can be done, given sufficient time and money.

In any case, if there is any desire to remove a portion of a bearing wall—any bearing wall—a structural engineer should be consulted.

4.3.2 Bracing

The main problem with bracing is that it might not be visible, because much bracing (in steel-framed buildings) is concealed. Particularly in high-rise buildings, much of the diagonal bracing that is installed to resist horizontal wind loads (mostly but not entirely) is hidden in core walls so as to keep the architectural design "clean," although it is not all that uncommon to see diagonal steel bracing crossing windows around the exterior of some steel-framed buildings.

This is only an issue because it is sometimes desirable to add a door way to access a core space from a new location or to add an opening in a core wall for some other reason. When the gypsum board (or whatever) is removed, a diagonal steel brace might be found running right through the proposed opening. What to do?

Call a structural engineer, who will review the situation to see if there is a way to modify the bracing to make it possible to add the opening. Of course, the cost of modifying the bracing should be compared to the cost of relocating the opening (assuming that is possible for the moment) so that the least cost solution is identified. (Few owners would be interested in paying to modify diagonal bracing if they have a choice.)

But this can happen in low-rise (even single-story) buildings too, so interior designers should not be surprised if such conflicts arise when adding openings to existing walls.

A similar situation occurs when trying to add an opening to a bearing wall too. At a minimum, a new opening in a bearing wall is going to require a structural lintel—a beam (or device) that spans the head of the opening to support the load above the opening. The size and complexity of the header will depend on the width of the opening and the construction of the wall. A 3′ 0″-wide opening in a stud wall would be simple; a 10′ 0″-wide opening in the 25″-thick solid wall mentioned previously would be much more complicated.

4.4 Structural systems

Buildings are built primarily of the following materials:

- Masonry (stone, brick, concrete masonry, structural tile, etc.)
- Soil (mud, adobe, etc.)
- Concrete
- Wood (heavy timber or light framing)
- Cast iron and steel
- Plastics
- Glass (glass block and sheets of glass)

Roofs can be made of any of these materials, except soil (which has no ability to span any significant distance). Floors can be made of any of these materials too.

Bearing walls can be made of any of these materials, except plastic and glass. Plastic or glass walls can be designed and built to support themselves as enclosure, but not to support roofs or floors. Obviously, soil and concrete bearing walls are going to be solid, but masonry bearing walls can be solid or hollow. (Concrete masonry units are usually hollow, unless grouted full for structural reasons, as are tile walls.)

Wood and steel bearing walls are usually hollow. Steel bearing walls would probably consist of light-gauge framing, similar to wood stud framing, because steel beam and column systems are framed systems and not bearing walls at all.

If a building has bearing walls, there will be roof and/or floor framing members of some kind (probably wood or steel, but concrete is possible too, especially if the bearing walls are concrete) that are supported by the bearing walls.

Can one tell by looking at a wall if it is a bearing wall? If there is visible framing that connects to the wall, yes. If the framing is not visible, then no. This is especially

Figure 4.4.A
Cast-in-place concrete beams and slabs

Figure 4.4.B
Cast-in-place concrete one-way joist

tricky in residential-style all-wood construction where virtually any wall can be a bearing wall; in that case, it is absolutely necessary to verify if framing members are supported by the wall (usually by removing gypsum board or other covering materials) in order to be sure.

Figure 4.4.C
Cast-in-place concrete two-way joists (waffle slab)

Figure 4.4.D
Unprotected prefabricated wood trusses

This is a crass generalization, but the older a building is, the more likely it is to have bearing walls. Buildings that have heavy masonry walls (even interior walls) should be assumed to have bearing walls, unless a qualified professional (structural engineer or architect) verifies that such is not the case.

Figure 4.4.E
Heavy timber

Figure 4.4.F
Protected open-web steel joists (bar joists)

Framed buildings, in wood, concrete, cast iron, and steel have been around for a long time, but it was common practice to use heavy masonry walls as bearing walls until the past 30 years or so, when it started to become more cost effective to use more framed systems.

Figure 4.4.G
Protected steel beams

Figure 4.4.H
Unprotected open-web steel joists

Figure 4.4.I
Unprotected steam beams and girder

Roofs and floors can be framed using concrete, wood, cast iron, and steel, or in combinations of these materials.

- If wood roof framing is exposed in the building, that means that the building is a Type V-B (or Type V-N in the past) building—unprotected combustible structure (see Figure 4.4.D).
- If there is wood framing but none of it is exposed, it is probably a Type V-A (or Type V-1-hr. in the past) building—protected combustible structure.
- If there is heavy timber framing, floor decks, and roof decks, it is a Type IV (or Type IV-HT in the past) building. For a Type IV building, the roof decking has to be at least 2″ thick and the floor decking has to be at least 3″ thick. If lightweight decks are used (plywood or similar), it is probably a Type V-B building that looks like a Type IV building (see Figure 4.4.E).
- If there is a combination of wood and steel framing, it is probably a Type III-A or III-B building, depending on the degree of protection.
- If the building has unprotected (no fire-proofing) steel framing that is exposed, it is probably a Type II-B (or Type II-N in the past) building (see Figures 4.4.H and 4.4.I.).
- If some of the exposed steel is protected, it may be a Type II-A building, or possibly a Type I-B building (see Figures 4.4.F and 4.4.G).
- If all of the exposed steel is protected, it is probably a Type I-A building, but it might be a Type I-B building (see Figures 4.4.F and 4.4.G).
- If all of the framing is concrete, it is probably a Type I-A building (but it could be a Type I-B) building (see Figures 4.4.A, 4.4.B, and 4.4.C).

The bottom line?

- If the building shell is an all non-combustible framed building (i.e. steel or concrete), then use only non-combustible interior framing.
- If the building shell is heavy timber, use heavy timber interior walls (not lightweight wood stud walls).
- If the building shell is known to be type III, use either non-combustible or combustible interior framing.
- If the building shell is protected or unprotected combustible material, use protected or unprotected combustible interior framing.

Summary

Avoid putting openings into bearing walls or walls where there might be diagonal bracing, if possible. (The author once worked on a pet store project where a preliminary space plan called for removing about 40′ 0″ of existing bearing wall. By re-working the space plan, he was able to keep most of the bearing wall in place, saving the owner more than $30,000.00 in the process. Owners like design changes like that.)

When in doubt, call in a structural engineer. But try not to be in doubt too often. Calling in a structural engineer can be a costly process, and who is going to pay the fee? If it is necessary to call in a structural engineer to evaluate something in an existing building, it is best to call in the engineer who designed the building in the first place, even if that is an out-of-town (or famous) engineer. Even if their rates are higher, it will be more cost effective for them to do the analysis because they already have all of the information for the building. Without knowing everything about a design, it can be difficult for a new engineer to come to a definitive conclusion. (There are structural engineers in most markets who specialize in historic buildings; in such cases, it is worth seeking out such an engineer because they are accustomed to dealing with the typical unknowns found in old buildings.)

Above all else, never assume that everything will be OK without structural checking. The potential problems that can result in a structural failure are truly stupendous and simply must be avoided.

Outcomes

4.2 Understanding that structural reinforcement might be required for unusually heavy or dense loads.

4.3 Understanding that structural modifications might be justifiable for certain situations.

4.4 Understanding various construction types based on structural systems—concrete, steel, wood, masonry, etc.

Chapter 5

What are architectural building systems?

Objectives

5.1 To understand walls
5.2 To understand doors
5.3 To understand door hardware
5.4 To understand door frames
5.5 To understand windows
5.6 To understand ceilings
5.7 To understand millwork
5.8 To understand vertical movement systems
5.9 To understand architectural documentation

5.1 Walls

5.1.1 What are walls?

Walls are assemblies of physical materials that are used to define spaces. They can be solid or hollow, short or tall, and transparent, translucent, or opaque. What is a door? A door is a way to get through a wall, which can take the form of a movable section of the wall itself or any of a wide variety of alternatives with or without frames. A doorway is simply an opening through a wall.

5.1.2 *Why do we build exterior walls?*

Mostly, we build exterior walls to protect ourselves from the weather: sun, wind, rain, heat, cold, sleet, snow, ice, etc. The techniques and materials that have been developed around the world over the past several thousand years—stone, brick, mud, rammed earth, adobe, ceramics, concrete, ice and snow, concrete unit masonry, stucco, wood, metal, glass, plastic, and fabric—came about in response to local climate conditions. No one has ever built a building in Moscow (Russia) using only fabric and it would be unheard of to see adobe in southern Alabama. But new all-glass buildings are being built in Dubai now, requiring massive investment in air-conditioning (for both installation and operation) due to the local very harsh, hot desert climate. To reiterate the sustainable design mantra, no one built glassy buildings in Dubai 100 years ago at all, let alone glassy high-rise buildings; it simply does not make sense. One can argue that sophisticated building technology, including "smart" glass that responds to the sun, can be used to make the extensive use of glass less problematic in harsh climates, but the point remains that simpler is better and less glass is better in most climate zones.

It is complicated and difficult to build simple buildings and more complicated and more difficult to build sophisticated buildings; simple designs are more likely to get built properly in any given situation so it only makes sense to keep everything as simple as possible.

5.1.2 *Why do we build interior walls?*

Reasons for building interior walls include:

- To provide physical separation (but not necessarily security), by separating one space from another with some kind of physical construction. Virtually any wall described in section 5.1.5 below can provide separation, if security and durability are not issues.
- To provide security, by separating one space from another with durable construction. For secure walls, tough materials are required: concrete, concrete masonry, brick, stone, plaster, cement board with ceramic tile, etc.
- To provide visual privacy, by separating one space from another with visually solid construction. Anything will work except clear glass.
- To provide acoustical privacy, by separating one space from another with acoustically effective construction (see Chapter 6 for more details).
- To provide surfaces to display artwork. Many different materials will work for this, but the hanging system is of key importance. If a **rail & cable**, or rail & hangers, system is used, the surface of the wall can be just about anything because nails or screws are not put into the wall to hold the displays. If nails or screws are used to secure artwork to the wall, easily repaired surfaces are required. Do not use stone, brick, cement plaster, or ceramic tile—unless there is a rail system of some type; do use wood or gypsum board (or something else that is reasonably soft and repairable).

- To provide surfaces to support wall-hung cabinets and similar items. Metal stud walls are not necessarily good choices for walls with heavy unbalanced loads (e.g. large overhead cabinets only on one side of the wall) because they are prone to vertical deflection (bending). That can be overcome with adequate blocking (wood or metal), but other types of walls might be more appropriate for heavy wall-mounted loads: thick masonry, wood studs, wood-reinforced metal studs, etc.
- To provide surfaces to support wall-mounted plumbing. As above, metal stud walls are prone to bending, so other choices might be more appropriate.
- To provide termination to a visual axis. Anything will do here—it is all about the design anyway.
- To provide direction for movement through a space or building. Anything will do here too—it is not about technical function; it is about design—the flow of people moving through or around a space.

Note that this list does not include "to support doors." Why not? Because we only need doors where we have walls that we need to pass through. No walls = no doors. Can doors be walls? Certainly, but that is a special case.

5.1.3 How permanent are interior walls?

Interior walls can be as permanent as we want them to be. Most interior walls are essentially permanent, meaning that they are not intended to be moved. However, there are "demountable" wall systems that are intended to be moved and then there are office cubicle furniture systems where the panels (which are a form of low wall—see 5.1.4 below) are intended to be moved around easily.

The main thing about moving walls, which applies primarily to demountable walls, is that they become more or less immovable if plumbing and or wiring is installed in them. Furniture systems panels, and some wall systems (DIRTT walls, specifically), are designed for internal wiring but not for plumbing. While it is relatively straightforward to plan facilities not to need plumbing in walls that are intended to be movable, it is not so simple to limit wiring.

5.1.4 How tall are interior walls?

Interior walls can be just about any height a designer can dream up—16″ (to sit on), 36″ (to lean on), 42″ (bar height), 48″, 60″, 6′0″, 7′0″, 8′0″, 8′6″, 9′0″, 10′0″, 12′0″ or more—literally the underside of the floor or the roof above is the limit.

Low walls are defined as walls that do not reach the ceiling. Such walls require special support because they have no bracing at the ceiling (or above), and that special support usually takes the form of a small **steel tube column** buried in the wall or a special lighter weight steel assembly that is designed just to support such a wall. This is necessary because a wall like this is usually very tall relative to its width at the base, which provides for an excellent lever arm (someone leaning against, or pushing on, the top of the wall)

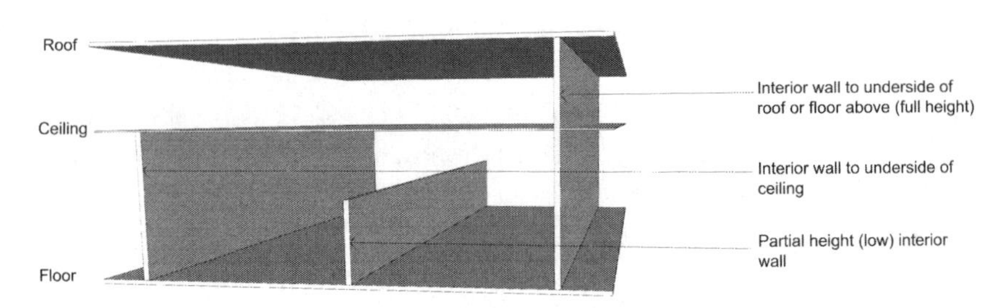

Figure 5.1.4.A
Wall heights

that can be used to dislodge the mounting at the floor. Low walls also require tops (which are usually unfinished in other types of walls), which can run the gamut from gypsum board wrapped around the top, finished, and painted, to metal trims, to elaborate and high-quality wood caps (with or without moldings and other special trim elements).

Ceiling-height walls are exactly that: walls that span from the floor to the underside of a ceiling (much more on ceilings in Section 5.4), whatever that ceiling might be. One of the most common standards in the commercial office world is a lightweight wall framed using metal (**cold-rolled** steel) studs and **gypsum board** that spans from the floor to the underside of a **lay-in ceiling**, sometimes using a special head track called an "eliminator track." (An eliminator style head track is attached to the ceiling and it has extra wide flanges on both sides to make finishing the gypsum board much easier; it is called an eliminator track because it eliminates the need for the gypsum board finishers to use tape.) Ceiling-height walls typically have very poor acoustical performance (more on that in Chapter 6).

Full-height walls are walls that span from the floor to the underside of the floor or roof above. If it is a roof above, such a structure usually slopes (sometimes just a little—$\frac{1}{8}$" per 12"—or sometimes a lot—12" per 12" or even more), so the wall needs to slope too. Also, when an interior wall is built full height, the top of wall condition (called the "head") must be designed to accommodate the structural **deflection** of the floor or roof above. Deflection is usually accommodated in metal studs by using an extra-deep (tall) head track with slightly short studs (shorter than the anticipated deflection) and with slots for fasteners (screws for metal studs or screws or nails for wood studs) or with no fasteners at all; this arrangement allows the head track to slide up and down the studs vertically so that the structure can deflect without crushing the wall. (The surfaces and finishes on such a wall are critical because they cannot extend all the way to the top of the wall, or they would be crushed under deflection too. In other words, a "top of wall gap" must be designed into the wall surfaces and finishes to allow for this inevitable movement.)

5.1.5 What are interior walls made of?

Numerous different materials have been used over the past hundreds of years for interior walls:

- *Solid stone.* This is rare, but possible. The only example known to the author was at the Upjohn Corporation headquarters building in Kalamazoo, Michigan, which was built in the mid-1960s; in that building, the interior walls consist of 6″-thick solid slabs of marble, from the floor to the underside of the ceiling. Using a substantial amount of this material should trigger a structural review, especially if it is located on an elevated (i.e. not on the ground) slab.

- *Stacked (or piled or joined) stone, with or without mortar.* This is the technique most commonly found all around the world in very old buildings. Using a substantial amount of this material should trigger a structural review, especially if it is located on an elevated (i.e. not on the ground) slab.
- ***Cast-in-place concrete***. Concrete is a mixture of water, sand (light aggregate), gravel (heavy, or coarse, aggregate), and Portland cement that is placed into forms wet. It is very rarely used for interior walls, although it is commonly used for exterior walls, especially for underground walls. Wiring and piping can be installed within solid concrete where necessary. Using a substantial amount of this material should trigger a structural review, especially if it is located on an elevated (i.e. not on the ground) slab.
- ***Pre-cast concrete***. Commonly used in parking garages and as exterior walls in some building types (warehouses, factories, large stores, even some churches), but rarely used for interior walls.
- *Solid* **concrete masonry units**. The hollows inside the walls are filled with grout after the wall is built. Using a substantial amount of this material should trigger a structural review, especially if it is located on an elevated (i.e. not on the ground) slab. The units are bonded together (end-to-end and top-to-bottom) using mortar, as in brick work.

www.betcosupreme.com (the web-site of Betco, Inc., an Oldcastle Company)
www.cement.org (the web-site of the Portland Cement Association)
www.swconcrete.com (the web-site of Headwaters Construction Materials)

- *Hollow concrete masonry units.* These are the same as the solid units, with mortar between units but without the grouting inside the units. Wiring and plumbing can be routed inside hollow CMU walls. Using a substantial amount of this material should trigger a structural review, especially if it is located on an elevated (i.e. not on the ground) slab.

www.betcosupreme.com (the web-site of Betco, Inc., an Oldcastle Company)
www.cement.org (the web-site of the Portland Cement Association)
www.swconcrete.com (the web-site of Headwaters Construction Materials)

- *Hollow structural tile.* This was a common system in some areas from the late 19th century up through the 1940s or so and it consists of hollow clay tiles that are stacked and **plastered**. Using a substantial amount of this material should trigger a structural review, especially if it is located on an elevated (i.e. not on the ground) slab.
- *Glazed structural tile.* Commonly used in schools in the 1950s and later, this is an updated version of the structural tile noted above, also using mortar. In this case, the tile has a fired enamel finish on one, or both, faces, which provides for one of the most durable finishes available for interior walls. It was commonly used as a

wainscot (4′0″ to 6′0″ high) on school corridor walls and it is seen occasionally today (where its high cost can be justified due to heavy wear-and-tear). Ironically, it is rarely used in schools any more. Using a substantial amount of this material should trigger a structural review, especially if it is located on an elevated (i.e. not on the ground) slab.

www.astraglaze.com (the web-site of Trenwyth)
www.elginbutler.com (the web-site of the Elgin Butler Company)

- *Wood studs*. Most commonly, 2 × 4 studs, usually at 16″ o.c. (on center) or sometimes at 24″ o.c. But wood stud walls can be built using 2 × 3, 2 × 6, 2 × 8, or even larger studs too. It is also possible to use studs sideways to make walls thinner, if the walls are not load bearing or very tall. Stud walls are usually hollow (although sometimes partially filled with insulation for acoustical reasons) and faced with some kind of opaque material. The most common facing is gypsum board, but numerous others products can be used too: plywood, MDF (medium density fiberboard), wheatgrass board, bamboo, cork, rubber, vinyl, solid wood planking, composite gypsum/paper boards, wood or metal **lath** and plaster (all three types of plaster), veneer plaster (a gypsum plaster finish coat over a special type of gypsum board—faster and more cost effective than traditional wet plaster), ceramic tile, stone, metal (solid, perforated, or mesh), plastics (especially 3Form, polycarbonate sheets similar to Polygal, and other composite products), plastic laminates, glass, glass tile, and fabric. For improved acoustical performance, **staggered studs** can be used as well as **double studs**.

  www.polygal.com (the web-site for Polygal)
  www.3-form.com (the web-site for 3-Form and 3-Form Chroma)

- *Metal studs*: usually steel and nearly always $1\frac{1}{2}$″ wide on the faces (similar to a wood 2 × 4). The most common depth is $3\frac{5}{8}$″ (again similar to a wood 2 × 4) and the most common spacing is 16″ o.c. although 24″ o.c. is not uncommon. Metal studs are also available in $1\frac{1}{2}$″, $2\frac{1}{2}$″, $5\frac{1}{2}$″ (non-standard), 6″, $7\frac{1}{2}$″, and $9\frac{1}{2}$″ depths, as well as in **gauges** from 26 through 12, where 26 is thin and 12 is very thick. The selection of gauge is dependent mostly on the height of the studs but also on load specifics; in other words, a wall with heavy cabinets on it will require stronger studs than a wall without the cabinets. And a tall wall will require a higher gauge than a short wall. Metal stud walls are also usually hollow (with the same exception for acoustical insulation noted above) and can be finished using all of the same methods noted under wood studs above.

  www.dietrichindustries.com (the web-site of Dietrich Metal Framing)
  www.steelnetwork.com (the web-site of The Steel Network, Inc.)

- *Solid wood*. Unusual but possible, using some arrangement of solid wood planks, probably in a **stile and rail** arrangement similar to doors.
- *Solid glass*. Solid glass blocks are available in 3″ thickness; extremely costly and bullet-proof in some applications. Using a substantial amount of this material should trigger a structural review, especially if it is located on an elevated (i.e. not on the

ground) slab. Also, is a solid glass wall really a wall? Or is it a window? See Section 5.3 below.

http://pittsburghcorning.com/specifications-and-details/cad/vistabrik-solid-glass-block.aspx

- *Hollow glass block.* Most glass block is hollow (most commonly $3\frac{5}{8}''$ thick with two faces each about $\frac{1}{4}''$ thick) and it is available in many different sizes, shapes, and styles (some nearly transparent but most translucent) and with or without internal inserts to block viewing. Also, is a hollow glass wall really a wall? Or is it a window? See Section 5.3 below.

  www.pittsburghcorning.com (the web-site of Pittsburgh Corning, a company that makes many different glass products, including both solid (as noted above) and hollow glass blocks)

- *Framed glass, using wood, steel, stainless steel, aluminum, etc.* Many systems are on the market for both exterior (usually called "storefront" or "curtainwall" systems) and interior glass walls, and they cover the gamut from very small and simple to large, heavy, and complicated. Exterior systems can be used in interior applications, but their weather-sealing features are unnecessary. Interior systems cannot be used on the exterior at all. Also, is a framed glass wall really a wall? Or is it a window? See Section 5.3 below.

  www.kawneer.com (the web-site for Kawneer, one of the leading storefront and curtain companies)
  www.nanawall.com (the web-site for Nana Wall, a company that makes folding glass wall systems)
  www.tubeliteinc.com (the web-site for Tubelite, a major competitor to Kawneer)
  www.usalum.com (the web-site for United States Aluminum, a major competitor to Kawneer)

- *"Frameless" glass.* It is possible to build walls using only heavy panes of glass and very minimal head and sill channels, which results in what appears to be a frameless wall. But, is a frameless glass wall really a wall? Or is it a window? See Section 5.3 below.

  www.avantisystemsusa.com (the web-site for Avanti Systems USA, a company that makes frameless glass doors (swinging and sliding) and frameless glass doors)

5.1.6 How thick are interior walls?

The most common wall in commercial construction is $4\frac{7}{8}''$ thick overall, consisting of two sheets of $\frac{5}{8}''$-thick gypsum board over $3\frac{5}{8}''$-thick metal studs. In residential construction, the most common thickness is $4\frac{1}{2}''$, consisting of two sheets of $\frac{1}{2}''$-thick gypsum board over 2 × 4 (nominal $3\frac{1}{2}''$-thick) wood studs.

Another common commercial system is $3\frac{3}{4}$"-thick: two $\frac{5}{8}$"-thick gypsum board sheets on $2\frac{1}{2}$"-thick metal studs.

But many other options are available, from $\frac{1}{2}$" solid glass to 8", 10", 12" or even thicker. Very thick walls usually result from double stud arrangements in acoustically sensitive situations.

For concrete masonry walls, standard thicknesses are $3\frac{5}{8}$", $5\frac{5}{8}$", $7\frac{5}{8}$", $9\frac{5}{8}$", and $11\frac{5}{8}$" (known nominally as 4", 6", 8", 10", and 12" block).

For solid concrete walls, 8" is as thin as it is likely to get if it is site formed and placed; 4" is practical for pre-cast concrete, but pre-cast concrete is very rarely used in interior walls.

It is very important to identify wall thickness early on in the space planning phase in order to avoid dimensional problems later on.

5.1.7 How are masonry (stacked stone, concrete masonry units, brick, and glazed structural tile) walls built?

To build an interior masonry wall, masons come in and lay out the locations of the walls on the floor slab and they start setting individual blocks in **mortar** beds (i.e. bed joints), filling in vertical joints (head joints) as they go. They can only lift the heavy blocks a few feet vertically, so they will erect scaffolding to finish the upper parts of the walls (raising the scaffolding as the walls get higher). As they move along vertically, they will **tool** the joints. A hand-held tool is used to compact the face of the mortar, making it smoother and more weather-resistant, and to shape the face of the joint. The most common shape is a concave joint, where the face of the joint is a small indented arc. Other joint types include flush joints (flat on the surface and not advisable), raked joints (raked out horizontally to emphasize the joints; OK for interior walls but not recommended for exterior joints that are exposed to weather), beaded joints (shallow raking with a projecting half-round bead in the middle of the joint), weathered joints (raked at the top, sloping to flush at the bottom), struck joints (raked at the bottom, sloping to flush at the top—not recommended for weather), and extruded joints (where the mortar protrudes beyond the face of the wall).

Tooling is done from the bottom up as the wall is built, and cleaning the wall (removing excess mortar, etc.) is done from the top down as the last step. If a wall is to be **grouted**, the grout is poured into each cell from the top of the wall until the wall is full.

When using brick or concrete masonry units, outside corners can be square or rounded (bullnose). That said, bullnose bricks are very rare and are likely to be seen only in older existing buildings.

Building unit masonry construction with mortar is a slow (and messy) process, but it does result in very tough walls that should be serviceable with minimal refinishing for many decades. From a sustainability point-of-view, masonry products (stone, sand, gravel, etc.) are low-**embodied energy** materials that are in plentiful supply, usually from a reasonably local source. Portland cement is a highly negative material from an environmental point-of-view (due to high energy use and polluting chemicals that are required),

Figure 5.1.7.A
Joint tooling

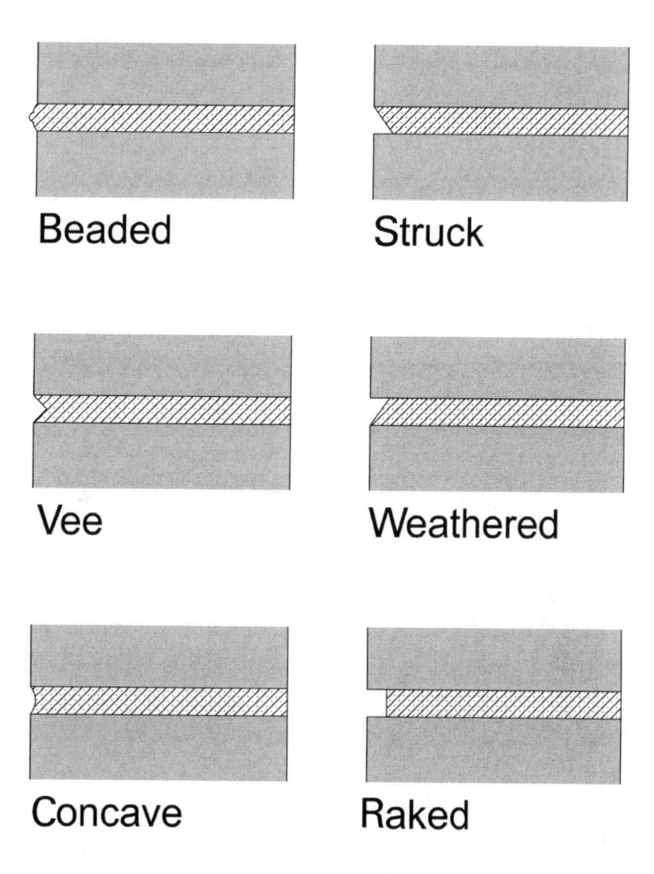

but it is used in relatively small quantities and the characteristics of the resulting construction make that a reasonable trade-off.

Brick and concrete masonry units are laid up in "bond" patterns, although the actual bonding is rarely relevant today. Bonding means connecting multiple layers of unit masonry together from layer to layer, horizontally; in brick, that is done by rotating some bricks 90 degrees so that they are half in one layer and half in the next. Few masonry installations today have more than one layer so this does not matter. For single-**wythe** masonry, the most common bond patterns are $\frac{1}{2}$ **running bond** and stack bond for both brick and concrete unit masonry and $\frac{1}{3}$ running bond for some sizes of bricks.

5.1.8 How are wood stud walls built?

Wood stud walls consist of sill plates, studs, and top plates. The sill plate is a flat stud that runs along the floor to form a base for the vertical studs. In panelized (pre-fabricated) walls, the studs are nailed through the sill plate from the bottom (before the wall is put into position). In site-built walls, the studs are **toe-nailed** to the sill plate. The top plate is usually a double plate, consisting of two flat studs running across the tops of the studs. The studs are either through-nailed (from the top) or toe-nailed (from the sides) to the first top plate, and the second top plate is through-nailed to the first top plate. Double top

Figure 5.1.7.B
$\frac{1}{2}$ running bond

Figure 5.1.7.C
$\frac{1}{3}$ running bond

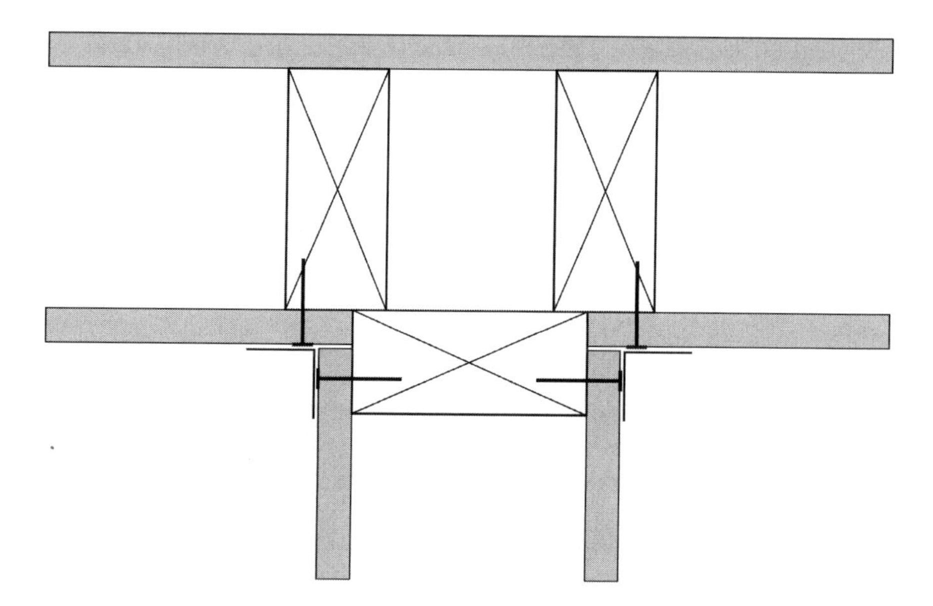

Figure 5.1.8.A
Wood stud wall lead

plates are especially important in bearing walls, because roof rafters or floor joints could land in between the studs; the double plate transfers the point load from the rafter or joist to the studs. In a non-bearing wall, a double top plate might not be necessary.

In stud walls, it is necessary to build "leads" in the corners to make it feasible to attach facing (gypsum board, plywood, etc.).

A lead consists of two studs at the end of an intersecting wall, with their inner edges aligned with the outer edges of the studs to provide nailing surfaces for the facing. (Double top plates make this issue simpler at ceilings too, which is why double top plates are usually used in non-bearing walls.) The wall that intersects a lead must begin with a stud that is corner-to-corner with both lead studs to provide nailing surfaces on the intersecting wall.

Wood is a renewable resource in the sense that we can always grow more of it, but if that growth is not planned, the supply might not match the demand and excessive destruction can occur. Most framing lumber comes from fast-growing softwood species such as pine, fir (especially Douglas Fir for high-quality framing lumber), and spruce grown mostly in the southeastern and northwestern United States; slow-growing hardwoods (such as maple, oak, ash, etc.) are unnecessary for most framing lumber and they are used primarily for furniture and trim.

Wood is available from certified forests that guarantee that each tree cut is replaced to ensure that the forest is never depleted. Such wood is more costly but using it is the right thing to do. See http://us.fsc.org/en-us, for more information about the Forestry Stewardship Council.

5.1.9 How are metal stud walls built?

The terminology and basic approach are the same as for wood studs, but the pieces are slightly different. Instead of a sill plate, in metal studs there is a runner track (usually a channel with 1″-high flat vertical faces). The studs fit into the runner channel and terminate at the top in a head track. A standard head track is an upside-down runner track. A head track for a full-height wall is extra-tall (2″ or even 3″ in some cases) and uses slotted fasteners (or no fasteners at all). And, as noted previously, walls that terminate at the underside of lay-in ceilings sometimes use eliminator tracks at their tops.

The sequencing is a little different too. In wood studs, in a site-built wall, the sill plate is put down, then the studs are put in, then the first top plate is put on, and then the second top plate is put on. In metal studs, the runner track goes down first, then the head track, and the studs go in last (this is also true with an eliminator track application). From a sustainability point-of-view, wood is good because it is renewable, although not necessarily rapidly renewable. Softwoods in the southeastern and northwestern US can grow to harvestable size in less than ten years; hardwoods tend to grow far more slowly, sometimes taking 100 years to reach maturity.

Even though steel is high-embodied energy material, it is also almost infinitely recyclable. With each recycling, the average embodied energy drops, and it can resemble a low-embodied energy material after it has been re-used five or six times. Today, nearly 90% of all steel used is recycled, so it can be an appropriate material for highly sustainable projects, for both heavy steel for structural framing and for light steel for interior (and sometimes exterior) walls.

www.dietrichindustries.com (the web-site of Dietrich Metal Framing)
www.steelnetwork.com (the web-site of The Steel Network, Inc.)

5.1.10 How are openings made in a stud walls?

In both wood and metal stud walls, openings require sills, heads, and **jambs**. Jambs are usually double vertical studs, but sills and heads have to be added. A non-structural head would consist of a sill plate or runner track spanning the opening (usually supported by shortened inner studs on both sides), with short cripple studs filling in above the opening. A sill would consist of a sill plate or head track with short cripple studs filling in below the sill. If structural headers or sills are required for some reason, a structural engineer or architect should be consulted. If the opening is for a door, there would be no sill.

5.1.11 How are walls demountable?

Demountable wall systems usually consist of some sort of metallic framing (exposed pre-finished steel or aluminum head and sill tracks, and aluminum or steel studs) and pre-finished wall panels (usually gypsum board factory-wrapped with vinyl wallcovering) that are clipped to the framing system with removable hardware. If someone wants to move the wall, the clips are removed, the panels are removed, the studs are taken out, the head and sill tracks are relocated and the process is reversed to reinstall the wall at its new location. That is all well and good in theory, but, in practice, few of these walls have ever been relocated. If a wall is 10 or 15 years old, with 10- or 15-year-old vinyl wallcovering on it, who wants to spend money to move that? And if conventional wiring has been added (and it often has been), an electrician is required for the dismantling and reassembly of the wall, making the process even more costly and awkward. In recent years, more sophisticated versions of these systems have hit the market, specifically DIRTT walls and several products from Steelcase (V.I.A., Post & Beam, Privacy Wall, and FlexForm Work Walls), which are more easily dismantled and relocated. Unsurprisingly, such systems are costly.

Systems furniture is different because it really is designed to be taken apart on a regular basis and reassembled in a new configuration or at a new location or both. And most of the systems include wiring provisions and replaceable panels (to avoid the 10- or 15-year-old vinyl wallcovering problem). Of course, such sophisticated products are very costly and most owners still do not move them around very much.

www.DIRTT.net
www.steelcase.com
www.warestar.com/material-handling-resources/brochures/Gravity-Lock-Demountable-Wall-Systems.pdf

5.1.12 Can walls move?

Sure, but it is not easy and they are really no longer walls if they actually move. Moving walls are called "doors," which takes us directly to Section 5.2.

5.2 Doors

5.2.1 Why do we need doors?

We need doors to allow us to get through the walls that we decided that we needed for physical separation, security, visual privacy, acoustical privacy, or for some other reason.

5.2.2 What is a door?

In its simplest form, a door is a section of wall that can be moved to allow access. But that is not what most doors look like, is it? Why not? Because designing and building sections of wall that actually move is too difficult and too costly for most situations. Part of the challenge with moving a section of a wall is that walls are designed to be static objects; that is, objects that do not move. So when you start moving around a chunk of stud wall (or some other kind of wall), unexpected things happen: sagging, warping, twisting, even breaking.

As result, two "standard" approaches have been developed for wood doors over the past few hundred years:

- Stile and rail doors
- Flush doors.

5.2.3 Stile and rail doors

Stile and rail doors are commonly used (especially in traditional design), and the most common form is the six-panel door. This is a door that has relatively wide vertical rails along both sides and in the middle, a wide lower cross rail (a stile), a somewhat narrower top stile, and two intermediate stiles to define six panels.

Note the six-panel door in the middle of Figure 5.2.3.A. The lower and middle panels are usually of roughly equal height and the top panel is usually quite short. The panels are inset, but the panel face is usually raised above the inset face, so this is called

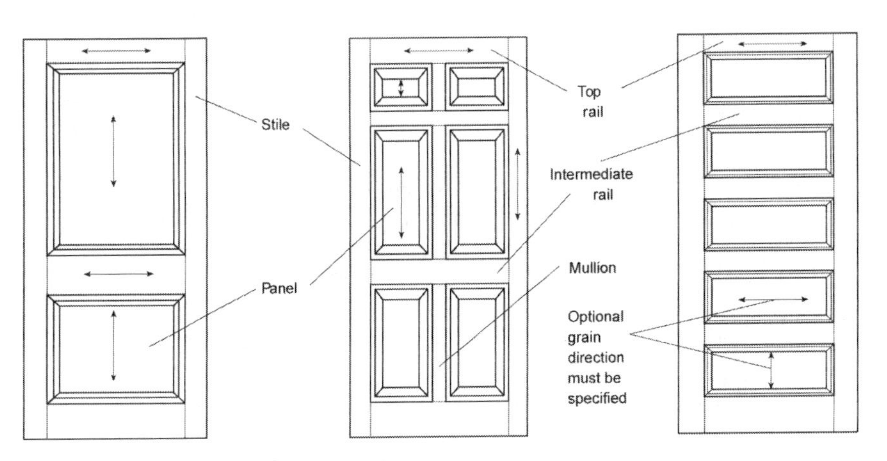

Figure 5.2.3.A
Stile and rail door layout (© Architectural Woodwork Institute, *QSI* 8th Edition, reproduced with permission)

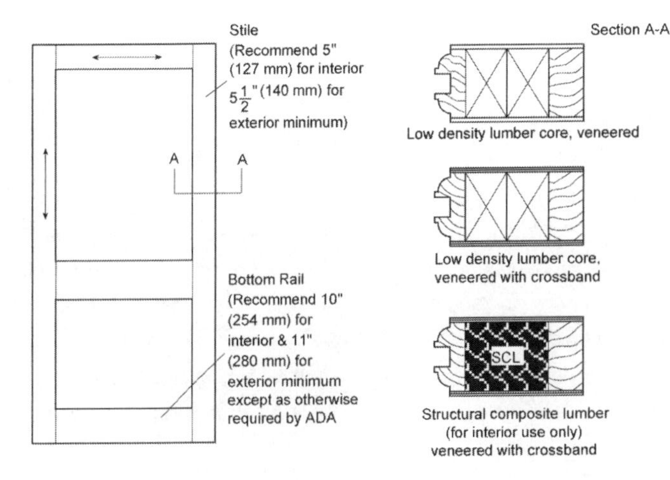

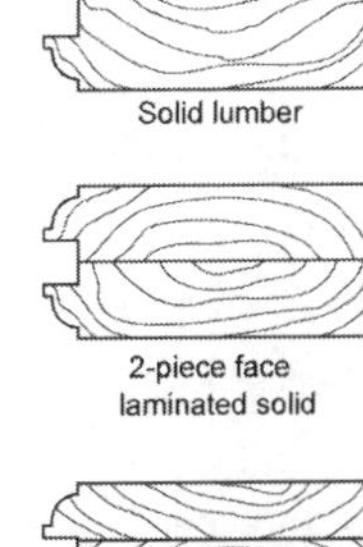

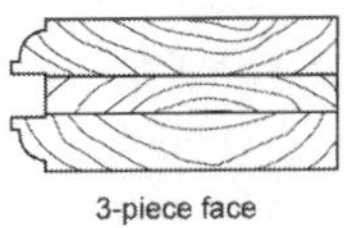

Figure 5.2.3.B
Stile and rail door construction (© Architectural Woodwork Institute, *QSI* 8th Edition, reproduced with permission)

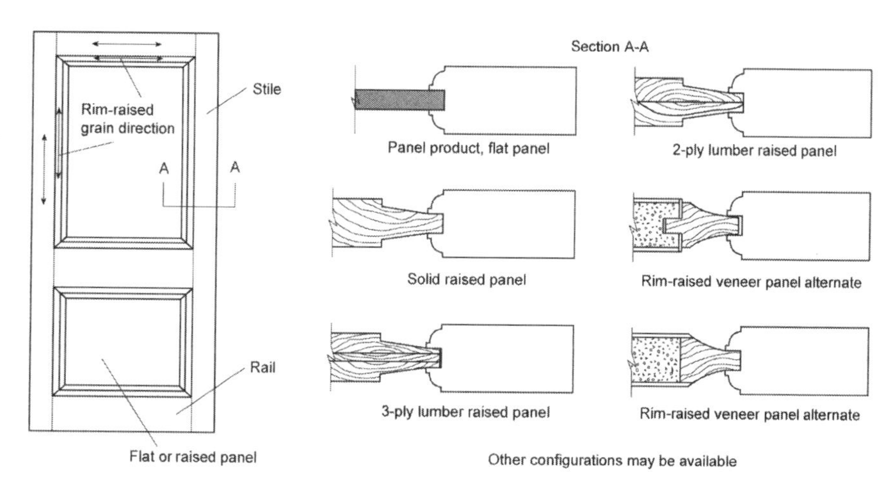

Figure 5.2.3.C
Stile and rail panel construction (© Architectural Woodwork Institute, *QSI* 8th Edition, reproduced with permission)

Stile
A
A
Rail
Section A-A
Flat bead stop
Moulded stop (one side applied)
Lipped moulding
Inset molding
Other configurations may be available

Figure 5.2.3.D
Panel and glass retention (© Architectural Woodwork Institute, *QSI* 8th Edition, reproduced with permission)

a "raised panel door." Most importantly, the panels are actually not attached within the frames of the rails and stiles, to allow for movement as the door expands and contracts with temperature and humidity changes. (Humidity changes in particular are tough on doors; if the door fits and operates properly under high humidity conditions, it is probably a little too small under cold and low humidity conditions; if it fits under cold and low humidity conditions, it probably sticks when it is hot and humid (anyone have a door like that?).) The stiles and rails are joined together using very tough and durable mortise and tenon or dowel style joinery, to keep the door square over time.

Stile and rail wood doors are seen in furniture and cabinetry as well as for both exterior and interior openings, in numerous different sizes and across a wide swath of quality levels. Traditionally, doors like this are made using solid wood (no veneer, no plastic, no metal, no insulation, etc.).

Multi-panel doors can have **lights** within one or more of the panels. Louvers for ventilation can be done too.

The most authoritative source available for door construction specifications and techniques is the Architectural Woodwork Institute (AWI) in Potomac Falls, Virginia (www.awinet.org). The AWI published (jointly with the Architectural Woodwork Manufacturers

Association of Canada and the Woodwork Institute) *Architectural Woodwork Standards, Edition 2*, on October 1, 2014, and this is by far the most complete and authoritative source for wood information that is available. It includes:

Section 1: Submittals
Section 2: Care & storage
Section 3: Lumber
Section 4: Sheet products
Section 5: Finishing
Section 6: Millwork
Section 7: Stairwork & rails
Section 8: Wall/ceiling surfacing & partitions
Section 9: Doors
Section 10: Casework
Section 11: Countertops
Section 12: Historic restoration work[1]

(www.awinet.org)

Section 9 "Doors" notes that stiles and rails can be constructed six different ways (see Figure 5.2.3.B):

1 Low density lumber core, veneered (where only the face skins are truly solid wood)
2 Low density lumber core, veneered with crossband
3 Structural composite lumber (for interior use only), veneered with crossband
4 Solid lumber (the old-fashioned method)
5 Two-piece laminated solid lumber (a way to use less costly smaller lumber)
6 Three-piece laminated solid lumber (another way to use less costly solid lumber)[2]

Six different panel types are defined (see Figure 5.2.3.C Panels):

1 Panel product, flat panel (no raised panels)
2 Raised, 2-ply solid lumber
3 Solid raised panel
4 Rim-raised veneer panel alternate (A)
5 3-ply lumber raised panel
6 Rim-raised veneer panel alternate (B)[3]

Four different panel or glass retention examples are provided (see Figure 5.2.3.C):

1 Flat bean stop (simple rectangular trim pieces on both sides)
2 Molded stop, one-side applied (here, a molding profile is cut into the opening on one side and a matching loose piece of trim is used on the other side)
3 Lipped molding (trim molding that projects beyond the face of the door)
4 Inset molding (applied moldings within the thickness of the door)[4]

(By the way, the standard door size in commercial work is 3′ 0″ wide by 7′ 0″ high and $1\frac{3}{4}$″ thick; the standard door size in residential work is 2′ 8″ wide (not entirely standard really and 3′ 0″-wide doors are becoming more common) by 6′ 8″ high and $1\frac{3}{8}$″ thick. This applies to all "standard" types of doors and not just wood doors.)

One of the disadvantages of exterior stile and rail doors is the lack of insulation, and it is now possible to get stamped vinyl and steel doors that appear to be multi-panel doors; in fact, they have single-piece face skins with concealed insulated cores of various types, and they probably are not stile and rail doors at all.

Sometimes, a designer does not want to have an inset, or raised, panel look, and a flatter looking door is preferred. Such a door can still be a stile and rail door, but it cannot be done easily with floating panels (as with the six-panel door described above). Instead, a door like that needs to be a rigid and stable assembly that will not warp or twist under typical ranges for temperature and humidity.

If a truly flat door is desired, it becomes a flush door.

www.etodoors.com (the web-site for ETO Doors, a company that makes many different types of solid wood doors)

www.solidhardwooddoors.com (the web-site for Allegheny Wood Works, Inc., a company that makes many different types of solid wood doors)

www.trustile.com (the web-site for TruStile Doors, LLC, a company that makes both residential and commercial stile and rail doors)

www.vtindustries.com (the web-site for VT Industries, Inc., a company that makes both stile and rail and flush doors)

5.2.4 Flush doors

A flush wood door has the stiles and rails concealed behind flat faces of some sort. The faces can be wood, plywood, wood veneer (from paint grade birch all the way up through exotic, rare, and extraordinarily beautiful species), plastic laminate, or even metal. A door like this will have narrow rails along both sides, a high stile across the bottom, a narrow stile across the top, and reinforcement panels (solid wood blocking) for hardware reinforcement (hinges, closer, lockset, etc.). The core of the door between the stiles and rails is made various different ways according to budget and preferences, from paper honeycomb materials to solid blocks of hardwood. Even the number of total plies (layers) in the door varies from three to seven, with seven being the highest quality and highest price.

Architectural Woodwork Standards, Edition 2, Section 9 "Doors," includes six details for **Dutch doors**;[5] five details for louvers,[6] and three details for **flashing**[7] for exterior doors. The AWI also publishes eight standard arrangements for hardware blocking[8] (see Figures 5.2.4.A, 5.2.4.B, 5.2.4.C, and 5.2.4.D).

The AWI defines the four basic flush wood door construction methods as follows:

"3-ply consists of a core with a plastic laminate face applied over both sides of the core."[9] It should be noted that this type of construction applies *only* to doors

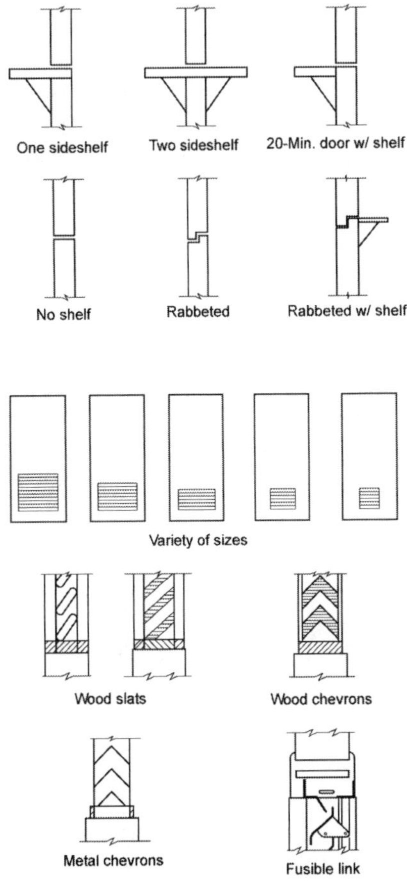

Figure 5.2.4.A
Dutch doors
(*Architectural Woodwork Standards*, 2nd edition, © AWI/AWMAC/WI 2014)

Figure 5.2.4.B
Louver options
(*Architectural Woodwork Standards*, 2nd edition, © AWI/AWMAC/WI 2014)

Generally, fusible link louvers installed in 45, 60, and 90 minute fire rated doors must comply with individual fire door authorities. Wood louvers are not allowed by NFPA 80 in fire rated doors. All doors must comply to accessibility requirements.

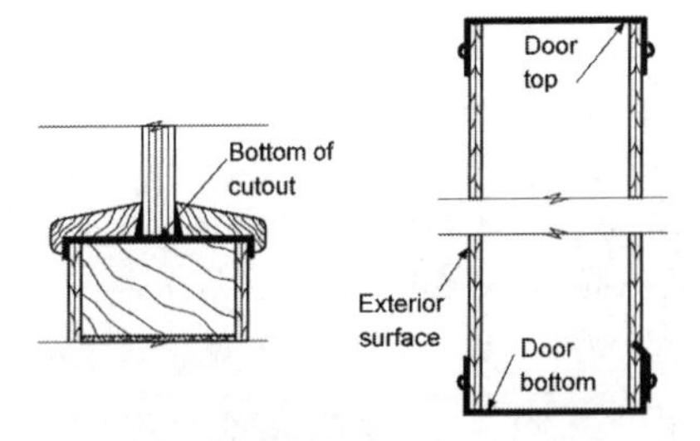

If the manufacturer is to flash the top of the door or the bottom edge of cutouts for exterior doors, it must be specified.

Figure 5.2.4.C
Flashings (*Architectural Woodwork Standards*, 2nd edition, © AWI/AWMAC/WI 2014)

For undercutting flexibility and specialized hardware applications, a number of internal blocking options are available from most manufacuterers. When blocking is required it is typically at particle core and fire resistant core doors. Options such as 5"(127 mm) top rail, 5" (127 mm) bottom rail, 5"x 18" (127 x 457 mm) lock blocks (may be one side only), $2\frac{1}{2}$" (64 mm) cross blocking are available, but there are other options are available. Consult your manufacturer early in the design process to determine requirements.

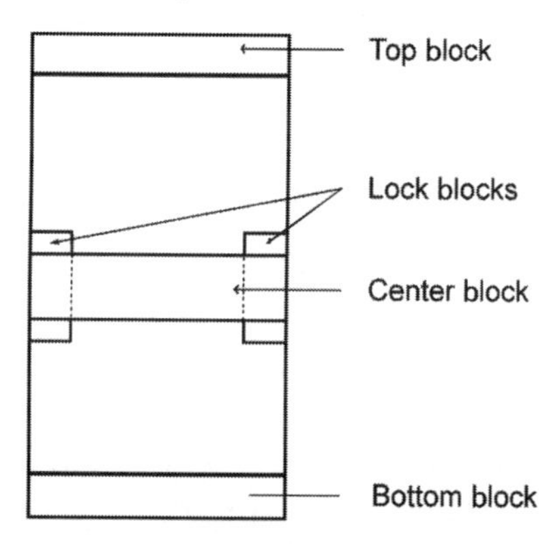

Hardware blocking, if desired, shall be specified from the following typical options:

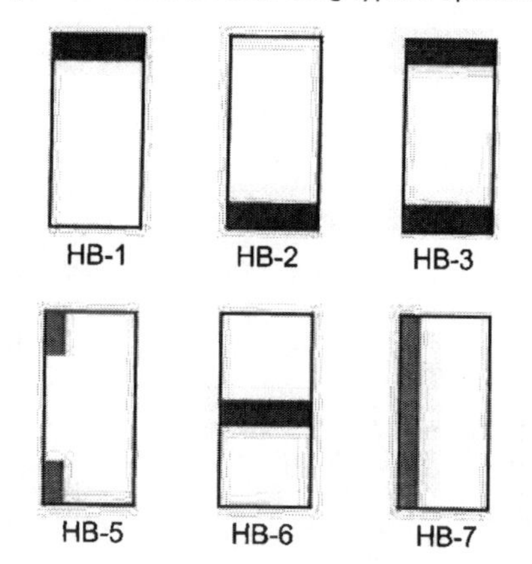

Top blocking may be full or partial width as required by its application.

Figure 5.2.4.D
Blocking options (*Architectural Woodwork Standards*, 2nd edition, © AWI/AWMAC/WI 2014)

with plastic laminate faces. The core can consist of particleboard, MDF, or agrifiber material.

With plastic laminate faces,"5-PLY consists of a wood veneer or composite cross band applied over the core before application of the face laminate."[10] It should be noted that this type of construction also applies *only* to doors with plastic laminate faces. The core can consist of particleboard, MDF, agrifiber material, or fire-resistant composite material.

With other faces, "5-ply consists of a center core on which is applied to each side a wood veneer or engineered composite cross band with a face veneer applied over the cross band."[11] It should be noted that this type of construction, and 7-ply below, applies to wood faced doors. The core can consist of particleboard, MDF, agrifiber material, staved lumber, structural composite lumber, or fire-resistant composite material. Staved lumber core doors are the highest quality and the most costly.

"7-ply consists of a center core on which is applied to each side 3-ply face skins."[12] The core can consist of particleboard, MDF, agrifiber material, staved lumber, structural composite lumber, or fire-resistant composite material. Staved lumber core doors are the highest quality and the most costly. 7-ply doors are also available with hollow cores (see Figure 5.2.4.E).

Lights can be added in virtually any conceivable size and arrangement, and can be filled in with plain glass, translucent glass, patterned glass, textured glass, patterned and textured glass, and even fire-rated ceramic materials that appear to be glass.

Flush doors are also available in steel and aluminum. Steel flush doors are called "hollow metal" because they are metal and hollow; clever, eh? Aluminum flush doors are a special type that has become quite rare in the past couple of decades; more on that in the next section.

Hollow metal doors can be some of the toughest doors available, depending on the gauge of the face panels, but they do require painting and future re-painting. They are also available with a wide variety of lights and/or louvers, just like wood doors.

www.mohawkdoors.com (the web-site for Mohawk Doors, a Masonite Company)
www.vtindustries.com (the web-site for VT Industries, Inc., a company that makes both stile and rail and flush doors)

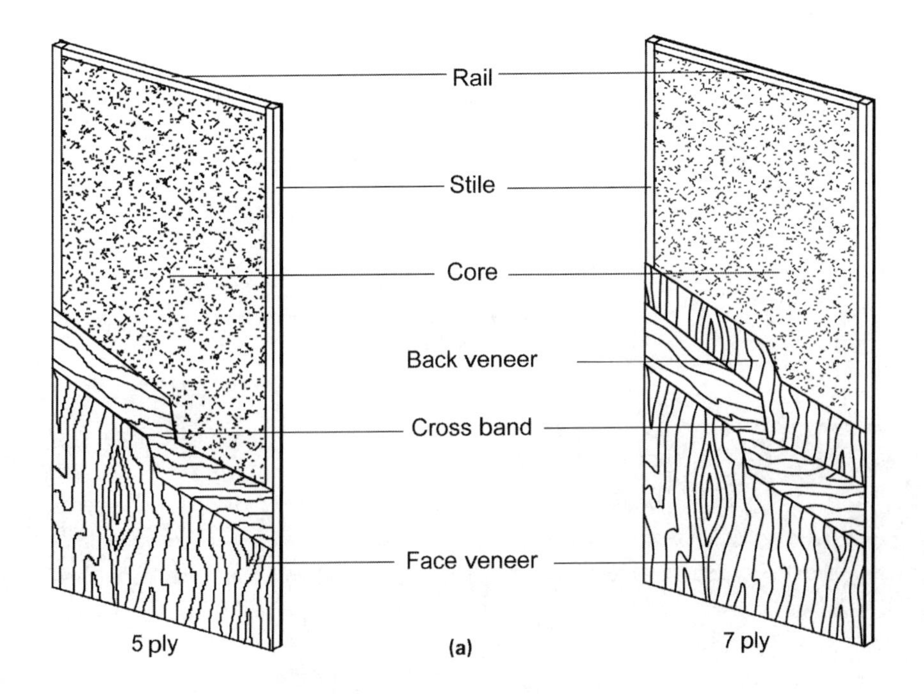

Figure 5.2.4.E
Door construction cut-away examples: (a) Wood veneer face with particleboard, MDF, or agrifiber core (PC-5/PC-7);

Figure 5.2.4.E (continued)
(b) HPDL face with particleboard, MDF, or agrifiber core (PC-HPDL-3/PC-HPDL-5);

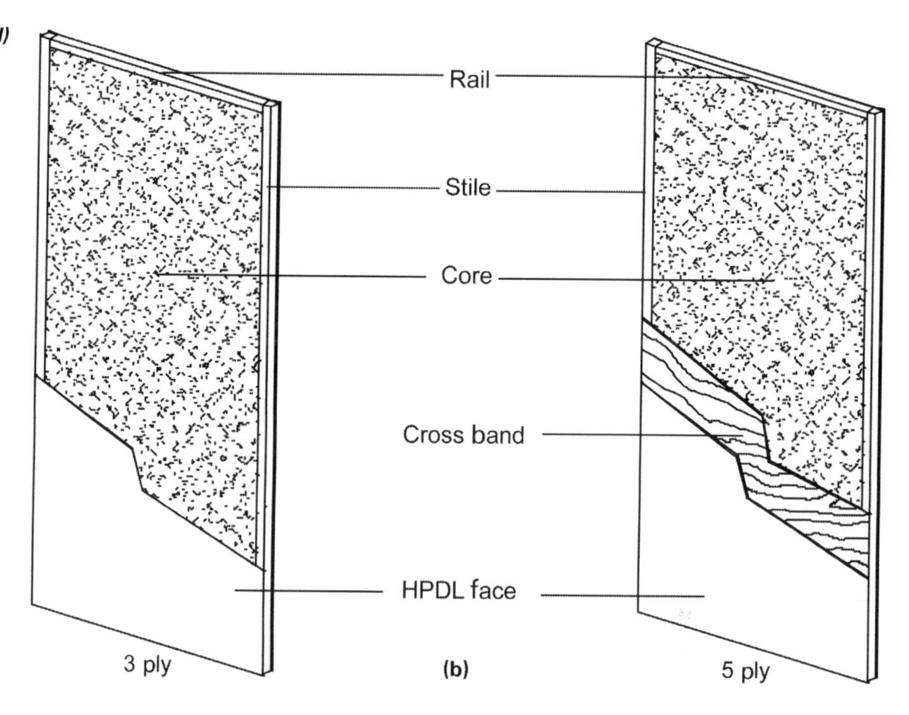

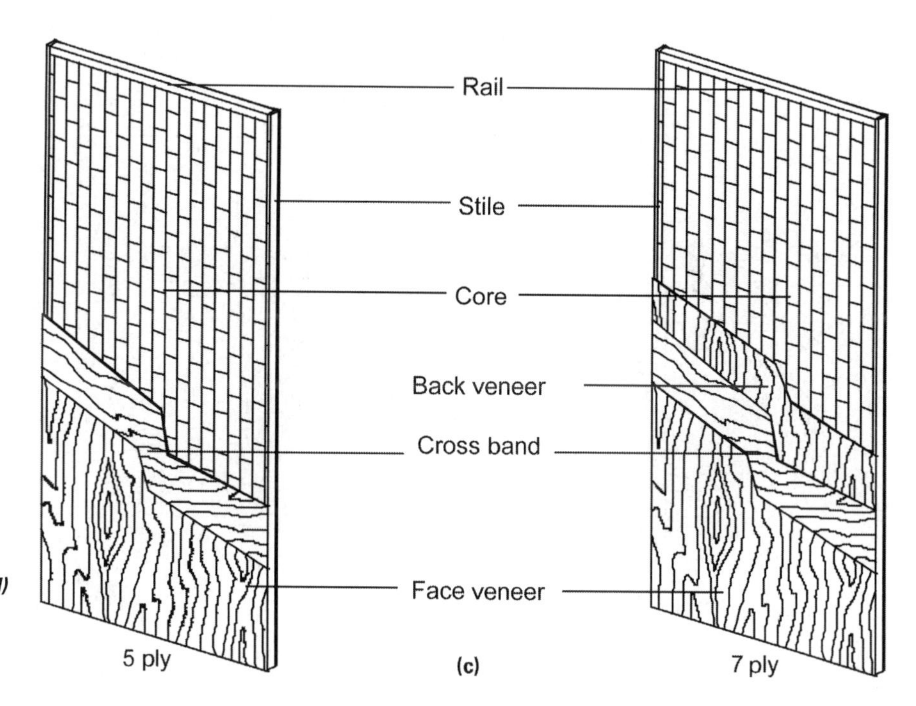

Figure 5.2.4.E (continued)
(c) Wood veneer face with staved lumber core (SLC-5/SLC-7);

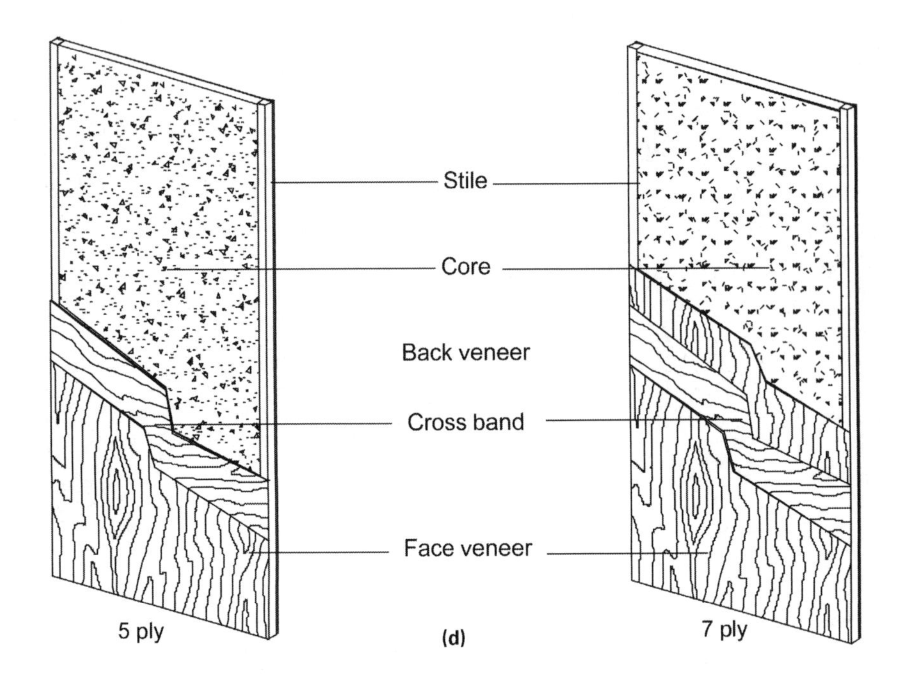

Figure 5.2.4.E (continued)
(d) Wood veneer face with structural composite lumber core (SCLC-5/SCLC-7);

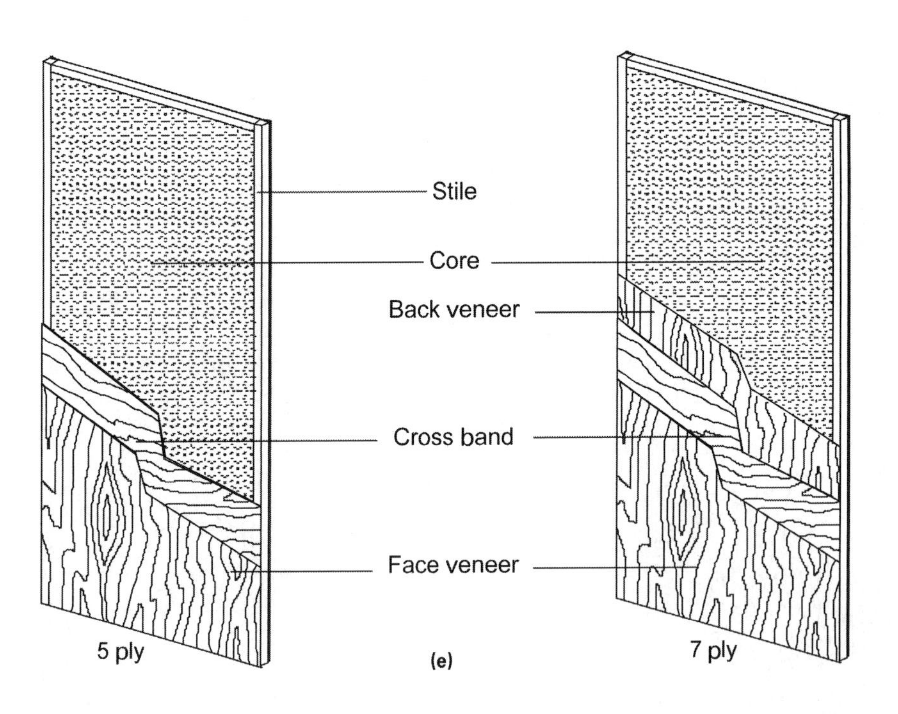

Figure 5.2.4.E (continued)
(e) Wood veneer face with fire-resistant composite core (FD-5/FD-7);

Figure 5.2.4.E (continued)
(f) HPDL with fire-resistant composite core (FD-HPDL);

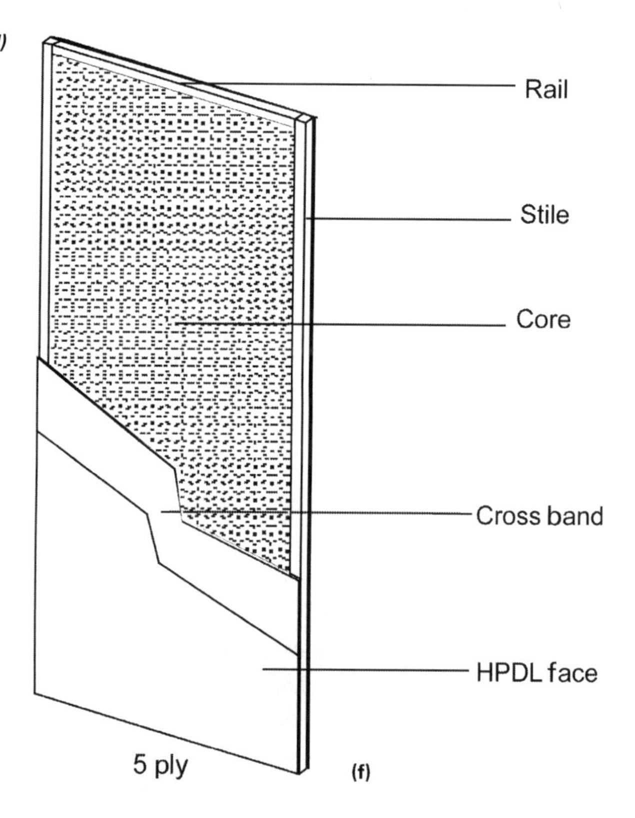

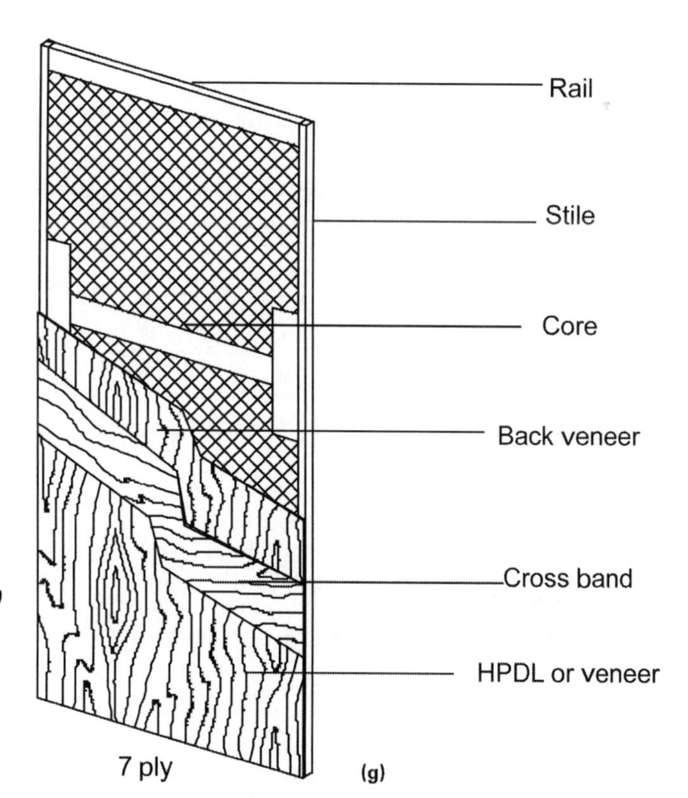

Figure 5.2.4.E (continued)
(g) Wood veneer/HPDL face with hollow core (HC-7).
Source: *Architectural Woodwork Standards,* 2nd edition, © AWI/AWMAC/WI 2014)

5.2.4.1 Wood door finishing

Wood finishing falls into two broad categories—clear and painted. Clear finishes are intended to show the graining and general character of the wood (with or without staining) and painted finishes are intended to hide the wood completely. Both types of finishes can be applied to both types of door—both stile and rail and flush types—in the field (by painters using brushes, rollers, or spray equipment) or in a shop or factory. Through no fault of their own, it is extraordinarily difficult for painters to achieve a high-quality finish in the dirty environment of a job-site, but field finishing is commonly done. Due to environmental regulations, high **VOC** finishes can be used only in controlled environments where workers have respirators and other protective equipment, so such products cannot be used in the field these days.

There are several basic categories of clear finishes: shellac, lacquer, varnish, oil, polyurethane, vinyl, polyester, and epoxy. Shellac, lacquer, varnish, and oil all have roots in natural products from the past (especially lacquers), but the others have all been created by humans through chemistry (and chemical engineering). Shellac, lacquer and varnish are available in the old-style organic type where secretions from beetles (lac beetles) are suspended in some type of oil; most of the oil evaporates during curing, leaving a protective coating on the object. New catalyzed lacquers and varnishes are available too, which are more consistent and more durable. Spar varnish is the old organic type of varnish that was developed for use on wood boats (a spar is a mast or beam on a sail boat), and the modern version is called "conversion varnish." The latter is one of the toughest factory finishes available for wood. There are a number of different polyurethanes on the market for various purposes, including **wet-curing** urethanes, which are the toughest finishes of all. Wet-curing urethane is used, with color, to paint airplanes and it is often used (without color) to finish wood floors in heavy traffic environments (restaurants and bars in particular).

Most of the clear finishes can be done as full (or high) gloss, semi-gloss, and satin (low) gloss, and some of them can be done as matte (no gloss) finishes too.

The AWI identifies 13 possible systems:

System 1: Lacquer, nitrocellose
System 2: Lacquer, precatalyzed
System 3: Lacquer, postcatalyzed
System 4: Latex acrylic, water-based
System 5: Varnish, conversion
System 6: Oil, synthetic penetrating (available in transparent only)
System 7: Vinyl, catalyzed
System 8: Acrylic cross linking, water-based
System 9: UV curable, acrylated epoxy, polyester or urethane
System 10: UV curable, water-based
System 11: Polyurethane, catalyzed
System 12: Polyurethane, water-based
System 13: Polyester, catalyzed[13]

How to choose? The AWI provides general performance information for each system in the following categories:

- General durability
- Repairability
- Abrasion resistance
- Finish clarity
- Yellowing in time
- Finish flexibility
- Moisture resistance
- Solvent resistance
- Stain resistance
- Heat resistance
- Household chemical resistance
- Build/solids
- Drying time[14]

Even more detailed information is provided for transparent and opaque finishes (separately), including resistance to staining by 24 different chemicals (from vinegar to nail polish remover to gasoline and vodka).[15]

The highest score for transparent finishes is 134 for System 9—UV curable, acrylated epoxy, polyester or urethane (the score is 133 for opaque finishes). But this system is not widely available among mill-working shops, so System 5—Varnish, conversion—is far more common. The score for System 5 is 129 for both transparent and opaque applications. Neither of these systems can be easily used in the field. The lowest scores for transparent finishes are 57 for Penetrating synthetic oil and 77 for Nitrocellulose lacquer. The reasons to use each system, and not to use each system, are provided in the right-hand column on the table.

Another factor that must be considered in wood finishing is the species of the wood itself and the desired final appearance. Some woods have open grain (ash, chestnut, elm, mahogany, red oak, white oak, according to the AWI[16]) and other woods have close (or closed) grain (birch, cherry, maple, and yellow poplar, according the AWI[17]). This does not mean that open grained woods cannot be filled and finished smoothly, but why would one use an open grain wood if a close grain look is the intention?

For purposes of appearance and long-term durability, all wood doors should be factory finished, using one of the more durable AWI finishing systems. Field-finished wood rarely looks as good and the field-applied finish does not last as long, so why use field finishing? Unfinished doors are usually available quickly (and they can be quickly finished in the field) but factory-finished doors have to be ordered well in advance (possibly two months or longer). It is vital to plan for the lead-time so that factory finishing can be done (www.awinet.org).

Finally, what about costs? Good quality commercial solid-core wood doors are expensive, running from a few hundred dollars each to $1,000.00/each or more. The most common wood door is 3′ 0″ wide by 7′ 0″ high with **rotary-sliced** birch veneer (rotary-sliced veneers have large and inconsistent grain figuring; **plain-sliced** or **quarter-sawn** is far more attractive—but more costly) on both sides and such a door, factory finished, could easily cost more than $350.00, not installed. Anything that changes—number of internal plies, facing quality, width, height, etc.—adds to the cost:

- Change to plain-sliced birch veneer? Add $75.00 (all costs are rough estimates and are for comparison purposes only)
- Change to plain-sliced cherry veneer? Add another $200.00
- Increase the height to 8′ 4″ (a standard in the Indianapolis office market for many years)? Add another $300.00
- Change to an exotic veneer? Add another $400.00 (or even more)

What would an 8′ 8″ high by 3′ 0″ wide solid wood door with quarter-sawn white oak veneer on both sides cost? Possibly more than $1,200.00.

It is usually less costly to improve the veneer on a door (a step or two—not into the realm of exotic woods) than to use a non-standard size (usually taller). In other words, it is probably more cost effective to use a good quality cherry veneer on 3′ 0″ × 7′ 0″ doors than it would be to use a plain-sliced birch veneer on 3′ 0″ × 8′ 0″ doors.

5.2.5 Aluminum doors

Aluminum doors, usually a combination of aluminum and glass, come in many different styles, and they are designed to be exterior doors. The broadest categories are: narrow stile, medium stile, and wide stile, which obviously describes the width of the vertical (and horizontal) framing members. Doors like this are usually filled in completely with glass, but there are other alternatives too (e.g. Fiberglass-Reinforced-Plastic (FRP) panels, either within the framing or over the whole door as a flush skin, or additional aluminum rails across the face of the door).

- Narrow stile aluminum doors will have stile width in the general range of 2″.
- Medium stile aluminum doors will have stile width in the general range of 4″.
- Wide stile aluminum doors will have stile width in the general range of 6″.

Sill rails tend to be a little larger than the stiles in narrow, medium, and wide stile doors. Head rails will usually be similar in width to the stiles.

Aluminum products like this are available in two basic types of finishes: anodized and painted. Anodizing is a chemical process that is used to bond a coating (colored or not) to the surface of aluminum, and it is extremely durable. The most common colors for anodizing are "clear," 'light bronze," "medium bronze," "dark bronze," and black. (Dark bronze has dominated the market for years, but it is hardly the most attractive color available.) But anodizing can be done in colors too, and some manufacturers might offer some colored options. Painting can be done in almost any color using wet paint or by powder coating; powder coating is probably more durable.

These doors are commonly used for the inner set of doors in a **vestibule**, mostly to match the outer doors. But these doors (and their matching wall/window framing systems) certainly can be used indoors in general.

www.kawneer.com (the web-site for Kawneer, one of the leading storefront and curtain companies)
www.nanawall.com (the web-site for Nana Wall, a company that makes folding glass wall systems)
www.tubeliteinc.com (the web-site for Tubelite, a major competitor to Kawneer)
www.usalum.com (the web-site for United States Aluminum, a major competitor to Kawneer)

5.3 Door hardware

Before talking about how to design and use door frames, it is necessary to summarize the vast subject of door hardware. Here are the basic categories:

- Hinges (including center pivot, offset pivot, knuckle (butts, olive, and piano), and concealed)
- Latches
- Locks (including cylindrical, dead-bolt, mortise, electric strikes, and electromagnetic)
- Closers (including spring hinges and hydraulic closers) and power operators
- Hold-opens
- Panic devices
- Mutes (or silencers)
- Thresholds
- Seals
- Remote stops (including floor, wall, and overhead mounted)
- Coordinating hardware (for pairs of doors in a single frame where there is an **astragal** to cover the vertical gap between the doors)

5.3.1 Hinges

Hinges are the most critical components for door hardware because it is the hinges that allow the door to move as intended. Many different types and sizes of hinges are available, most of which are used for exotic purposes only. The main types are:

- Center pivots
- Offset pivots
- Knuckle (including butt, olive, and piano)
- Concealed

www.erbutler.com (the web-site for E.R. Butler & Company)

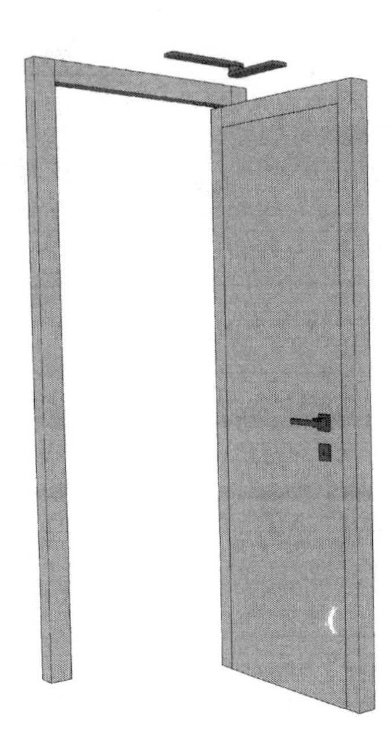

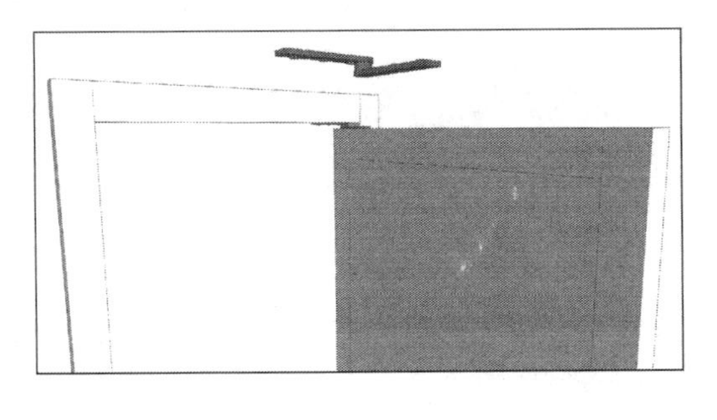

Figure 5.3.1.1
Center pivot hinge

5.3.1.1 Center pivots

The most basic hinge is a center pivot.

In this situation, some sort of pin (anything from a sharpened nail to a complicated and carefully made shaft) rotates in some sort of cup; sophisticated center pivot hinges for heavy-duty doors use bearings in the hinges, but that is not absolutely required. If these hinges are installed along the center line of the door, they will swing 360 degrees unless remote stops are provided, so they are good for doors that swing in and out. However, the dual-swinging nature does make it awkward to install stops and virtually impossible to install acoustical seals, so dual-swinging doors on center pivot hinges should not be used in acoustically sensitive environments. Center pivots are also challenging to install because it is so difficult to align the top and bottom hinges. These hinges require no jamb at all, but it is necessary to have a sill (or floor) and a door head (or ceiling) for mounting the hinges themselves. The main advantage to using center pivots is that they are usually concealed, and take nothing away from the design.

5.3.1.2 Offset pivots

Offset pivot hinges have a mechanism that is similar to center pivots, but they are designed for doors that swing in one direction only, which is done by moving the pivot center to outside the face of the door.

Each offset pivot hinge has two arms, either both horizontal or one horizontal and one vertical, according to the weight of the door and the installation conditions. For typical doors, there are usually two offset pivots: one at the sill and one at the head. The sill and head pivots will have arms that attach to the bottom and top of the door (and sometimes on the back face), respectively, and separate arms that attach to the floor or frame head (or ceiling). Sometimes, the "frame" arms mount vertically along the face

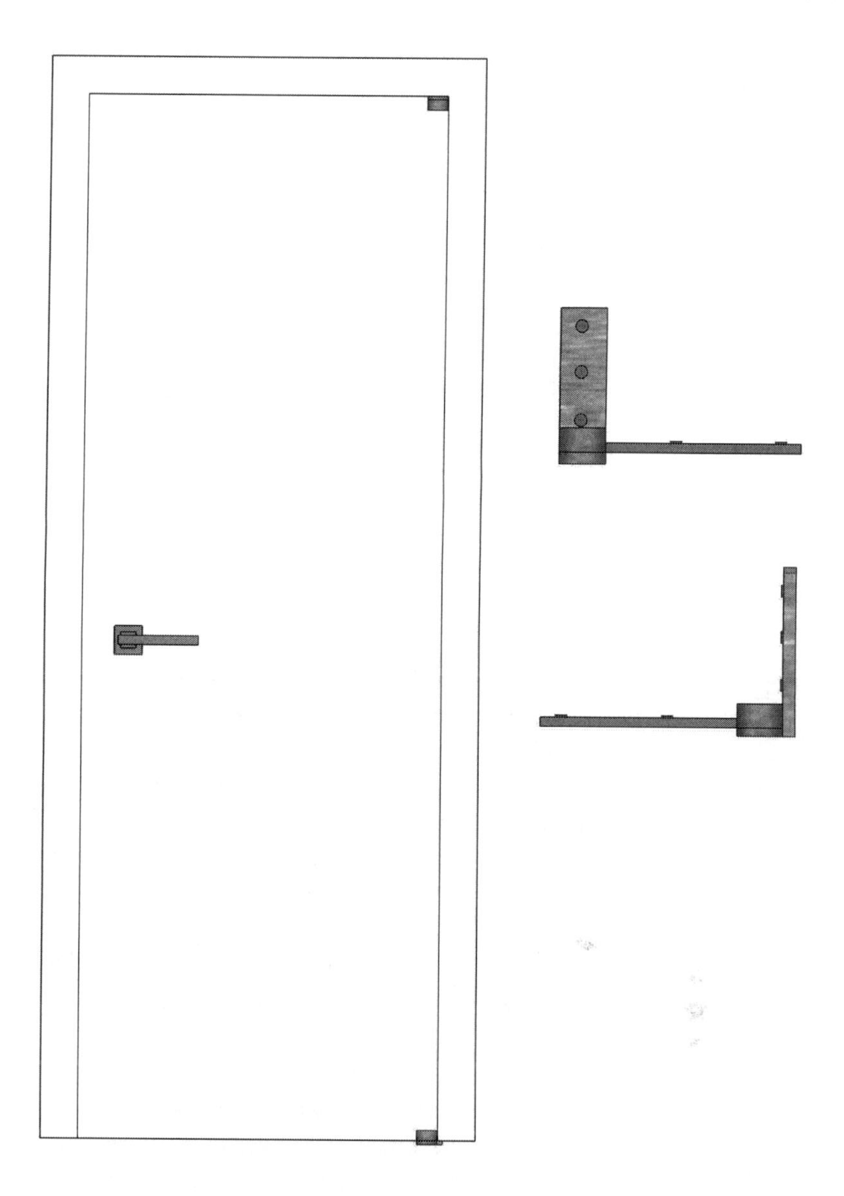

Figure 5.3.1.2
Offset pivot hinge

of the frame. For very large or unusually heavy doors, there could be one or more intermediate pivots, which are designed to be mounted to the edge of the door and face of the frame, at the middle or third points of the door's edge. The main advantage to using offset pivots is that they are visible only at the top and bottom of the door (unless there are intermediate pivots), again taking little away from the design. But as in center pivots, alignment can be difficult, so we see relatively few offset pivot hinges in the field.

5.3.1.3 Knuckles

Knuckle hinges are a broad classification that includes any hinge with two parts that join together (as interlocking one's fingers) with knuckles and which are held together with some kind of pin. The most common sub-classifications are: butt, olive, and piano.

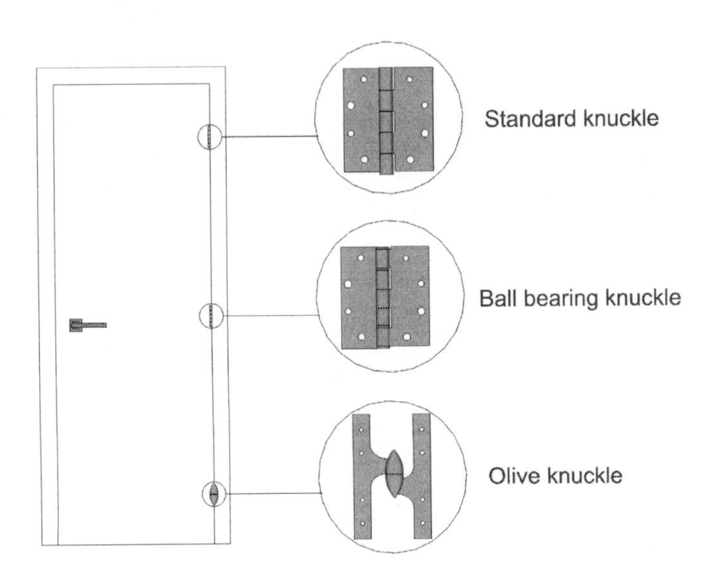

Figure 5.3.1.3.A
Knuckle hinges

Butt hinges are the most common hinges, and they are used extensively in commercial work and almost exclusively in residential work. The name simply means that the open "barrels" of the hinge butt together when the hinge is assembled. They are rated by size: 2″ × 2″, 3″ × 3″, $3\frac{1}{2}$″ × $3\frac{1}{2}$″, 4″ × 4″, 5″ × 5″, etc.; they are available in standard and ball bearing configurations; and they are available with standard, hospital, and non-removable pins. (A standard pin has a button at the top that holds it in the hinge. A hospital pin has a sloping top on an extra-tall button to minimize dust collection, and a non-removable pin can be removed only with special tools for high security applications.) Ball bearing hinges are far more durable (and usually much quieter), but they are also substantially more costly. The most common materials for hinges are brass (sometimes chrome-plated) and steel (and some stainless steel). Brass is used almost exclusively in residential work, but steel is more common in commercial work (especially if the hinges have ball bearings). Brass is used because the soft metal works better as a bearing surface where the moving parts of the hinges rub together. (Many years ago, high-quality hinges were often made from bronze, a costly but high-quality and very durable alloy of copper and other metals; such hinges are rare today. The author has seen such hinges that were installed in 1876 for very large doors and which operate perfectly today.)

All such hinges come in three different mounting arrangements: surface, half-surface, and recessed (Figure 5.3.1.3.B). This means that the plate on either side of the hinge (one plate is attached to the door edge and the other plate is attached to the door frame jamb) is either on the surface or recessed into the surface. A surface hinge has both of the hinge plates mounted on the surface; a recessed hinge has both of the hinge plates recessed, and a half-surface hinge has one of each (either way too—recessed in the frame and surface on the door or recessed in the door and surface on the frame). By far, most hinges are fully recessed these days, in both the commercial and residential areas.

The size and number of the hinges used depends on the size and weight of the door and the extent of the owner's budget. More hinges cost more money, but larger

Figure 5.3.1.3.B
Surface, half-surface, and recessed knuckle hinges

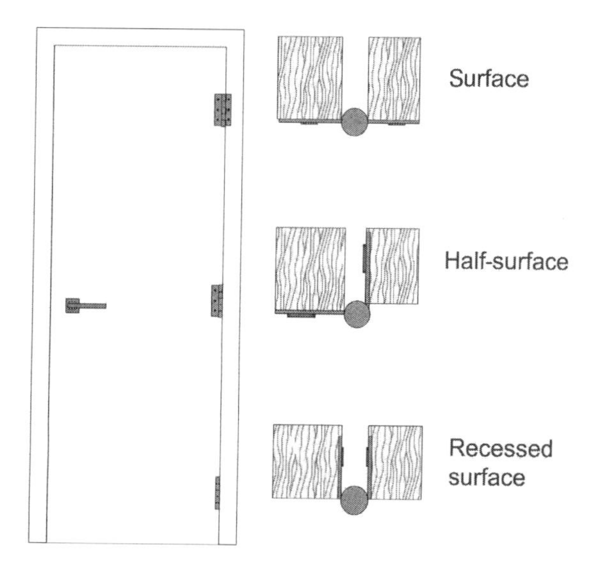

(and better) hinges cost more money too. It is still common to see residential doors mounted with two smallish butt hinges, but nearly all commercial doors are mounted using three larger hinges, often ball bearing hinges. Very heavy or very large commercial doors could use four hinges. (In the commercial world, hinges come in pairs, so a door with three hinges has $1\frac{1}{2}$ pairs, not three hinges.) It is easy to spot butt hinges on a door because the hinge barrels stick out from the edge of the door on the hinge side; not all that attractive, but the standard of the industry.

Olive hinges are special knuckle hinges that have decorative barrels (instead of the fairly crude looking barrels of a butt hinge); however, they are quite rare.

In the past, really heavy doors and very high use doors (entrance doors at hotels and airports, for example) usually used pivot hinges, but that changed recently when the heavy-duty nylon-bearing piano hinge entered the market. Piano hinges are very long butt hinges that were developed to attach the lid of a piano (to minimize twisting and binding), and manufacturers now make heavy-duty versions (with nylon bearings) for heavy use doors. Heavy-duty piano hinges are easy to spot too because they usually have continuous aluminum covers along the full height of the door; if you see a door with what looks like an aluminum plate running the full height on one side or the other, that is probably a heavy-duty piano hinge. (The cover protects the hinge from physical damage and helps to keep out the dirt too.) Piano hinges are also used in some furniture pieces.

5.3.1.4 Concealed

Concealed hinges come in two basic types: (1) concealed when the door is closed but quite visible when the door is open (very common European style cabinet hinges are the best example of this type; they are invisible when the cabinet doors are closed but they are large, bulky, and obvious when the doors are open); and (2) concealed when the door is closed but not obvious when the door is open. A company called Soss pioneered the latter type, making a wide range of styles and sizes for cabinet doors all the way up through full-sized commercial doors. Recently, a company called Simonswerk has introduced the Tectus line of hinges that takes this style to new heights of beauty, strength, and durability. The hallmark of the latter type of concealed hinge is that its mechanism is installed within pockets in the edge of the door and the jamb of the frame. When the door is open, the knuckles can be seen, but it is a far more attractive look than typical butt type hinges.

5.3.2 Latches

Latches are used to hold doors in the closed position. They come in many forms, from a wood cross-bar on a barn door, to a hook and eye on a shed door, a touch-latch on a cabinet door, a roller latch on a cabinet door, and to a retractable latch bolt for a typical door. The retractable latch bolt is standard for both residential and commercial doors, and it is usually operated by a turning handle of some kind (a door knob in the old days or a lever handle for better accessibility today). The bolt is spring loaded and it has a curved front edge that causes the bolt to retract (push in) when it hits the strike plate on the jamb of the door frame. The strike plate has a hole in it that is shaped to allowed the latch bolt to spring back, where the vertical face at the back of the bolt holds the door closed against the hole in the strike plate. Turning the handle (on either side of the door) pulls back the latch bolt and allows the door to open. The industry-wide term for a latch bolt and two handles is a "passage set," which distinguishes it from a "lock set." True passage sets cannot be locked.

When double doors are used, it is sometimes desirable to keep one of the two doors closed most of the time. In a case like that, vertical latch bolts are often used, in one of two forms: (1) concealed and (2) surface. Concealed latch bolts are built into the edge of the door and are usually operated by flipping levers or by simply pushing up or down on the ends of the bolts (via a depression carved into the door edge). Surface latch bolts are installed on the face of the door; they are quite ugly, so they are usually used only for industrial doors. Latching can also be done (for single or double doors) with head and sill bolts, operated by a panic device (see 5.3.6 below). In this case, there are vertical rods that extend up and down from the panic device to the bolts; these rods and bolts can be surface-mounted (ugly but easy) or concealed inside hollow doors. Some doors might use only head bolts, which would eliminate half of the rods and bolts.

5.3.3 Locks

Locks are used to keep doors secured in the closed position, openable with a key, core, swipe card, proximity card, etc. They also take many forms, from a padlock on a hasp on the shed, to a dead-bolt on the front door at home, to cylindrical locksets (where the locking mechanism is located within the cylinder that connects the two handles together), and to mortise locksets (where the mechanism of the lock is located within the edge of the door in a large pocket, or mortise). The two of these that are critical for this discussion are cylindrical and mortise locksets.

Just as in passage sets, a lockset is a combination of two handles and a locking mechanism. Cylindrical locksets are lower in cost but less durable because the whole mechanism has to fit into a 1″-diameter (roughly) cylinder that is only about 3″ long; the case for a mortise lockset is about 7″ high, 4″ deep, and 1″ thick, which provides far more space for heavier duty springs and other components. Naturally, mortise locksets are considerably more costly and are usually only seen in very heavy use environments (airports, hospitals, schools, maybe).

Both cylindrical and mortise locksets are defined by lock "function," meaning how and when it is either locked or unlocked. "Passage set" is sometimes listed as a lock function, but it is not really a lock at all; it is just a latch. The most common lockset functions are:

- Privacy (usually used on small restrooms and sometimes on bedrooms): locking this lock is accomplished by pushing in or turning a button on the end of the inside knob or handle and there is no conventional key. The outside knob or handle will have a small hole in it that provides access to unlocking with a special kind of simple key (sometimes a toothpick will work).
- Classroom: this is locked, or unlocked, by key from the outside but it can always be unlocked by a turn of the handle from the inside. This is to prevent locking someone into a room accidentally.
- Storeroom: unlockable only with a key from the outside; always unlocked from the inside.
- Entrance: unlockable with a key from outside; lockable from the inside by pushing in or turning a button. This is what is usually seen on residential entrance doors.

Dead-bolts are separate devices when using cylindrical locksets but they are integral to mortise locksets. A dead-bolt is simply a heavy-duty square bolt (as compared to the curved latch bolt that can be unlocked with a credit card or other flexible strip) that provides greater security. Dead-bolts are operated by key-only from the outside or by key or thumb-turn from the inside. Dead-bolts are commonly used for entrance doors to stores, restaurants, etc. to provide for maximum security, but it is often required to post a sign reading "This door is to remain unlocked during occupied hours" if using such a dead-bolt.

Electric strikes are used to provide remote release of locked doors. If you have ever entered a building (or a space) where you had to wait to be "buzzed" in, this might have been an electric strike. If you heard an obvious clunking noise just before you could open the door, you heard the electric strike operating. This device is a heavy-duty latch bolt (almost a dead-bolt) that is retracted by an electromagnet when someone pushes a button to active the magnet. Because it is located in the door jamb, it is necessary to get wiring to the device. They are somewhat prone to mechanical failure.

Electromagnetic locks are similar to electric strikes except that there are no moving parts at all. An electromagnetic lock is usually mounted to the door head, hanging down below the top of the door. It contacts a large steel plate that is strongly bolted to the door, and holds that plate when power is applied to activate the electromagnet. These devices can be so strong that someone could tear the door apart before breaking the holding force of the electromagnet (partly due to the large area of the magnet—usually about 9″ wide by 3″ high or so). These devices (and electric strikes) can be operated by remote buttons, by sensors built directly into the lock housings (electromagnetic locks only), by remotely mounted sensors, or by timers.

Here are a number of the most widely available hardware manufacturers:

Best Access Systems, a division of Stanley (www.bestaccess.com), a company that makes conventional and electronic locking systems
Corbin/Russwin (www.corbinrusswin.com), a 160-year-old hardware company
Kwikset (www.kwikset.com), a residential grade of hardware
Omnia Industries (www.omniaindustries.com)
Schlage (www.schlage.com)
Yale (www.yalecommercial.com)

5.3.4 Closers

Automatic door closers (not closures, please) are required for fire-rated doors and are often used for other doors where the desire is to keep the door closed under normal circumstances. For relatively light weight (and light duty) doors, special butt hinges are available that have interior springs to close the door automatically; these should be avoided in commercial projects.

The standard type of closer is a hydraulic device (that moves hydraulic fluid from one chamber to another to move an arm that is connected to the door), which is usually mounted at the head of a door. Unlike electromagnetic locks, the closer body can be mounted to the frame or the door, according to preference and functionality. Fifty years ago, these devices were usually in bronze cases that were bulbous looking; today, they are usually hidden by simple plastic or metal covers. All such closers have adjustable closing speed and closing force, so it is not necessary to put up with a slamming door just because it has a closer on it. Some closers will "hold open" at pre-set positions as well (usually 90 degrees), but such hold-open features are usually not code-compliant for fire-rated doors. Special closers with integral smoke detection are available too; these closers hold doors in the open position, closing them if smoke is sensed.

For doors with center pivot hinges, closers are available that can be recessed into the floor or ceiling. Floor closers used to be very common for heavy use entrance doors, but they have faded from favor over the past 20 or 30 years or so. Some center pivot closers can "hold open" at 90 degrees too.

For residential doors (especially exterior storm doors), lighter duty devices are made. For very light doors, one such device could be mounted near the middle of the door (height), but for heavier doors, two of these closers might be required (one at the top and one at the bottom).

www.lcnclosers.com (the web-site for LCN Closers, a division of Ingersoll Rand)
www.rixson.com (the web-site for Rixson)
www.stanleydoorclosers.com (the web-site for the door closer division of Stanley)

5.3.5 Door operators

Many building entrance doors—both swinging and sliding—are powered operated, which means that they open by themselves when triggered by a sensor or a switch. This has long been true at grocery store entrances and exits where carts are commonly used. In the past, the sensors for these operators were in mats on the ground or floor that sensed when a cart or person put weight on the pad; today, such sensors are nearly always mounted to the head rail above the door and they are more reliable. That said though, automatic swinging doors are prone to mechanical problems and everyone reading this has probably encountered an automatic door that either would not open automatically or which was tied open for that reason.

Power operators are also available for horizontal sliding doors and for overhead rolling and coiling doors (think garage doors for the latter). For overhead doors, there is even a category for "super-fast" vertical doors that open and close in a second or two; such doors are usually used in large-scale refrigerated warehouses (especially freezers).

All power operated doors can be automatic, meaning that they open when approached, or they can be operated from switches of various types. Power operated doors at building entrances are used mostly for accessibility reasons.

5.3.6 Hold-opens

As noted under closers, some closers have built-in hold-open capabilities (including the closers with integral smoke detectors). But separate mechanical devices can be added to hold-open doors too. These take the forms of: (1) overhead arm devices (usually installed in conjunction with a non-hold-open closer); (2) wall- or floor-mounted mechanical devices (including rubber wedges); and (3) electromagnetic devices that are controlled by the fire alarm system. Overhead arm devices are not often used, unless they serve multiple purposes. Wall- or floor-mounted latches, hooks, spring-loaded devices, and even rubber wedges can all be used *if* the door is question is not fire-rated. Electromagnetic devices connected to the fire alarm are the reverse of an electric strike; when power is on (from the fire alarm system), the device will hold the door open. A signal from the fire alarm system turns the power off, which will allow the door to close. These devices can be floor- or wall-mounted too, with special plates mounted to the doors.

5.3.7 Panic devices

Panic devices, often called "panic bars," are required in most large occupancies and some smaller assembly occupancies in order to prevent individuals rushing toward exit doors from getting trapped by a latched (or locked) door. The idea is very simple—if someone simply pushes on a panic device, it unlatches (and unlocks) the door and the door opens, allowing exiting. The earliest versions of these devices were actual bars that spanned across the width of door, positioned so that almost any pressure would push the bar down and open the door. More recent versions are more likely to look like flat plates (more attractive), but they function in exactly the same way.

There is a problem with panic devices though. It is always possible to open the door from the inside, so that it is always possible to defeat security to let someone into a secure facility. Many stores add separate alarms to such doors so that they will know if someone opens one of these doors; the separate alarms make a local noise only (they are not connected to the fire alarm) and their main purpose is deterrence. Many older schools (and other facilities) make it a habit to chain the panic devices so that they will not work when the building is unoccupied, but that is dangerous (and possibly illegal) and should not be done. This security issue can be a problem in many different types of facilities (usually facilities where there is an admission fee to get in): theaters, concert halls, arenas, YMCAs (and similar facilities), museums, swimming pools, etc.

www.vonduprin.com (the web-site for Von Duprin, a division of Ingersoll Rand, which is the leading manufacturer of panic devices)

5.3.8 Mutes (or silencers)

When using hollow metal door frames (see 5.4 below), a door slam can be quite loud, so it is standard in the industry to provide vinyl mutes, also called silencers, along the strike jamb of the frame. (Standard hollow metal frames have pre-drilled holes just for these devices.) They do not eliminate door closing noise, but they do make it quieter. Similar devices (vinyl or felt pads) can be used with wood and aluminum frames too.

5.3.9 Thresholds

Thresholds are required for exterior doors, but they are used for interior doors only under special circumstances (e.g. all-glass doors that latch into the threshold, to resolve floor transition issues, to separate wet and dry areas, etc.). Thresholds are usually aluminum and they are available in numerous shapes and sizes; for accessible facilities, the maximum height of a threshold is $\frac{1}{2}$" (which can be very challenging). Thresholds are available with flat tops; flat tops with vinyl seals; stepped tops; stepped tops with vinyl seals; sloping tops; and sloping tops with vinyl seals, and in many different widths. One simply finds the appropriate product for the specific situation.

5.3.10 Seals

Seals are available for weather applications (mostly in exterior doors) and for acoustical applications. The most common seals are vinyl seals on the tops of thresholds (see 5.3.8 above) and vinyl seals along the jambs and head, all for weather protection. The same seals can be used as well for minimal acoustical performance. If a seal will not work in a threshold for some reason, a device called an "automatic door bottom" can be used in its place. An automatic door bottom is a mechanical device that is either recessed into the bottom edge of the door or mounted on the inside face of the door along the bottom edge. It has an operating rod that contacts the hinge jamb of the door frame when the door closes, which causes a mechanism to lower a seal down to the floor (or threshold). They can be highly effective, although they are costly. Automatic door bottoms only work well with thresholds or hard-surface flooring (they cannot seal to carpet or rugs very well). For more demanding acoustical applications, higher grades of head, jamb, and sill seals are available, up to and including seals that inflate when the door closes. The latter are exotic, costly, and to be avoided if possible. (When high acoustical performance is required for doors, it is often necessary to use double doors. See Chapter 6 for more information.)

5.3.11 Remote stops

Doors swing around and can easily crash into walls, wall trim, furnishings, appliances, etc., so it is often necessary to have mechanical stops installed to minimize the potential damage. These devices are often wall-mounted (usually at the level of the lockset handle); floor-mounted (to contact the door itself), or even ceiling mounted (which the author did once for some frameless center-pivoted doors). Many different styles and types are available in a number of different materials from numerous manufacturers.

5.3.12 Coordinating hardware

If two fire-rated doors are used in a single fire-rated frame, the gap between the meeting edges of the doors cannot be left open when they are in the closed position, so one of the two doors has to have an astragal plate to cover up that gap. If these doors are both closing (by automatic closers), only the door without the plate can close first—otherwise, the doors will not close properly at all, let alone being latched and sealed as required for fire-rated doors. So additional hardware is required that will coordinate the closing of the two doors to make sure that the plateless door closes first—that is coordinating hardware.

5.4 Door frames

5.4.1 Do we need door frames?

Whether we need door frames depends on the door and the desired design effect, but it is not always necessary to use conventional door frames for working doors.

5.4.2 What are door frames for?

Simply put, the purpose of door frames is to support doors, so that they can move as required, and provide whatever features are necessary: security, physical privacy, visual privacy, acoustical privacy, etc. A door frame holds the door in position and it provides locations and means for mounting door hardware.

5.4.3 What are door frames made of?

Door frames can be made of wood, fire-retardant wood, steel (usually hollow metal), aluminum, fiberglass, and plastic. Bronze has been used for doors, door frames (and windows) in the past, but the extremely high cost makes it very unlikely to be used again in the future.

Generally speaking, the door frame should be more durable than the door itself, mostly because it is usually much simpler to replace a door than it is to replace a door frame. So it makes sense to hang wood doors in hollow metal frames, but it does not make sense to hang hollow metal doors in wood frames. In the commercial world, most doors hang in hollow metal frames (of one type or another), although wood frames, and no frames, are used on occasion. In the residential world, wood door frames are far more common, especially in one- and two-family dwellings.

Can a hollow metal door hang in a wood or aluminum frame? Sure, but it is not a very good idea.

Can a wood door hang in a steel or aluminum frame? Sure.

Can an aluminum door hang in a wood or hollow frame? Sure, but an aluminum frame would be more appropriate.

5.4.3.1 Wood door frames

Wood frames are the oldest and simplest type, sometimes consisting of little more than three boards around the sides and top of an opening. A door frame head (the portion across the top of the opening) is really only required when there is head-mounted hardware (such as a door closer or a special head-mounted electromagnetic lock), but they are usually provided anyway, for alignment purposes if for nothing else. (In the residential world, pre-hung wood doors are very common. This consists of a door, on hinges, in a complete frame (with loose trim pieces) with a strap or board across the bottom. After the unit is installed and secured, the bottom strap or board is removed.) One jamb provides a mounting location for hinges (usually) and the other jamb provides a mounting location for a latch, lock, or both. If one-directional hinges are used, a stop is necessary to prevent damage to the hinges by allowing the door to over-swing in the wrong direction. (Stops also make it simpler to latch and/or lock doors.)

The most commonly used wood frame consists of 12 pieces: two frame jambs, one frame head, two jamb stops, one head stop, four jamb trims, and two head trims. Why are the stops separate? And do we really need all of that trim?

The stops are usually separate because it is the cheapest and easiest way to do it. But that does not mean that there are not other ways to go about it. Having a separately applied stop gives the finished frame the appearance of a "double-rabbetted" frame. A rabbet in woodworking is a cut-away corner, so the applied stop makes it look like both corners were cut away from a thicker piece of wood to create the stop. What if you, as designer, do not like double-rabbetted frames? Then design a single-rabbeted frame, where the stop becomes an integral piece and it is created by rabbetting one corner out of a larger piece. (It would not be wise to pretend that such custom wood frames would be inexpensive.) This single-rabbet look can be done by using an extra wide applied stop too, of course (although that will be obvious unless the trim covers up the joint between the frame and the stop). Most wood frames are of the **wrap-around** style, which means that the frame is as wide as the wall is thick and trim is applied to the wall surface.

Why do we use trim? To make life easier for workmen. The trim covers up gaps between the frame and the wall facings (often quite large) and makes the finished opening look more decorated (even if the decoration is minimalistic). Can you do wood door frames without trim? Yes, but doing less work (by omitting all the trim) is going to cost a great deal more, because it will require workmen to do things they are not used to: finishing the edges of the gypsum board (or whatever) tight to the door frame, being much more careful, etc.

5.4.3.2 Steel door frames

Steel door frames are available in four basic types:

- Welded hollow metal
- Knock-down hollow metal
- Cold-rolled pre-finished steel
- Solid steel

Each of these will be covered separately, but it can be said that the last option is rare. It is offered by a few companies who still make old-style solid steel factory-type windows, and, while such frames can be elegant (meaning narrow and unobtrusive), they are very costly.

5.4.3.2.1 Welded hollow metal

For institutional (schools, hospitals, correctional facilities, etc.) and commercial projects, this is the most common type of door frame because it is the toughest. Welded hollow metal frames have standard $\frac{1}{2}''$ offset (the distance from the wall to the face of the frame, unless the frame is inset) and 2″ face width (sometimes 4″ width for heads of frames in masonry walls to match up to coursing) and they are available in a range of metal gauges. But most frames like this use heavy metal gauges for toughness (16 gauge is not uncommon).

If a frame like this is installed in a wrap-around version in a typical commercial stud wall that is $4\frac{7}{8}''$ thick, the overall frame width will be $5\frac{7}{8}''$ ($4\frac{7}{8}'' + 2 \times \frac{1}{2}''$). Given that a standard commercial door is $1\frac{3}{4}''$ thick, the offset distance to the integral stop is $1\frac{7}{8}''$. If the frame is symmetrical (not necessarily required), the stop will be $5\frac{7}{8}'' - 1\frac{7}{8}'' - 1\frac{7}{8}'' = 2\frac{1}{8}''$ wide. If a $1\frac{1}{2}''$-wide stop is desired, the frame will be asymmetrical and the offset on the non-door side will be $5\frac{7}{8}'' - 1\frac{7}{8}'' - 1\frac{1}{2}'' = 2\frac{1}{2}''$. The double-rabbet look, with the stop expressed on both sides, is most common, but single-rabbet frames can be done as well. As the frame gets wider for thicker walls, the stop of the non-door side offset, or both, would get wider. These frames are also done as inset frames, especially in masonry walls, which means that the frame is completely inside the wall opening. This protects the frame and can make for a more attractive installation, although bullnose corners are usually required for masonry (brick, concrete masonry, or tile) to make sure that corners do not chip off over time.

A frame like this comes to the job-site in one piece, meaning that the head rail is welded to both jambs; the frame is also already prepped for the required hardware package. The frame pieces are mitered at joining corners, which is where the welds are done, but the welds are ground smooth and the joints are rarely visible in the completed frame. This does mean that each frame is custom made. That does not mean that these frames are high cost or unaffordable, but it does mean that they are **long-lead** items, taking as long as six to eight weeks to obtain. If a schedule is very fast moving, this delay can be a problem.

Welded frames are installed in walls as the walls are built (whether they are stud walls or masonry walls) because the frame is anchored directly to the wall; in masonry applications, the jambs of the frame are usually filled with cement grout to make the installation even more secure. The only trick to the installation is to make sure that the door side of the frame is on the door side of the wall. This type of frame also comes to the job-site primed for field finishing and there are no other options.

Finally, it is easy to incorporate special features into these frames, such as single or double sidelights, adjacent windows, separate individual windows, walls of windows, etc.

5.4.3.2.2 Knock-down hollow metal

Knock-down (KD) hollow metal frames are made from the same parts as welded frames, but they are not welded into complete frames in a factory. Instead, a typical KD frame would arrive on the job-site in three pieces: two jambs and a head, already prepared for hardware. In this case, the installer has to make sure to put the door side on the door side of the wall and to get the strike side on the strike side of the opening. KD frames are anchored to the walls using large screws through both jambs (there are usually no anchors in the head). KD frames are usually done as wrap-around frames in stud walls, but inset frames can be done too. In masonry walls, inset KD frames are usually required. Any wall thickness can be accommodated with an inset frame.

Sidelights and separate windows can be done with KD frames too, but it is usually necessary to build framing between openings for support. KD frames also come factory-primed and require field finishing. Lead-time is much shorter than for welded frames.

5.4.3.2.3 Cold-rolled pre-finished steel

These frames are similar to KD frames, wrap-around only, with a few key differences.

First, the trim pieces are separate for this type of frame, which gives the frame a different profile. A base channel is installed at the head and both jambs, which overlaps the wall framing. That base channel includes clips for attaching the trim pieces, which snap-on. The trim piece is usually about $1\frac{1}{2}$" wide and about $\frac{3}{8}$" high, for a slimmer profile than a KD frame, but it snaps on about $\frac{1}{4}$" back from the frame channel corner, which gives an extra corner.

Figure 5.4.3.2.3
Pre-finished door frame

Second, they are pre-finished which means that they are painted (usually with baked-on enamel) in the factory and require no finishing in the field.

Third, they are stock products (as long as the hardware preparation that is needed meets the standards) and they are readily available, usually in a matter of days rather than weeks for KD frames. The standard sizes cover most typical wall thicknesses ($3\frac{3}{4}$″ and $4\frac{7}{8}$″ especially) but odd wall thicknesses could be tricky.

Frames like this can be used for sidelights and windows, but framing is usually required as for KD frames.

One of the advantages to this type of frame is how easy it is to install a wood trim. Adding wood trim to a KD or welded frame requires either large trim (to rabbet out a large enough pocket to cover 2″ × $\frac{1}{2}$″) or a special frame installation to have no $\frac{1}{2}$″ projection, either of which can be costly. To add wood trim to a cold-rolled frame, one simply throws away (or recycles) the trim pieces and the clips and installs wood right over the base channel.

This type of frame is very commonly used for fast-moving projects where there simply is not time to wait for welded or even KD frames.

5.4.3.2.4 Solid steel

As noted previously, the solid steel door frame is a sub-set of the old solid steel window frame systems. These systems use steel channels and angles, about $\frac{1}{8}$″ thick, to form sashes and frames with profiles (widths) that are narrower than any other type of window or door frame. Such widths are generally around $1\frac{1}{4}$″, which is quite small. Some of these products do not offer thermal break properties, which are required under modern energy codes for exterior openings, so they have limited application to commercial work. But if one wants a very small door frame, this is a good way to get there.

www.hopeswindows.com (this company also offers bronze windows and doors)

5.4.3.3 Aluminum door frames

Aluminum door frames are a sub-set of either of two systems: storefront or curtainwall. Storefront is an aluminum and glass framing system that is most commonly used for single-story storefronts, but it is also used extensively in relatively small-scale expanses of glass in commercial and institutional buildings. (Today, such systems are limited to 12′ 0″ in height, for wind-loading reasons, but the author's 1964-vintage two-story office building has a storefront that is 18′ 0″ tall.) For higher openings than 12′ 0″ or for all-glass installations, it is necessary to use curtainwall instead of storefront. The difference is that curtainwall systems have the ability to withstand whatever loads might be necessary; if it is not obvious, curtainwall systems are considerably more costly than storefront systems, so architects are encouraged to use storefront when they can.

Either system uses a kit-of-parts kind of approach, with aluminum extrusions for vertical mullions and horizontal muntins, in all sorts of combinations. The sizes of these extrusions vary from roughly $1\frac{1}{2}$″ × 4″ to 3″ × 9″; they can accommodate all sorts

Figure 5.4.3.3
Old storefront with narrow stile doors

of different glass panels (single-glazed for interior applications or double-glazed for most exterior applications) and solid panels (usually called spandrel panels); and they can even use butt-jointed glass where there is no framing member between the sections of glass (although sometimes there is such a member behind the glass). Anodizing is the most common way to finish these extrusions, from plain (no color), to bronze tones (light, medium, and dark), to black, and even in some colors. Powder-coated paint finishes are possible too, in a wide variety of colors.

The door frames in these systems are simply the standard extrusions prepped for door hardware—hinges, latch plates, etc.—so the look can be very sleek, especially if using a narrow stile aluminum door or possibly all-glass doors.

5.5 Windows

Interior windows can be done in numerous different ways, from simple custom wood frames, simple channels, hollow metal, cold-rolled steel, aluminum storefront, or even by using exterior windows (plastic, wood, or metal) in interior walls. The latter should be avoided because exterior windows have features that complicate interior installation, especially sloping exterior sills that are required to shed water, but they can be used if the complications are accounted for in the design.

5.5.1 Wood windows

Wood interior windows are most easily done by framing an opening with wood, adding applied glass stops to one side, installing glass, and applying wood glass stops to the other side to hold the glass in position. This assembly can then be trimmed in whatever manner is consistent with the rest of the design. But applied stops on both sides can be crude, so this can be done with a single-rabbet approach, with a fixed stop on one side and removable stop on the other side; the removal stop can be hidden by trim so that it looks as though there are no removable stops in the finished installation. These techniques work well with relatively small panes of glass; large expanses of glass might require different techniques. In any case, there is usually a need for structural framing between large groupings of small windows or between large windows; this might take the form of studs concealed inside the trim (or even steel in extreme cases).

It is common to use these techniques to build up full-custom wood-and-glass walls and screens in commercial and institutional projects.

If the look of old-style divided light windows is desired, or if rounded heads are desired, it might not be feasible to have the installation custom built due to cost, so this might be an area where it would make sense to install standard wood exterior windows (which are available with round heads, etc.) in an interior wall. Just remember to detail accordingly and be careful about wall thickness, among other things.

5.5.2 Steel windows

As noted under door frames, welded hollow metal, KD hollow metal, and cold-rolled pre-finished steel systems can be used to make interior windows, door sidelights, and even walls of windows. One does have to take framing requirements into account though and it might be necessary to hide wood or metal studs, or even a small steel column, into these systems to make them stable.

5.6 Ceilings

Ceilings are simply surfaces over our heads, and they cover the gamut from materials adhered to structural materials (usually concrete but sometimes steel) to materials

Figure 5.5.2
Cold-rolled door frame with sidelight

suspended from structure (often plaster, gypsum board, wood, metal, etc.) to special suspended ceiling systems (concealed **splines**, and lay-in systems of many types). For much of history, the ceiling in a room would run from wall-to-wall, but, in recent years, partial ceilings—usually called "clouds"—have become very popular. It is also not uncommon to see multiple ceilings in some spaces.

When partial ceilings are relatively small as compared to the spaces where they are located, they are sometimes called "bulkheads." The term "soffit" is also used but that is more common outside the US; in the US, a soffit is usually an exterior ceiling. Bulkheads can be framed in wood or metal and covered (or not) in metal, plastic, wood, glass, gypsum board—really anything is possible. Bulkheads can be very costly though because they are labor-intensive to frame and to cover.

In the residential world, "tray ceilings" are common, where the framing is done in two, or more, levels and the gypsum board follows the framing; in commercial work, we might just call this a bulkhead around the perimeter.

Bulkheads are usually used to transition between differing ceiling heights, to cover the tops of cabinetry (to avoid dust build up or just for the look), to hide systems (like ductwork or piping), or to achieve the overall design look. Is a 3′ 0″-thick wall with a fireplace and an entertainment system recessed into it a wall? Or walls between and a bulkhead above? No one can say with certainty.

Very small bulkheads, like ones used to hide piping, can be done with pre-manufactured parts too, in metal, plastics, and composites.

But back to the ceilings themselves. As noted above, there are only two general categories: applied and suspended.

5.6.1 Applied ceilings

In the residential world, where most construction is done using wood framing that is covered by gypsum board, most ceilings are gypsum board applied to the wood framing. This is less common in the commercial world, but it is not rare; often such ceilings are required to be fire-rated in the commercial world though which can be challenging for lighting (see Chapter 8).

Before the 1950s, most applied ceilings would have been Portland cement plaster (or lime plaster before the middle of the 19th century) on wood or expanded metal lath. In residential construction, the lath is likely to be narrow wood boards, about $\frac{3}{8}$″ apart, nailed directly to the wood ceiling joists or wood rafters; in commercial construction, the lath is likely to be metal attached directly to the same type of framing. The plaster would have been done in three coats—scratch coat, brown coat, and finish coat—adding up to roughly $\frac{7}{8}$″ in thickness, and the surface would probably have been very smooth for final painting or wallpapering. This system can be highly durable, although there is a tendency for the plaster to break away from wood lath over time. Also, it is very difficult to put nails into plaster like this.

Beginning in the 1950s, wet plastering was used less and less and "drywall" was used more and more. "Drywall" consists of gypsum board sheets (solid gypsum between outer layers of heavy paper) installed on the framing, and then wet-taping and finishing joints and nail or screw heads. Drywalling is much faster than plastering, but it is also inherently less durable, although this is less of an issue on ceilings than for walls.

From the middle of the 19th century through the 1920s or so, it was common to apply decorative tin sheets over the plaster (or sometimes directly to wood framing). Unfortunately, much of that material has been damaged as later renovations have occurred.

Today, applied ceilings are really only seen in wood framed buildings where a "residential look" is desired.

5.6.2 Suspended ceilings

Suspended ceilings are also divided into two basic types—"hard" and "accessible." Hard suspended ceilings refers to gypsum board, plaster (still used in some situations), metal, or wood and accessible suspended ceilings refers to everything else.

5.6.2.1 Hard suspended ceilings

Hard suspended ceilings are a challenge in commercial and institutional projects because mechanical and electrical equipment above such ceilings must be accessible for service and/or replacement; this means that **access panels** are required in hard ceilings in such circumstances. Generally speaking, access panels are to be avoided because they disrupt the continuity of the surface (no matter how well they are built) and continuity of surface is a major design reason for using a ceiling like this; sometimes, it is feasible to keep mechanical and electrical equipment out of the area, but not always.

Suspended plaster ceilings are usually done with a system of small steel channels, wires, and metal lath. First, $1\frac{1}{2}$"-high steel channels are suspended at 4′0″ o.c. Second, $\frac{3}{4}$" steel channels are installed perpendicular to the $1\frac{1}{2}$" channels at 2′0″ o.c., tied with wires. Third, the metal lath is wired to the small channels, and, fourth, the plaster is

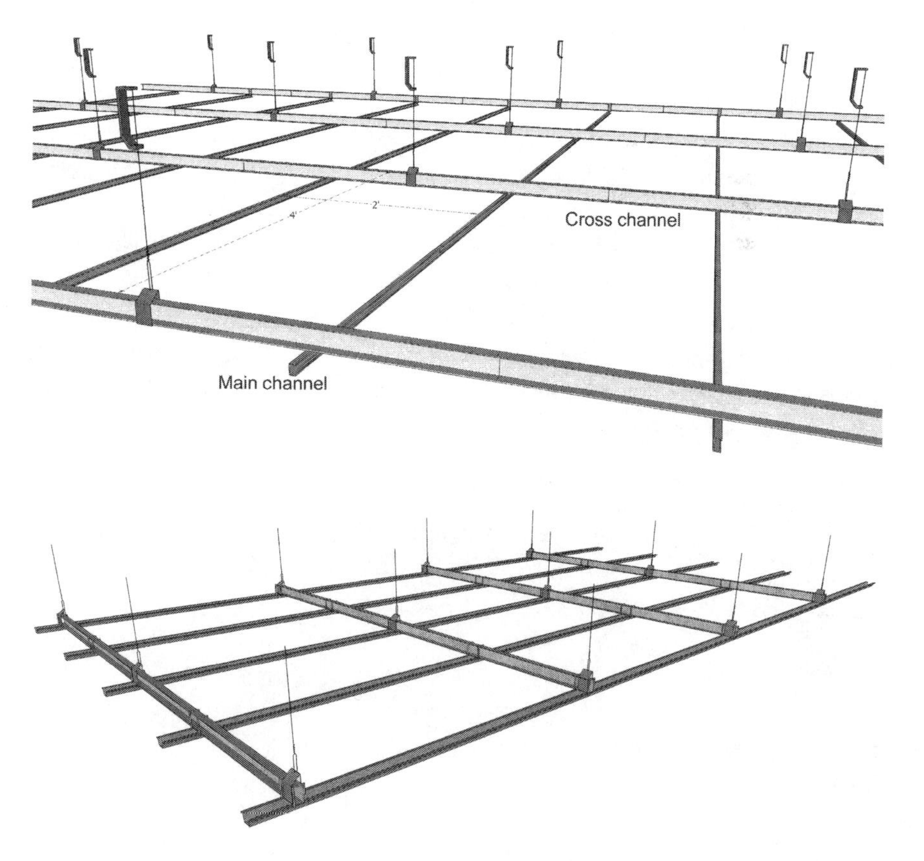

Figure 5.6.2.1.A
Black iron channel suspension

Figure 5.6.2.1.B
Black iron and furring channel suspension

applied to the lath. This system is still used for exterior soffits and for shower room ceilings due to its durability.

Suspended gypsum board ceilings can be done with the $1\frac{1}{2}''$ channels too, with $\frac{7}{8}''$ high "hat channels" at 2′ 0″ o.c. for screws to attach the gypsum board, but a special form of 2 × 4 grid is used as well. The special grid has extra wide flanges on the tees so that it is easy to attach the gypsum board; the main tees are stronger (gypsum board is heavier than most lay-in panels), and more wires are used to hang the main tees.

When acoustic materials started to become available around the middle of the 20th century, these were often bonded (glued) directly to plaster or gypsum board. The most common form was 12 × 12 tiles with uniformly spaced holes (at roughly 1″ o.c.); this is seen all over schools and some offices from the 1940s through the 1960s. At a glance, a ceiling like this might look like it is suspended, but in all likelihood it is not.

5.6.2.2 Accessible suspended ceilings

Because there is so much concealed wiring, piping, and ductwork in most commercial and institutional buildings, accessible ceilings are very useful. Accessible simply means that special tools are not required to gain access to the space above the ceiling.

Accessible ceilings initially took two forms: concealed framing and exposed framing. The concealed framing system, called "concealed spline," is still available but it is rarely used today. In a concealed spline suspended ceiling, the surface of the ceiling is completely flat and there are only hair-line joints between 12 × 12 tiles; the tiles are not usually the same perforated tiles mentioned above but they could be. These ceilings are accessed by using a special tool (really just a hook) to pull down the first panel; once there is an opening, it is much easier to remove additional tiles to gain access. But it is very difficult to put them back together and previous attempts at doing so are usually apparent in the ceiling. This system is costly too.

The main alternative to concealed spline is the typical exposed grid system. The first version of this used a $\frac{15}{16}''$ wide snap-together grid to make 2′ × 4′ cells that can be infilled with ceiling panels (called "pads"), lights, HVAC diffusers, or return air grills. The system can be done in 2′ × 2′ cells as well, which performs better but costs more. Many different pads are available—the least costly and most common is called "omni-fissured" where there are holes and streaks of holes in a random pattern over the face of the pad.

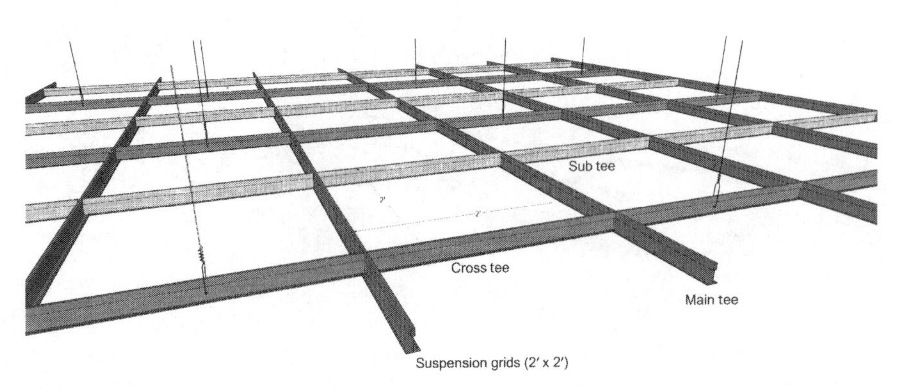

Figure 5.6.2.2.A
Suspended ceiling grids

Other finishes include: pin-perforated (which resembles a randomized version of the old 12 × 12 tiles); textured; vinyl-faced; and even wood and metal looks. Most of the pads are made from mineral fiber (a gray mat core) but some are made from fiberglass (a fibrous yellow core). The main advantage to this type of ceiling is that it provides the best acoustic absorption possible relative to cost; putting in a ceiling like this goes a long way to making most spaces acoustically functional.

After some years, architects who use a 5′0″ building planning module (common in large high-rise buildings) complained that the 2′ ceiling module cannot line up with their column grids (which is apparent from outside at night), so a new version of this system was developed which makes 20″ × 60″ cells (or 20″ × 20″ cells). But 60″-long light fixtures never caught on, so these ceilings have either 6″ ceiling pads or metal plates at the ends of the light fixtures. This entire system never really caught on either and it is rare today.

The next complaint was that the grid itself is too big, so a narrower grid was developed that is $\frac{9}{16}$″ wide.

But the real complaint about lay-in ceilings is that they are not flat: the pads sit on top of the thin grid flanges. That is only $\frac{1}{16}$″ or so, but it is visually obvious in a space. There are two solutions to this issue: "tegular" pads and slotted grids.

Tegular pads have rabetted edges so that the suface of the pad drops $\frac{1}{4}$″ or so below the face of the grid; the idea is to de-emphasize that there is a snap-together grid by emphasizing the geometry of that same grid. As long as there is a wall-to-wall ceiling within each room, this can work, but if walls stop at the ceiling grid, it is awkward to do

Figure 5.6.2.2.B

Tees and pads: (a) $\frac{15}{16}$″ lay-in grid; (b) $\frac{9}{16}$″ lay-in grid; (c) $\frac{1}{4}$″ tegular lay-in grid; (d) $\frac{1}{8}$″ tegular lay-in grid.

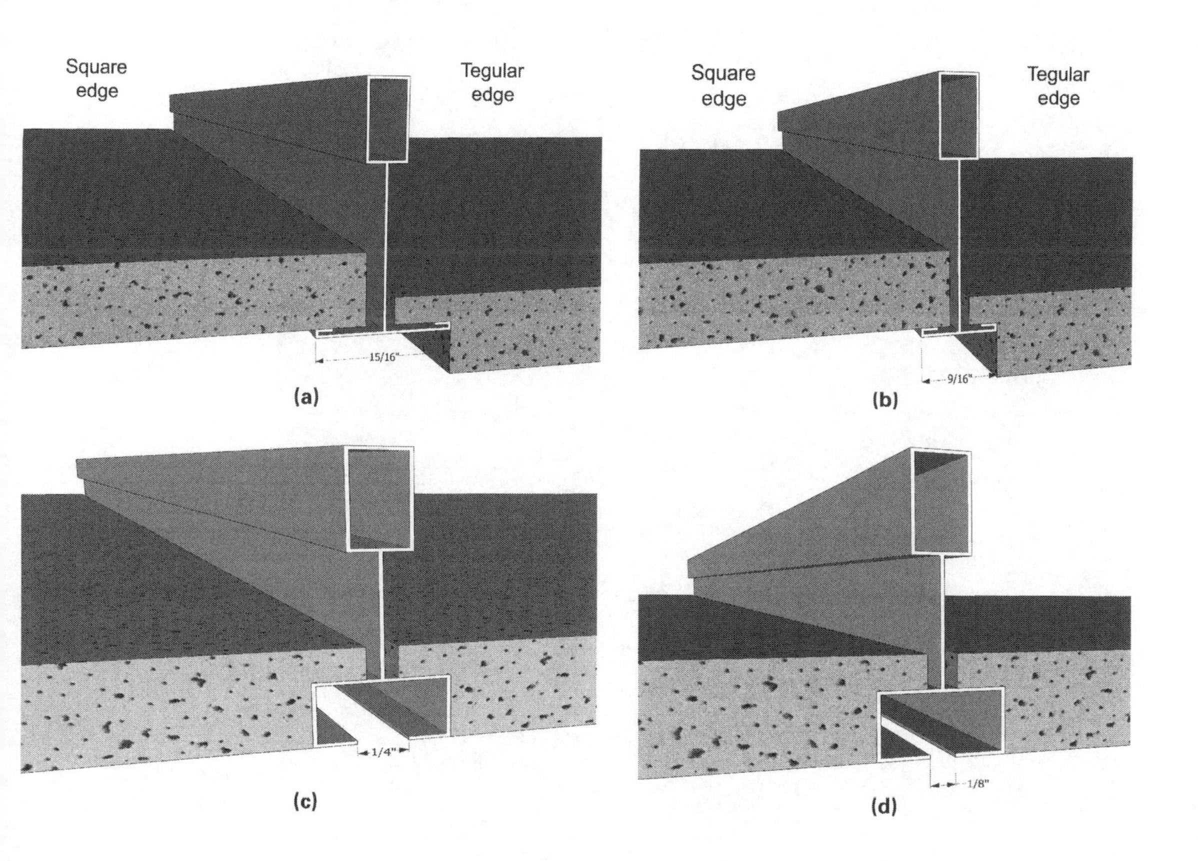

all of the necessary cutting to make this look right. (In addition to cutting a pad to size, the workman also has to rabbet the cut edge, which is not easily done.) As a result of these challenges, the author has never been an advocate for tegular ceiling pads, even though that is what is in his personal office (the ceiling was already there when the building was purchased).

Slotted grids are narrow grids ($\frac{9}{16}$″ wide) that are about $\frac{3}{8}$″ high, so that they form a small "box" with a slot in the bottom (narrow or wide slot). In this system, the pads are rabetted as well (to different dimensions than tegular pads, so that they are *not* interchangeable) but the result is a smooth ceiling, where the surface of the pad lines

Figure 5.6.2.2.C
Old tegular ceiling

up with the surface of the grid. This is the author's preferred system because it can approach the look of a hard ceiling (if using slightly perforated mostly smooth pads) while being fully accessible and less costly—a win win. Of course, slotted grids are slightly more costly, as are 2′ × 2′ grids as compared to 2′ × 4′ grids.

5.6.2.3 Clouds

Clouds can be gypsum board, wood, metal, plaster, plastic, etc. They can be almost any size and almost any shape, and there can be multiple levels of clouds in a single space, which can vary in shape and material too. The greatest challenge to clouds is how to finish the edges.

In lay-in systems, the manufacturers offer standard edge trims that can be square (and in various heights, commonly 4″ and 6″) and even knife-edged. Custom square-edge trims can be done by sheet metal contractors too.

The main issue about the edge of the cloud is whether or not it needs to hide something. Obviously, a knife-edge trim will hide nothing that is above the thickness of the grid framing, which might be OK if there is nothing to see (unlikely though—there go those pesky mechanical and electrical guys again). If there is a desire for up-lighting around the edge of a cloud, even if it is just LED tape that is only $\frac{1}{8}$″ high, some kind of concealing edge is advisable; the more there is to hide, the higher the edge should probably be.

5.7 Millwork

Millwork is another vast subject unto itself and millwork design is one of the most interesting, exciting, and gratifying aspects of interior design. But there is much to learn. On the practical side, it would be good to get to know a local millworker well enough to be able to spend some time in the shop; to see how things are done and how things really go together. Also, it is vital that we designers listen to millworkers; they know more about how to make moldings, casework, and furniture than we probably ever will and we should never forget that.

That said, the sub-categories in millwork include: standing and running trim (including custom moldings); wall paneling; casegoods (from standard kitchen cabinets to custom reception desks); doors and frames; stairways and railings; and furniture. But before delving into each sub-category, we should first talk about wood—solid, real wood—wood veneer, and wood alternatives.

The wood that we use in buildings, no matter what we use it for, is nothing more or less than the partially dried remains of a tree. The tree is cut down; branches are removed; the bark is removed (usually) and the trunk is cut into pieces of useful size (we hope). Larger branches might be cut into useable pieces as well. Then the wood is dried because freshly cut wood is far too wet to use for anything. Wood is dried in two ways—in kilns and in the open air (although sheltered from rain and snow). The best way is to let wood air-dry but that is slow and costly, so kiln drying is used almost exclusively.

Wood that is used for framing is dried less than wood that is intended to be used for furniture, flooring, etc.

In the residential world, lots of wood is still used both for framing the buildings (using mostly relatively low-quality softwood, usually spruce, pine, or fir) and for door frames and trim inside the finished building (probably pine but possibly oak, although oak seems to be out of fashion at the moment, or cherry or other high-end species). Solid wood might be used for exterior trim; in doors, and maybe even in floors. It would be very rare to see solid wood used in residential casework.

In the commercial/institutional world, less wood is used. There is still some framing for wood buildings and still some interior trim (though much less), as well as some exterior trim, some doors, and maybe even some floors. And, again, it would be very rare to see solid wood in commercial casework.

The use of wood has changed dramatically over the years. The same building, a church, where the author has seen the 1876 bronze hinges has extensive interior solid walnut trim—the entire sanctuary in fact—and solid wood cabinets in the basement kitchen. There is solid wood trim all over the building, including 10′-tall solid walnut doors. When the building was built (1873–76) walnut trees ran rampant in central Indiana, so the wood was easy to come by. Today, such trees are so valuable that those who own them need security to prevent theft, and, even at that, large boards simply are not available because there are no large trees. *All* of the large trees are gone. There will be more someday, but it takes hundreds of years to grow a large walnut tree in central Indiana.

A relatively new type of wood has become popular in recent years, and that is what is called reclaimed wood—sometimes from old barns but also from the bottoms of the Great Lakes. Even though wood that gets wet, then dries, gets wet again, dries again, etc. rots, wood that it always wet does not rot. So there is an industry that recovers logs from the bottoms of lakes (notably the Great Lakes), some of which might have been there for hundreds of years, dries out the logs for the last time and then uses the wood for furniture and building projects. It can be quite beautiful and it is usually very costly. Re-used barn wood is generally somewhat rough and weathered looking but it is somewhat less costly, although it can also be very beautiful. It is certainly better to re-use this wood than to burn it or put it in a landfill, but it is material that is not "readily renewable," unlike new pine wood that grows very quickly. In other words, once all of the logs have been pulled out of the lakes, there will not be any more for a very long time, if ever. And once the last old barn is gone, there will not be any more of that wood either.

And then there is certified sustainably grown wood, as mentioned previously. Such wood is grown in such a way as to assure no forest depletion, so it is far better for the environment and the planet overall—the right thing to do. See http://us.fsc.org/en-us, for more information about the Forestry Stewardship Council.

So what should a designer do if an owner wants to have beautiful wood in a project? Tell them no because the best trees are gone? Tell them that only wood from fast-growing trees can be used? Use no wood at all? Use veneer?

What is veneer?

Veneer is very thin wood that is peeled or sliced from a log to achieve maximum beauty and to spread that beauty across a wide area. Veneer might be only

0.05″ thick (less than $\frac{1}{32}$″), so a single 1″ board sliced into 20 veneer leaves can cover 20 times the surface that the single board would cover. But veneer, being so thin, is easily damaged and must be protected. Veneers are usually used by adhering (gluing) them to some kind of solid and stable substrate, most commonly high-density particle board these days. This material, although made from wood scraps, is dimensionally stable (so that it will not warp and twist easily) and it forms a good substrate for veneer. Veneer could be applied to lesser quality solid wood, plywood, and other materials as well. If veneer is applied correctly and detailed well, it can appear to be solid wood for all intents and purposes. Figure 5.7 shows four common veneer cuts.

Back to the what to do question—what should one do to use wood in a project in an environmentally sensitive manner? First, use solid wood only when there really is no other good option. Do the design work so that there are other options and use certified renewable wood if available (and affordable). In other words, do not design a complex, heavily shaped molding that can be done only with solid cherry; instead, design something simpler that can be made from a less scarce or more easily replenished material. Design the project so that a beautiful and rare wood is an accent; so that only a small amount of it is needed; so that it truly shines. Second, use veneers to the greatest extent feasible, especially for wall paneling, fronts of reception desks, cabinet fronts, table tops, etc. Third, use more easily replenished soft woods where you can; these generally work best when painted but there is nothing wrong with painted wood. And, last, do not use an imported material that appears to be both more available and less costly to substitute for a rare local material; deforestation somewhere else in the world

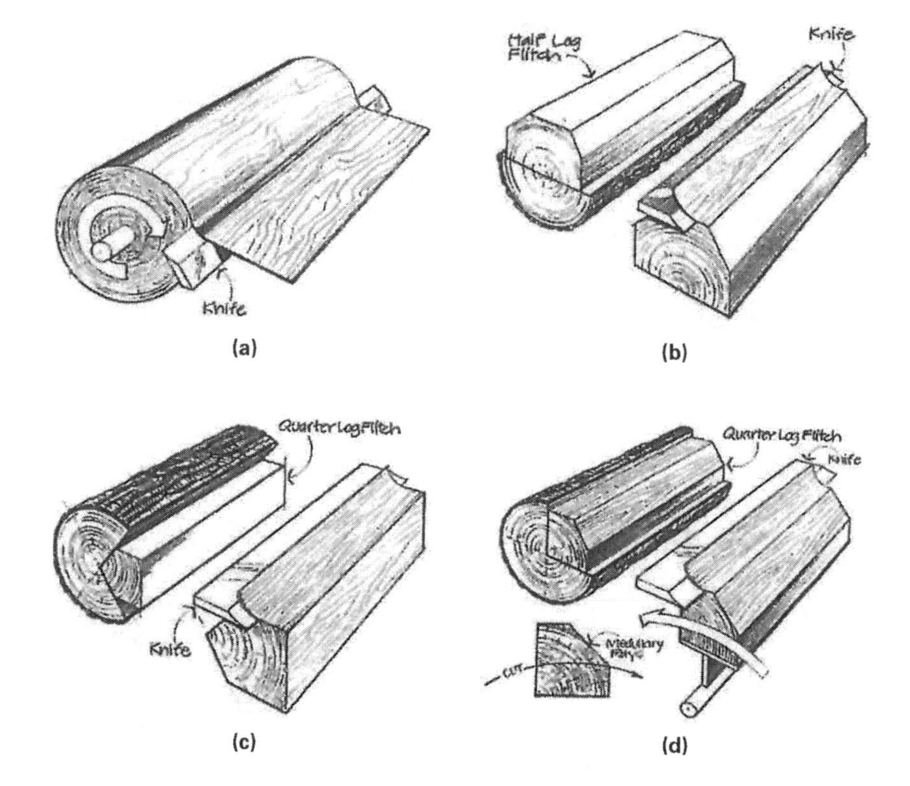

Figure 5.7
Four common veneer cuts: (a) Rotary slicing; (b) plain slicing; (c) quarter slicing; (d) rift slicing.
Source: *Architectural Woodwork Standards*, 2nd edition, © AWI/AWMAC/WI 2014

is still a problem and should be avoided, especially to wrap someone's conference room in faux walnut. That owner probably does not need walnut wall paneling at all.

5.7.1 Wood species

The first decision to make when selecting a wood species is to determine if the preference is for hardwood or softwood. Softwoods, such as fir, pine, and spruce, are commonly used for structural framing but less commonly used for finish work, the primary exception being low-quality trim that is made from pine. Most decorative wood work is done using one or more of the many hardwoods that are available on the market, especially in veneer form.

Hardwoods come from deciduous trees (usually) and softwoods come from coniferous trees (usually); also, most hardwoods are actually hard (except balsa) and most softwoods are relatively soft (except yew). If hardwood is selected, it is necessary to determine if open grain or close, or closed, grain is desired; sometimes, these are also called "coarse" and "smooth," respectively. Open grain woods have visible spaces—pores—between the wood fibers, and examples include ash, elm, and oak. Close, or closed grain woods also have pores but they are much smaller and not apparent to the naked eye. Examples of close grain woods include birch, cherry, maple, and yellow poplar. When a wood like this is stained, the staining leaves the open pores open so the look does not change. Adding a clear finish can leave the pores open, if it is a light finish, or filled in, if it is a heavy finish. But, as noted previously, why would one want to finish an open grained wood in such a way as to make it look like it is not open grained?

Many exotic species are available as well, mostly as veneer, but sometimes in solid boards. Common exotic species (which can come from all over the world) include:

- Anigre and figured anigre
- Aspen
- Basswood
- Birch
- Birdseye maple (and various other maples)
- Bloodwood
- Bubinga and figured bubinga
- Butternut
- American chestnut
- Cocobolo
- Ebony (various varieties)
- Hickory
- Holly
- Honey locust
- Koa
- Lacewood
- Mahogany (many varieties)
- Olivewood
- Padauk
- Purpleheart
- Sapele
- Teak
- Wenge
- Zebrawood

5.7.2 Solid wood or veneer?

The choice between solid wood and veneer, as described previously, mostly has to do with the application: moldings generally do not work with veneers and there is usually no reason to use solid wood for a table top or a wall panel. But availability plays a major role here and each designer should make an effort to determine what might and might not be reasonably available for a given project. Stocks of exotic solid woods are usually limited in most locations, although the larger a city gets, the more likely it is to see a larger selection. Exotic veneers tend to be stocked only by a few national distributors.

If one wants to use an exotic veneer, it is important to see the veneer in person, even if that means that samples have to be shipped in from the suppliers (at the designer's cost). Veneers are grouped into "**flitches**," which come from a single log, with sequential layers of individual sheets, or leaves, of veneer. One can have a distributor ship in two or three leaves from the flitch for selection purposes. The extent of the project matters too because the supplier might not want to sell just a few leaves from a flitch for a small project; on the other hand, for a large project, more than one flitch might be required so it will be necessary to see samples from more than one for matching purposes. The length of the flitch can be important too; if 8′-high wall panels are needed,

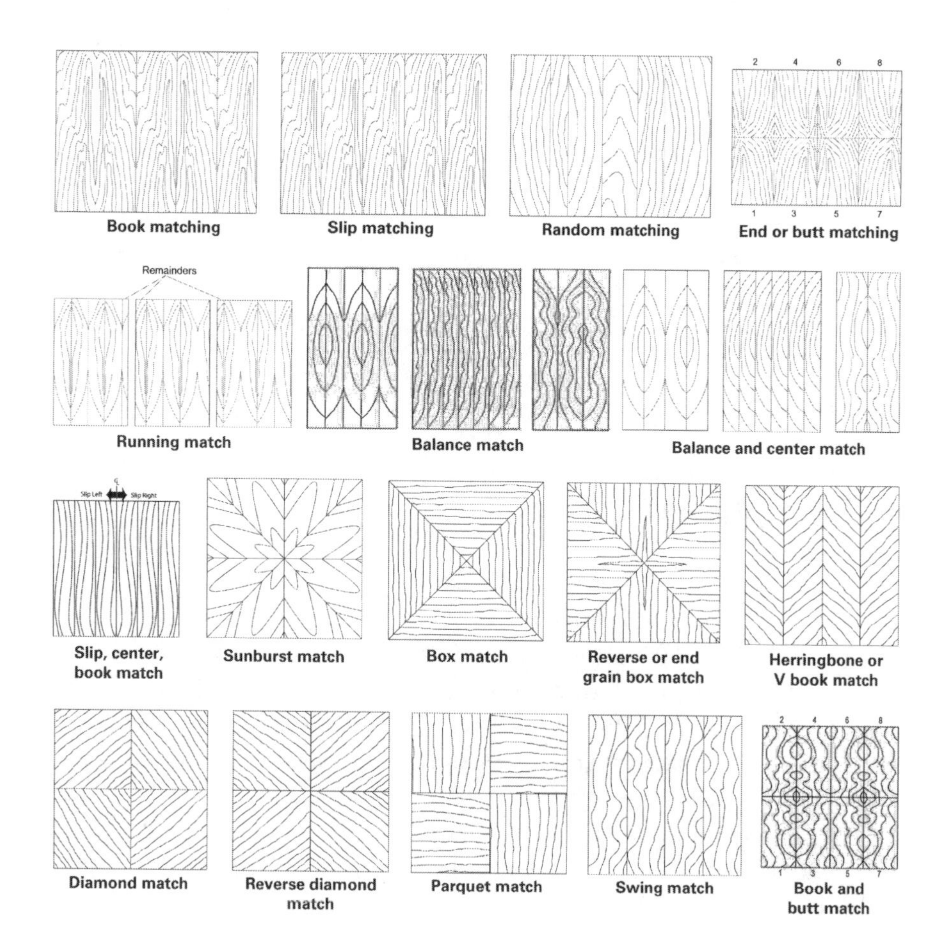

Figure 5.7.2
Veneer matching patterns.
Source: *Architectural Woodwork Standards,* 2nd edition, © AWI/AWMAC/WI 2014

the flitch will need to be a little longer. It is possible to leave the flitch selection to the millworker, but it is best for the designer to get involved if the look is highly critical.

5.7.3 Finishing

Millwork finishing can easily be the downfall of a beautiful project. The least costly approach is always to have painters finish wood in the field, whether the finish is unstained clear, stained clear, or painted, but it must be remembered that job-sites are usually dirty, dusty especially, and it is nearly impossible for a painter to achieve a high-quality finish under those conditions. Also, the most durable finish products—such a most catalyzed finishes—cannot be used in the field due to environment and health risks, so field-applied finishes tend to be less durable. If a design is going to include substantial millwork that is a focal point, it is very important to have such work finished in the shop, or factory, where finishes are applied in a controlled environment spray booth. Such finishes can be comparable to factory furniture finishes—really beautiful in other words. Pre-finished millwork must be installed very carefully in the field to avoid damaging the finishes, and that includes using putty to hide nail holes, etc. When millwork is pre-finished, it is usually installed by the millworker's crews and not by field carpenters. Field carpenters can install it, but the millworker will probably have to come back to do the putty work (color matching is very critical).

The finish systems that are used are the same as those used for wood doors, which were explained under 5.2.4.1 previously.

In rare cases, it might be possible to create something that resembles a spray booth in the field, by enclosing a work space that is kept clean and dust-free, but it would still be impossible to use the best finish materials, unless the workmen use respirators as in factory spray booth work.

Overall, the best results are nearly always achieved by having the millworkers finish the work in the shop, and that includes standing and running trim, wall panels, casegoods, furniture, etc.—everything, in other words.

5.7.4 Standing and running trim

Standing and running trim includes baseboards, chair-rails, crown molds, door casings, low wall caps, wall end trims, and any other form of linear wood used on walls and/or ceilings. Many standard shapes are available in various hardwoods, some softwoods, and even plastic (for curving applications). Wood can be used on curving surfaces too, either by sawing small grooves (kerfs) in the back of the piece to allow it to bend, by steaming the wood, or a combination of both. Needless to say, doing curving woodwork is very labor-intensive and very costly.

Many designers like to design custom moldings that relate directly to a design. Is this practical? Mostly, that depends on quantity. Having a custom molding made means that a millworker has to have a special knife made for the machine that cuts the molding from a solid board; having the knife made could be in the $1,000.00 range. So if one were having only 100 lf of trim made, the cost of the knife would be $10.00/lf, which would probably not be affordable. But if there were 1,000 lf of molding, the knife cost would be only $1.00/lf, which could well be affordable. These costs are in addition to the cost of the raw wood and labor, plus finishing and installation, of course. Standing and running trim can easily cost $15.00/lf, installed, more if the wood is exotic or if the moldings are unusually large. That might not sound like a large number, but the quantities of such items can be large. For example, the author's office space is roughly 4,200 sf (for a staff of 13). But the perimeter of all of the interior spaces is roughly 970 lf (not including the restrooms and the storage room), so the cost for wood base throughout could easily be $12,000.00 to $16,000.00, or about $3/sf to $4/sf. The actual renovation budget for the space was less than $25.00/sf, so we did not buy wood base throughout. We did have custom-made base installed when we changed the flooring in the first floor public lobby; it is simple 1″ × 6″ solid poplar.

See Figure 5.7.4.A for an example of a custom molding sample (painted poplar) from a project in the early 1990s and see Figure 5.7.4.B for the wood base installation in the author's building lobby.

Figure 5.7.4.A
Custom molding

Figure 5.7.4.B
Custom wood base and casing

5.7.5 Wall paneling

Over the long history of human design and construction, many approaches have been used to create wood walls. The simplest technique is to attach individual boards, vertically or horizontally, to framing or some other surface, with butt joints, overlapping joints, or tongue-and-groove joints. Beyond that, simple or decorative **battens** (small wood trim boards) can be applied over the butt joints. If larger continuous surfaces are desired beyond what can be done with individual boards (18″ or less, even when large trees were still plentiful), veneers are usually applied to plywood, MDF, or some other stable substrate.

When veneers are used, individual panels can be quite large (as large as can be handled, moved into the space, and installed) because the narrow veneer flitch is "matched" in various different ways across the width of the larger panel. When large panels are used, grooves can be added to applied features to provide a lot of depth to the surface. Multiple species of solid wood and veneers can be combined in endless patterns.

In the middle of the 20th century, what can only be called "faux" paneling became popular in both residential and commercial projects. (The author's 1958 house and his 1964 office building both still have some of this material in place, although it is either painted over or covered over.) Faux paneling is some sort of panel, similar to MDF at the high end of the quality range and little more than compressed paper at the low end, usually $\frac{1}{8}$" (low) to $\frac{1}{4}$" (high) thick with a veneer (high) or paper or vinyl (low—with an image of wood grain) face and with vertical grooves, with variable spacing. The grooves are usually a dark color to contrast with the finished surface. This kind of material is much lower in cost than solid wood paneling and faster to install, and it was used very extensively for several decades (it was probably the most common for residential renovations, where it was used to cover up bad plaster, drywall, or concrete masonry in basements, and "dress up" a space). But it has become a blight on the industry, where now even the highest quality versions look dated and low quality; the low-quality versions have mostly disappeared, simply because they fell apart.

Wall paneling in its traditional form was usually full height, meaning from floor to ceiling, but partial-height paneling—usually called wainscotting—is common as well. Wainscotting usually ends at the top in a chair rail (if low enough to experience bumping from chair backs—hence the term "chair rail") or some other type of wood molding (running trim) if higher.

Wood paneling has also been used on ceilings, although this is less common.

Today, when wood paneling is done, whether on walls or ceilings, it is usually done with veneered panels with various trim techniques to achieve a particular look. But it is important to keep in mind that lining the walls of a room—any room—with full-height wood is going to be very costly and should be carefully considered. See Figure 5.7.2 for various veneer matching patterns.

5.7.6 Cabinetry

Cabinetry, also called "case goods", is yet another vast subject in itself. There are huge industrial cabinet manufacturers, from very low quality (paint, melamine, and particleboard only) to very high quality (even some solid wood, but high-quality finishes, beautiful hardware, etc.) with every imaginable level in between. These products are sold through retail stores, showrooms, and even designers' offices in rare cases.

Virtually any conceivable look, from raised panels in cherry to sleek and smooth ultra-polished lacquer can be had in these standardized products as well as extensive options for sizes and types.

Traditionally, kitchen cabinets in the US are designed to have countertops at +36" above finished floow (aff). "Standard" countertops are $1\frac{1}{2}$" thick, so standard base cabinets are $34\frac{1}{2}$" high. Standard countertops are 25" deep and are intended for a 1" overhang in the front, so standard base cabinets are 24" deep. Upper cabinets are most commonly 12" deep, but greater depth is possible in some products lines, and height varies from 18" to 42", usually on a 3" module. Cabinet widths are also usually on a 3" module, so standard widths include 12", 15", 18", 21", 24", 27", 30", 33", 36", 39", 42", 45", and 48". (Not all of the sizes are available in all product lines.)

Cabinets like these are made using two basic techniques: framed and frameless. Framed cabinets have solid sides, back, and floor, and a "frame" around the front to support the door. Frameless cabinets are the same except that there is no front frame and the doors are hung on the sides of the cabinet instead. There are two significant differences here: first, the two styles look different because the frame will be at least partly visible in framed cabinets, no matter how the doors are designed and installed, and, second, the frameless style is inherently less durable because the door hardware is attached to non-solid material (particle board, MDF, etc.) and the long-term performance tends to be compromised. Framed cabinets can still have a sleek look, just not as simple and planar as frameless cabinets. (The author's 1958 decidedly mid-century modern house has its original kitchen cabinets, which are framed style but very sleek. The drawers have shaped fronts and the bottoms of the fronts are shaped to act as pulls so there are no visible drawer pulls; this cannot be done with frameless cabinets. Despite having been painted several times, these cabinets are mostly in very good condition after nearly 60 years.)

But all of this has been about what we call "stock" cabinets; cabinets that are pre-made to standard sizes and configurations in factories and installed in groups in buildings. There is a completely separate world of all-custom cabinets too.

In the commercial world, frameless all-plastic laminate cabinets dominate in kitchenettes, break rooms, work rooms, file rooms, etc. These are commonly custom made though, at least in part because nearly everything in commercial facilities has to be accessible these days and accessible countertops can be no higher than 34" (not 36") and the horizontal reach cannot exceed 24" (not 25").

The custom cabinet world is divided into three distinct areas. First, there are custom versions of the factory cabinets for residential projects. These can be very high quality but they can be made to *any* dimensions and configurations that might be needed. Second, as noted above, custom-built frameless plastic laminate (usually) faced cabinets dominate in the commercial world of utility spaces. And, third, full-custom cabinetry is used in the commercial world for breakfronts and AV cabinets in conference rooms, floating shelves, book shelves, wall-units, furniture in offices, special workplace desks (nurses stations, labs, etc.) reception desks and on and on. The latter category knows almost no bounds and anything can happen. Such casegoods might include glass, metal, stone, and even concrete, as well as basic wood, wood veneer, and plastic laminate.

For finishing, as noted previously, it is better to do less millwork well than to do more millwork badly. Among other things, this means that shop-finishing is required, especially for dramatic pieces or areas that are under special scrutiny—walls of elevator lobbies and reception areas, cabinetry in major conference rooms, etc. (The author had a law firm project many years ago where the interior designer wanted cherry crown mold, wood base, and over-sized cherry doors throughout the space, but so much high-quality wood was simply not affordable within the budget. So decisions were made to use standard 7′ 0″-high cherry doors (with high-quality trim) and to make most of the crown mold and wood base in painted poplar. The space still looked very good and it was done within the budget.)

5.7.7 Furniture

True custom furniture is unusual in both residential and commercial projects, especially for chairs. As it turns out, it is risky to design, manufacture, and sell chairs in the US because there are serious liability concerns due to potential injuries if something breaks, so it is next to impossible to convince a local millworker to make a full-custom chair. It is feasible to do everything else though.

The author had two experiences in the past that involved custom furniture design. (For the purposes of this discussion, a specially made slab desk in an office is not really "furniture.") Both of these projects were designed in 1989. In the first case, the project was a small Japanese restaurant and the design included clear-finished poplar throughout, including frames for modified Shoji screens, the screens themselves, bench seats, steps, shoe boxes, wall paneling, table tops (using some walnut), the check-in desk, a special box for a very special Geisha doll, and the chairs. A full-custom chair was designed but it was not possible to have it made. The millworker on the project had a design for an oak sled-base chair and the author was allowed to modify that by raising the bottom rails to a few inches above the floor and using solid poplar instead of oak. The results were impressive, with beautiful extensive millwork in the space.

The second project was a large accounting firm office and two special pieces were needed, one for the reception area and one in a bare-looking empty hallway niche. The author designed a breakfront of sorts, to be made from **rift-sawn** white oak (very costly, but it was needed to match the veneers in the space) and marble (for the top) for the reception area, and a long narrow table with a base shelf using the same wood and marble for the niche. In 1989, these two pieces were quoted to cost about $12,000.00, so a decision was made to omit the marble top from the table, saving more than $2,000.00. The finished pieces were hand-built by a local craftsman, using solid wood, they were quite beautiful, and they were still very costly at nearly $10,000.00. Since this time, the author designed and built his own coffee and end tables (using solid cherry, maple, and walnut with figured anigre veneer tops) for his living room and designed and had built a TV stand (using white oak and walnut) for his family room. The living rooms tables are now 23 years old and doing well. The TV stand is gone.

5.8 Vertical movement systems

5.8.1 *Stairs*

Stairs are needed to walk from one floor level to another, whether that is two steps or a full story; single steps are trip hazards and should be avoided whenever possible.

To design a stair, the geometry must be clearly understood. Also, the rules for residential and commercial stairs are significantly different. In both cases though, it is necessary to calculate the riser height (and to select a **tread** size) to make the geometry work out.

For residential stairs, the International Residential Code requires the following:

- Risers $\leq 7\frac{3}{4}''$ with the difference between two risers being no more than $\frac{3}{8}''$.
- Tread depth $\geq 10''$ and the greatest difference between two treads is $\frac{3}{8}''$.
- Stair and landing width is $\geq 36''$.
- Risers are allowed to slope up to 30 degrees from vertical.
- Open risers are allowed as long as a 4″ diameter sphere cannot pass through.

For commercial stairs, the International Building Code requires the following:

- Risers $\geq 4''$ and $\leq 7''$ with the difference between the largest and smallest risers being no more than 3/8″.
- Tread depth $\geq 11''$ and the greatest difference between two treads is $\frac{3}{8}''$
- **Nosings** $\leq 1\frac{1}{4}''$.
- Stair and landing width is $\geq 44''$ (unless occupant load is less than 50).
- Risers are allowed to slope up to 30 degrees from vertical.
- Open risers are allowed as long as a 4″ diameter sphere cannot pass through, but only for stairs that are *not* part of the accessible route.

So how to calculate a stair?

We will do a residential example first, for a simple straight run stair. Let us say that a residence has a 9′ 0″ ceiling on the first floor and that the framing of the second floor is 1′ 3″ thick. That makes the overall rise of the stairway 9′ 0″ + 1′ 3″ = 10′ 3″. For calculations like this, it is simpler to work in inches and not feet and inches, so we will convert 10′ 3″ to inches, or 123″.

Step 1: Divide the overall rise by the maximum riser height: $123'' \div 7.75'' = 15.9$

Step 2: Round up the results of step 1 to the nearest whole number to get the actual number of risers: $15.9 \approx 16$

Step 3: Divide the overall rise by the actual number of risers to get the actual riser height: $123'' \div 16 = 7.69''$. (This is a rare calculation in design and construction where precision is needed. Rounding this off to 7.7″ would result in an error of $16 \times 0.01 = 0.23''$, almost a quarter of an inch.)

Step 4: In any stairway, the number of risers is one more than the number of treads (there is a riser at the bottom at the first tread and an extra one at the far end of last tread), so this stair will have 16 – 1 = 15 treads.

Step 5: If the treads are 10″ long, the overall length of the stair will be 15 × 10″ = 150″ (= 12′ 6″). If the stair is 3′ 0″ wide, it will take up 12′ 6″ × 3′ 0″ = 37.5 square feet, not counting top and bottom landings.

Here is a simple commercial example. In this case, the overall rise of the stair will be 15′ 0″ or 180″ (a fairly typical floor-to-floor height in commercial buildings, although it varies quite a lot).

Step 1: Divide the overall rise by the maximum riser height: 180″ ÷ 7.0″ = 25.7

Step 2: Round up the results of step 1 to the nearest whole number to get the actual number of risers: 25.7 ≈ 26

Step 3: Divide the overall rise by the actual number of risers to get the actual riser height: 180″ ÷ 26 = 6.92″. (This is a rare calculation in design and construction where precision is needed. Rounding this off to 6.9″ would result in an error of 26 × 0.02 = 0.52″, approximately half an inch.)

Step 4: In any stairway, the number of risers is one more than the number of treads (there is a riser at the bottom at the first tread and an extra one at the far end of the last tread), so this stair will have 26 – 1 = 25 treads.

Step 5: If the treads are 11″ long, the overall length of the stair will be 25 × 11″ = 275″ (= 22′ 11″. If the stair is 3′ 8″ wide, it will take up 22′ 11″ × 3′ 8″ = 84.03 square feet, not counting top and bottom landings.

Obviously, the larger tread depth and lower slope of a commercial stair cause it to take up more space.

If a stair turns a corner, or doubles back (a switch-back stair), this gets more complicated because landings have to be taken into account. The easiest way to think about this is to imagine a straight run stair with a landing in it; in this case, it is easy to see that the landing would not add any risers and it is essentially just an elongated or widened tread, so the impact on the size of the stair would be the difference between the required landing width and the required tread width. For a residential stair, that would be 36″ – 10″ = 26″ longer; for a commercial stair, it would be 44″ – 11″ = 33″ longer. This is all still true if the landing is in a corner or it serves both up and down stairs, as in a switch-back stair.

Dealing with landings is really an issue of handling the stair **stringers**. A stair stringer is the framing member that runs along the sides of the stair to support the treads. In residential stairs, stringers are most commonly made from wood 2″ × 12″s and they can be exposed along the sides of the steps or concealed under the treads. In commercial stairs, stringers are most commonly steel channels or heavy steel plates, although steel or aluminum tubes and wood are sometimes used as well. It is possible to build stairs using only glass and small metal connectors, but these are generally only seen in Apple stores in large cities.

All of this gets very interesting when stringers turn corners at landings. The author does not like to see stringers "step up" at corners (he thinks that it is unsightly and that it makes for ugly and complicated handrails and guards), which is necessary unless the landing extends as the first tread in the next run going up; nevertheless, it is possible to build stairs with stringers that step up.

In any case, it is highly unlikely that an interior designer would actually design a stair anyway (except possibly in a one- or two-family dwelling) because an architect or professional engineer would be required to do the technical design. The purpose of this discussion is to point out how to figure out roughly how much space a stair will need.

As far as guards (required by code, for class I buildings, along any edge that is more than 30″ above an adjacent surface) and handrails (required by code as well, with requirements varying with stair width) go, that is all about the look again. Today, there are excellent options for guards from almost glass-only (supported only by a heavy channel along the bottom) to cables, rails, and solid walls and handrails that include wood, plastic, and various metals. It is highly advisable for students and practitioners to pay attention to stairs in the field, noting what they are made of and how they are put together (at least so far as that is possible).

5.8.2 Elevators, lifts, and escalators

All elevators and lifts are rated by their weight-carrying capacity, which is directly related to the car size. Small car = low capacity and large car = high capacity. Typical elevator capacities start at 2,500 pounds; hospital elevators that are large enough to move patients in their beds are usually 6,000 pounds; over-sized elevators (usually in industrial facilities) can be 10,000 pounds, 15,000 pounds, or even larger. A residential elevator might be rated for only 800 pounds or so.

An interior designer would never design an elevator. Architects design elevators in the sense that they establish locations, capacities, car sizes, hoistway sizes, machine room sizes, penthouse sizes, etc. but the elevator manufacturers actually design the machines and their controls. Nevertheless, it is conceivable that a new elevator, lift, or even escalator could be added in an existing building and could therefore impact space planning, hence the need to understand the basics of elevators, lifts, and escalators.

5.8.2.1 Traditional traction elevators

The oldest form of elevator is called a traction elevator because the mechanism uses cables to raise and lower the car; the cables are moved by large wheels attached to powerful DC electric motors using friction (called "traction" in this case). This type of elevator was subject to catastrophic failure until a very famous inventor named Elisha Otis came along. Even though it is commonly understood that Mr. Otis invented the elevator, this is not the case. Instead, he designed—perfected really—safety brakes that would engage if the cables (called ropes) break; this prevented falls to the bottom and the resulting mayhem. In traction elevators, counterweights are necessary to minimize the horsepower that is required to move the car; the car goes up while the counterweight goes down and vice versa. (This is commonly seen in action movies where people are seen in elevator shafts, dodging moving cars and counterweights; this never happens in reality of course because it is obviously highly dangerous.)

Traction elevators are still required for some mid-rise and all high-rise buildings and they can move very fast in very tall buildings. (The fastest elevators move at velocities approaching 1,500 feet per minute, which equates to about 17 miles per hour—vertically!) Traction elevators are the most costly option though, so other types are used when feasible.

In very tall buildings, only one or two elevators will run the full height of the building (usually) and other elevators run to groups of floors; this is done to minimize travel time. These grouped elevators can be stacked on top of one another; for example, three elevators might run from 1–20 with another group running from 21–40 and another group running from 41–60, all in the same hoistway. (Hoistway is the correct term for an elevator shaft.)

For planning purposes, traction elevators are space efficient because they take up no additional floor space beyond the hoistway except at the top. The drive machines sit directly on top of the hoistways, so it is necessary to have a large penthouse on top of the elevator hoistways with this type of elevator.

5.8.2.2 Traditional hydraulic elevators

Hydraulic elevators are much newer. With a standard hydraulic elevator, there is a large hydraulic cylinder attached to the bottom of the car, which extends from a casing to raise the car and drops into the casing to lower the car, using hydraulic pressure from a pump and fluid reservoir. Elevators like this are noisy and slow and they are limited to a vertical rise of 60 feet or so. This does mean in a standard hydraulic elevator that the hydraulic cylinder under the car is more than 60 feet long if the elevator can rise 60 feet (so the casing is more than 60 feet below the bottom of the car as well). Holes like this are drilled before the building is built because it is usually not practical to get the drilling machine to the site later in the process.

There are alternatives to fixed and one-piece cylinders, which include single telescoping cylinders (the casing is much shorter and the cylinder actually extends) and multiple wall-mounted telescoping cylinders. The latter requires no drilling under the elevator at all. Of course, these last two options are most costly and are not often used.

Hydraulic elevators do require more horsepower than traction elevators, primarily because there are no counterweights; 25 horsepower is low; 60 horsepower is not uncommon. Motors of these sizes require large power circuits and the wiring can be very costly.

The main issue to remember about standard hydraulic elevators is that they require what is called a "machine room" at the lowest level directly adjacent to the hoistway to simplify running the hydraulic piping from the machine to the cylinder. It is possible to separate the machine room from the hoistway horizontally, but the farther away it is, the more complicated and costly it gets. A typical elevator machine room varies in size according to the weight capacity of the elevator, but 10′ × 12′ would be a typical size.

5.8.2.3 Machine room-less elevators

This is the newest category of elevators, and it encompasses many different sub-types. Mostly, these are simplified traction elevators (meaning that they are cable-driven) with small machines that are located inside the hoistway, usually at the top (but not always). The main advantages to this type are: (1) there is no machine room and (2) they require less power (back to counterweights again).

These elevators are very popular for smallish elevators for mid-rise applications (again, 60 vertical feet or so) but they are more costly than conventional hydraulic elevators, so they often get changed at the last minute to reduce costs. (It must be noted that some of the savings in the elevator itself will be used up by the higher wiring cost for the less costly traditional hydraulic elevator; also, space will be lost for the latter to accommodate the machine room.)

5.8.2.4 Residential elevators

This type of elevator can be either cable or hydraulic in a small hoistway, usually with a small closet at the lowest level for the drive machine. The cars are very small, usually only 3′ × 5′ or so, with low load capacity, slow speed, and low overall rise height, but they are not costly and they are seen in some single family dwellings. This type of elevator can only be used in commercial applications if fully accessible cars are not required.

5.8.2.5 Lifts

This machine is often called a "wheelchair lift" because its main purpose is to lift someone in a wheelchair no more than a single story (often limited to 12 feet vertical) and usually without a hoistway. They are frequently located outdoors where they are used to provide access to building entrances not at grade, but they can be used for indoor level transitions too. Given that they are used to move wheelchairs vertically, their platforms are usually 5′ × 5′ and the overall unit is usually about 5.5′ × 6.5′ in plan. Dumbwaiters are another common form of lift, except that dumbwaiters move trays or carts—never people.

5.8.2.6 Escalators

Escalators are moving stairs, and they are basically a rotating large loop that forms steps when it comes up from below. They are commonly seen in large building lobbies, multi-level malls, airports, hotels, sports arenas, casinos, etc. They usually run in straight lines but it is possible to use curving escalators (at extremely high cost). The mechanism is a machine in a "pit" at the foot of the escalator, which must be planned into the structure very carefully. It would be highly unusual to add an escalator inside an existing building, but anything is possible.

5.9 Architectural documentation

It is far beyond the scope of this book to cover the detailed methods for documenting architecture and interior design for construction, but a few basic comments might be useful. First, like much else in the design world, this is changing rapidly, primarily due to the use of Building Information Management (BIM), which often takes the form of Autodesk's software REVIT. This is causing massive change in documentation because contractors can have direct access to 3-dimensional data that was created by the design team, which, at least in theory, should simplify the construction process. But things are not quite as simple as they seem.

Second, let us go back in time a bit to look at how things were done with earlier technologies. Logically, drawings have long been used (for hundreds, really thousands, of years) to help workmen understand what the architect wanted to be built, but prior to the late 19th century, there was no practical way to make a full-sized copy of a drawing. Prior to that time, the common way to do such drawings was to draw on water-color boards using ruling pens and to use water-color washes where color would be useful. These drawings can be quite beautiful and they are rare by definition. (The author had the good fortune to work in a firm that was founded in 1854 and they had records of construction from the 1870s. In those days, the architect took the drawings to the job-site every day and then took them back to his office every evening.)

But what are these drawings? How do they work?

Architects, interior designers, and even engineers work by imagining solutions to complicated 3-dimensional problems, and then, somehow, they have to find a way to help workmen on a job-site understand how to build this imagined solution. So over a period of many years, a specialized form of drawing—called technical drawing—was developed. This type of drawing cut the proposed building horizontally at key points (usually the foundation, each floor level, and above the roof, all looking downward) and vertically around the perimeter and maybe all the way through. The results of these cuts are what are called "plans" and "sections." Additional drawings were done of the exterior walls, which are called "elevations" and sometimes more detailed drawings (at larger scale) were done of portions of the plans, sections, or elevations, to describe elements more clearly; these latter drawings are called "details." Clearly, these drawings are not the building itself; they are simply a representation of something that is intended. There is nothing real here at all.

When blueprinting (white lines on dark blue paper—a relatively permanent process) was invented late in the 19th century, it became possible to make copies of these large drawings, so architects started drawing on flexible sheets of linen instead of on rigid boards (it also made it easier to move the original drawings, which could now be rolled up). Later, another faster, and less costly printing process was invented, called Diazo printing, which uses an ultraviolet light and an ammonia-based development process to make copies with blue lines on white paper (a temporary process—they will fade away to nothing in bright light). To lower costs, original drawings could also be done on vellum (high-quality paper) and then on mylar (very tough plastic, which required special plastic leads or ink).

In the 1970s, computers began to be used to produce technical drawings and software (such as AutoCad from Autodesk and Microstation from Bentley Systems) now completely dominates the production of technical drawings. But the basic approach to the drawings was the same: plans, sections, elevations, and details.

By the 1980s, 3-D software was being developed and that has now matured into current BIM packages that are very powerful and practical to use.

But all of this is still a representation of something that is intended and it is not something that is real. An elaborate 3-D model is still not the actual building, right? How could it be?

This matters because it speaks to the mind-set of those who produce this data (whether 2-D or 3-D), the interior designers, architects, and engineers of the AEI industry. The critical element in a design office is not the "what" that is shown in the drawing or model; it is the "why" that explains that what, if that makes sense. Showing anything in a drawing, or putting it in a model, is pointless if there is no understood "why" behind that information.

Students are often taught standardized ways to produce technical drawings, sometimes to the point of being told that there is a "right" and a "wrong" way. Offices also have "standards" for graphic output of various types. But there is no right or wrong here. There are no laws that govern the production of plans and sections; there are no universal industry-wide standards (although there are many commonly held "typical" approaches). The author has been drafting technical documents in offices since 1981 and he has been supervising the production by others since 1989 and it has become clear that the standards, even if one accepts that there are some, do not matter. What does matter is clarity, plain and simple. If a drawing or a model tells a workman something that is clear and that gets the work done, that drawing or model has done its job (whether it meets a someone's "standard" or not). If a drawing or model confuses the workman, it has not. That is all the drawings and models are for: period. They have no intrinsic value of their own. Very well done technical drawings can be beautiful, even artistic, but art is not part of the equation here. Contractors do not care if the drawings are pretty; they only care that they can understand their meaning.

The fear with BIM projects is that the designers who create the models might not fully grasp the why part of this. The software makes it easy to get a floor connected to an exterior wall, or a countertop connected to an interior partition. But it does not really matter that the wall connects to the exterior wall or that the countertop connects to the partition; what matters is *how* those connections are made, and the worry is that it might be too easy for that *how* to get lost. It will remain paramount, at least in the author's opinion, for designers to remain focused on the why at all times; the how is important but it means little, or nothing at all, without the why.

On a more mundane note, is often taught that technical drawings require dimension strings—strings of numbers that add up across a plan (usually)—so that workmen know how large everything is supposed to be. But there is a problem with dimensioning of technical drawings, which is that the likelihood of error is forgotten. If a group of students is handed measuring tapes and asked to measure—very carefully—every wall in a typical studio space and record the results, a set of numbers will be generated. If the same group re-measures the same space six months later (without reference to the first set of numbers) the new numbers will not exactly match the first numbers. This is true because no one is perfect and no one can measure anything perfectly. If the numbers vary by $\frac{1}{8}$″ or so, that is excellent and there should be no problem. But what if someone measured 15′ 0″ one time and 14′ 11″ the next time? Does that matter?

It might. If the room is a partner office in a law firm with built-in furniture that requires 15′ 0″ clear and the built room is 14′ 11″ clear, someone is going to get in a lot of trouble. (Things like this happen every day, unfortunately.) How can this be avoided? By not using complete dimension strings, especially when working in existing buildings (as is the nature of interior design). In any space plan, some spaces are more critical, dimensionally, than others: in some corporate environments, it really matters that offices for staff of equal rank be exactly equal in size; the law partner's office might be critical due to furniture; corridor widths might be critical for code reasons. But the dimensions of storage rooms, break rooms, even conference rooms might not be critical.

When a project is built in the field, a "layout" person marks the proposed wall locations on the existing floor, usually using chalk lines (maybe paint lines). The person who does this layout is usually experienced but this person has no knowledge of the "why" behind any of the dimensions; there is no way for this person to know that the 15′ 0″ number for the law partner's office is critical unless someone says so. So say so, somehow. That can be done by noting "hold" next to a critical dimension. Also, do not use complete dimension strings that run all the way across a space or a building; instead, leave out the least important dimension (maybe that storage room?) and then, if there is an error, it will be in a storage room, or some other non-critical space, where no one notices or cares. If two adjacent spaces need to be the same size and they are locked between two existing points (columns, perhaps), just put on the drawing that they are to be "equal"; that is a perfect legitimate dimension. These techniques all get to the "why" so that we, as designers, give information to those in the field in a way that helps them to understand the why too (even though they might not care) or to control inevitable error (workmen cannot measure perfectly either) so that it is not damaging to the client or the designer. Of course, if the contractors access the 3-D model directly, this type of control is lost.

Ultimately, what is needed should be drawn, or modeled, and what is not needed should not be drawn or modeled. That sounds simple enough, but it is really quite challenging in practice. Just try to remember the "why": why do we need this line, that tag, dimension, etc.? If you do not know why, you probably do not need it at all.

Summary

The objective of this chapter is to provide a solid basis in technical information for walls, doors, door hardware, door frames, windows, ceilings, millwork, vertical movement systems, and architectural documentation, and it has been a whirlwind tour. There is no need to be concerned with remembering all of this right away. Over time, and with practice (literal practice), much of this will become familiar. But sometimes, things become too familiar and projects become rote: we always use $3\frac{5}{8}$″ 18 ga. metal studs @ 16″ o.c., this type of door frame, or that door construction, these hinges, that drawer pull, catalyzed lacquer on rift-sawn white oak ... whatever. Part of the idea of this chapter is to provide a relatively concise summary that broadens the scope of what can be, rather than what usually is, as inelegant as that sounds. It is really easy to get into habits as a designer, and this chapter (the whole book, really) hopes to push against that tendency.

Outcomes

5.1 Understanding that walls encompass various heights (partial, to-ceiling, and full-height), can be hollow or solid, thin or thick, and can be constructed using many different materials.
5.2 Understanding that doors can be various sizes, thicknesses, and made from many different materials.
5.3 Understanding that door hardware is highly complex, consisting of hinges, latches, locks, closers, and various accessories.
5.4 Understanding that door frames might or might not be required, and can be made from wood, aluminum, and steel.
5.5 Understanding that interior windows can be nearly any size, can be nearly frameless, and can be framed using wood, aluminum, and steel.
5.6 Understanding that ceilings can be full or partial height, high or low, accessible or inaccessible, smooth or rough, and made from many different materials.
5.7 Understanding that millwork includes standing and running trim, wood doors and frames, and casework (cabinetry) and that finishing (whether painted or clear) is critical to design and long-term performance.
5.8 Understanding that vertical movement systems include stairs (wood, steel, aluminum, and even glass), elevators (conventional traction, conventional hydraulic, machine roomless, and residential), lifts, and escalators.

Notes

1 AWI, *Architectural Woodwork Standards, 2nd Edition*, pages 14–15
2 Ibid., page 248
3 Ibid., page 249
4 Ibid.
5 Ibid., page 245
6 Ibid.

7 Ibid.
8 Ibid., page 255
9 Ibid., page 243
10 Ibid.
11 Ibid.
12 Ibid.
13 Ibid., page 115
14 Ibid., page 113
15 Ibid., page 114
16 Ibid., page 53
17 Ibid.

Chapter 6

What are architectural acoustics?

Objectives

6.1 Introduction
6.2 To understand basic acoustical theory
6.3 To understand basic sound absorption
6.4 To understand basic room acoustics
6.5 To understand basic sound isolation
6.6 To understand basic mechanical systems sound isolation
6.7 To understand basic speech privacy
6.8 To understand basic electronic sound systems and home theaters

6.1 Introduction

Noise problems are very common in many buildings—from noisy neighbors in apartment buildings to loud mechanical systems in offices, or from being overheard in the doctor's office to not being able to understand speech in a large space. Both too loud and too quiet can be serious problems in the wrong situation. The detailed technical design of treatments and systems for performance venues (stage theaters, movie theaters, concert halls, opera houses, arenas, large conference centers, etc.) is far beyond the scope of this book (and beyond the scope of interior design practice), but interior designers can, and should, design according to sound (no pun intended) acoustical principles while also avoiding typical problems. There are several basic issues to be covered:

- *Basic theory*: what sound is and why it affects spaces and occupants the way that it does.
- *Sound absorption*: using materials to absorb sound to affect reverberation (excessive reverberation is experienced as "echoes") and noise reduction.
- *Room acoustics*: shaping rooms for acoustical purposes.
- *Sound isolation*: how to keep sound from one space from causing problems in another space.
- *Mechanical systems sound isolation*: all about vibrations.
- *Speech privacy*, especially in open office environments.
- *Electronic sound systems*, especially in home theaters.

6.2 Basic theory

Sound is vibration that is transmitted through **elastic materials** at a speed that varies with the density of the material. Sound travels at the highest speeds through very dense solid materials and at the slowest speed through very light gaseous materials (such as air). Here are some typical speeds:

- Speed of sound through solid steel: 16,000 feet per second (fps)
- Speed of sound through solid concrete: 12,000 fps
- Speed of sound through water: 4,800 fps
- Speed of sound through air (at sea level): 1,130 fps[1]

Swimmers are usually familiar with this because they know that sounds move readily through water. So why don't we use water, or concrete, or steel to move sound around? Because **fidelity** tends to decrease as density increases. Also, our ears are usually in air, so we are most interested in sound propagated in air.

The form of these vibrations is roughly a sine wave, with half of each cycle above normal pressure and half of each cycle below normal pressure. Sound is thought of mostly from the point of view of "cycles per second," also known as hertz (in honor of the physicist Heinrich Hertz, who did pioneering work in electromagnetism), which represents the number of complete cycles (both halves of the wave) in one second of time. For high numbers, kilohertz is often used in lieu of hertz; 1,000 hertz (Hz) is equal to 1.0 kilohertz (kHz) (see Figure 6.2).

This figure shows a 0.1-second time period along the horizontal axis and sound pressure along the vertical axis. Six complete cycles are shown, which means that this is a 60 Hz tone (6 cycles ÷ 0.1 seconds = 60 Hz). The time period, or the time for a single complete cycle is (1 ÷ 60 Hz = 0.017 seconds). The amplitude—the height of the curve both above and below the center line—represents the loudness of the sound.

When pure tones are produced, they actually look like this on a graph. Pure tones are produced only by tone generators, although some pipe organ pipes produce

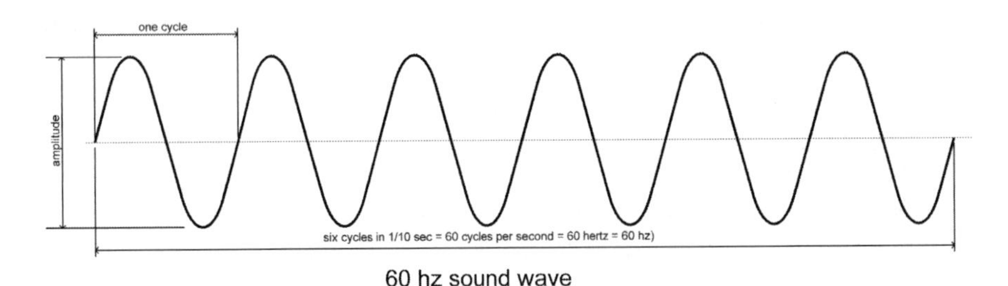

Figure 6.2 **Frequency, period, and amplitude in sound**

nearly pure tones, so most sound graphs do not look like this. Instead, most sounds are composite sounds that are made up of numerous different frequencies and amplitudes occurring simultaneously—when this is harmonious, we often call the sound "music." (Some music is not particularly harmonious, but that is irrelevant to the discussion at hand.)

If a sound is travelling through air at 1,130 fps and it has a frequency of 60 Hz, the wavelength is 1,130 fps ÷ 60 Hz = 18.83 feet. If the sound has a frequency of 600 Hz, the wavelength is 1.883 feet, and if the sound has a frequency of 6,000 Hz (or 6 kHz), the wavelength is 0.1883 feet (or about 2.26 inches). Wavelengths are of great importance in music and architectural acoustics alike. In music, wavelength (or frequency) is perceived as pitch—is a note high or low? High notes have large frequencies with short wavelengths and low notes have low frequencies with long wavelengths.

Young human ears can hear frequencies from 20 Hz to 20,000 Hz (20 kHz); old human ears (and old in this context means more than 20 or so) can hear frequencies from 20 Hz to about 4,000 Hz (4 kHz). This loss of high frequency hearing—called presbycusis—is normal and it happens to everyone.

The wavelength for 20 Hz is 1,130 fps ÷ 20 Hz = 56.5 feet and the wavelength for 20 kHz is 1,130 fps ÷ 20,000 Hz = 0.057 feet, or about 0.68 inches. At 4,000 Hz, the wavelength is 1,130 fps ÷ 4,000 Hz = 0.283 feet, or about 3.40 inches.

It should come as no surprise that sound waves having a wavelength of 3.40 inches behave differently, especially in enclosed spaces, than sound waves having a wavelength of 56.5 feet. This difference is one of the keys to understanding architectural acoustics.

At this point, it is worth noting that the speed of light in air is 186,000 miles per second, or 982,080,000 feet per second, or roughly 869,100 times faster than the speed of sound! This is why one sees mouths moving sooner than one hears the sound from someone talking at a great distance; if the listener is 1,500 feet away from a speaker (or singer or other sound source), it would take 1.3 seconds for the sound to reach the listener but it would take only 0.0000015 seconds for the light to reach the listener.

Another important effect worth noting is that air can absorb high frequencies (say above 2,000 Hz or so) but it cannot absorb low frequencies. This is the reason that thunder is heard as a low pitched sound although a nearby lightning strike includes sounds of many frequencies; it is actually the same sound, but when it is far away, the air absorbs much of the high frequency component, leaving only the low frequency "rumbling."

Within the overall range of hearing response (20 Hz to 4,000 Hz, or 20 Hz to 20,000 Hz for the very young), the typical human voice covers the range from 125 Hz to about 8,000 Hz (the high frequency sound are most sibilants—the hissing sounds associated mostly with saying "ess"); a piano can cover the range from 31.5 Hz to 4,000 Hz. (Middle "C," for musical readers, is defined as 250 Hz and "A above middle C," as used for tuning in orchestras, is usually set at 440 Hz.) In music, a doubling of frequency represents an octave, which is divided into eight notes. That means that "C above middle C" would be 250 Hz × 2 = 500 Hz and the C above that would be 1,000 Hz (1 kHz). It all makes sense. Typical women's voices are centered higher in the overall voice range than typical men's voices, although not all women's voices are necessarily "higher" than all men's voices. Men's voices exhibit a wider overall range as well, from rather high to very low.

Across the full hearing frequency response, the sensitivity is not the same. In other words, we hear some frequencies better than others, with maximum sensitivity occurring at 4,000 Hz (4 kHz). Minimum sensitivity occurs at the lowest frequencies, with the highest frequencies being roughly in the middle.

What about loudness? Sound power is measured in watts, but keep in mind that a watt of sound is a large amount while a watt of light is a relatively small amount. Sound intensity (loudness) is measured in watts per square meter (W/m^2). A really loud sound is produced at a level of $10 W/m^2$ (or $10^1 W/m^2$, or 10 to the power of $1 W/m^2$); the quietest sound that can be heard is at a level of $0.000000000001 W/m^2$ or $10^{-12} W/m^2$. This vast range makes calculations awkward at best, so a system has been devised to reduce this range to a manageable scale by applying logarithms (a mathematical technique that compresses the range by powers of 10). The resulting scale ranges from 0 to 140 and the unit is the decibel (denoted dB). (The Bel is a more general unit in physics that was named in honor of Alexander Graham Bell, who was a pioneer in sound research before he became famous for inventing the telephone.)

How loud is:

140 decibels (dB)? A jet engine 75 feet away.
120 dB? Some rock concerts and some thunder, if close by.
100 dB? Crowd noise at a football game (sometimes louder) or a car horn close by.
80 dB? A noisy cafeteria.
50 dB? Typical office.
30 dB? Quiet residence
10 dB? Rustling of leaves or breathing

Here are some general ranges[2]:

- Painful and dangerous: 126–140 dB
- Deafening: 104–124 dB
- Very loud: 82–102 dB
- Loud: 62–80 dB
- Moderate: 42–62 dB

- Faint: 22–40 dB
- Very faint: 0–20 dB

If the background noise in a classroom is 40 dB and someone turns on a sound system that produces 80 dB, is the resulting loudness 120 dB? No, because the logarithm conversion process makes it inaccurate to add dB. Instead, it is necessary to convert back to W/m^2, do the math, and convert back to the logarithms.

Here is how the math works. To find dB for a given loudness, use

$$L_1 = 10 \log I/I_o$$

where:

L_1 = loudness
I = intensity in W/m^2
I_o = reference sound intensity (always 10^{-12}W/m^2).

Let us use 0.00045W/m^2 for this first example.

$L_1 = 10 \log I/I_o$
$L_1 = 10 \log 0.00045/10^{-12}$
$L_1 = 10 \log 4.5 \times 10^{-4}/10^{-12}$
$L_1 = 10 \log (4.5 \times 10^{-4} \times 10^{12})$
$L_1 = 10 \log 4.5 \times 10^{8}$
$L_1 = 10\ (8.65)$
$L_1 = 86.5$ dB

Using this same equation from the first step to add 40 dB to 80 dB,

40 dB $= 10 \log I_{40}/I_o$
4 dB $= \log I_{40}/I_o$
antilog 4 dB $= \text{antilog}(\log I_{40}/I_o)$
10,000 $= I_{40}/I_o$
10,000 $= I_{40}/10^{-12}$
$10{,}000 \times 10^{-12} = I_{40}$
$10^5 \times 10^{-12} = I_{40}$
$10^{-7} = I_{40}$

Using the same equation again, from the second step,

80 dB $= 10 \log I_{80}/I_o$
8 dB $= \log I_{80}/I_o$
antilog 8 dB $= \text{antilog}(\log I_{80}/I_o)$
100,000,000 $= I_{80}/I_o$

$10^{8} = I_{80}/10^{-12}$
$10^{8} \times 10^{-12} = I_{80}$
$10^{-4} = I_{80}$

Now, from the third step,

$I_{40} + I_{80} = I_{120}$

So

$10^{-7} + 10^{-4} = 0.000001 + 0.0001 = 0.001001 = I_{120}$

Back to the equation again, to get the final answer:

$L_{120} = 10 \log I_{120}/I_{o}$
$L_{120} = 10 \log 0.001001/10^{-12}$
$L_{120} = 10 \log 100{,}100{,}000$
$L_{120} = 10\ (8)$
$L_{120} = 80\,dB$

So 80 db + 40 db = 80 dB! That is correct because what we are doing is adding a very small number to another very small number. There is another, much simpler, way to add these figures together directly in dB form:

When two dB values differ by	**Add the following dB to the higher value**
0 or 1	3
2 or 3	2
4 to 8	1
9 or more	0

Using this method results in 80 dB + 40 dB = 80 dB, just as before.
So the sum of 80 dB + 80 dB = 83 dB.
And the sum of 40 dB + 50 dB + 60 dB + 65 dB

= (40 dB + 50 dB) + (60 dB + 65 dB)
= 50 dB + 66 dB
= 66 dB

This technique works in any order, so

40 dB + 60 dB + 50 dB + 65 dB
= (40 dB + 60 dB) + (50 dB + 65 dB)
= 50 dB + 66 dB
= 66 dB

It should be noted that doubling of sound intensity is always equivalent to 3 dB.

6.3 Sound absorption

Sound in a free area (outdoors) radiates in a spherical form (in all directions) and its intensity falls off with the square of the distance (twice as far = one quarter the loudness, three times as far = $\frac{1}{9}$ times as loud, and four times as far = $\frac{1}{16}$ times as loud ...). When sound is propagated inside a space, there is still fall-off near the source, but the sound level in the space actually *increases*, simply due to the enclosure. This is referred to as "sound build-up." Sound trapped in a space can have three consequences:

1. problems outside the space due to leakage, sound transfer, etc.;
2. excessive reverberation, causing poor speech intelligibility;
3. excessive noise—just too loud.

The methodology to deal with these issues was developed by Professor Wallace Sabine at Harvard University, beginning in 1895. At the time, there was a relatively new building—the Fogg Art Museum—on the Harvard campus that had a semi-circular tiered lecture room at one end. But no one was able to use the room because it had an echo problem, it was too noisy, and no one could understand a speaker (poor speech intelligibility). Professor Sabine measured a sound persistence time of 5.5 seconds; given that the average English-speaking person can say more than 15 syllables in 5 seconds, it is no surprise that speakers could not be understood.

The room consisted of all hard surfaces: plaster walls and ceilings, hard flooring, and hard furniture. Professor Sabine determined that adding materials that absorb sound might reduce the problems, and he conducted an extensive series of experiments using hundreds of 3″-thick hair-fiber stuffed seat cushions from a nearby theater. Ultimately, the unit of acoustical absorption became the "Sabin" (pronounced Saybin) and the formula to calculate reverberation became

$$T_R = (0.049 \times V) \div \text{total absorption (in Sabins)}$$

where

T_R = reverberation time, in seconds (defined as the length of time that it takes the sound level to drop by 60 dB, or to one millionth of its original intensity)
V = room volume, in cubic feet
0.049 is a constant that adjusts for English units; a different figure is used for the metric system
total absorption = all absorption due to all materials in the room

This calculation is frequency dependent because materials are not equally absorbent at all frequencies. This means that the outcome will vary from one frequency to another, so it is necessary to do this calculation at all six of the standard frequencies that have been determined: 125 Hz, 250 Hz, 500 Hz, 1,000 Hz, 2,000 Hz, and 4,000 Hz.

Material	Area (sf)	125 hz Coeff. ∂	125 hz Absorb. (Sabins)	250 hz Coeff. ∂	250 hz Absorb. (Sabins)	500 hz Coeff. ∂	500 hz Absorb. (Sabins)	1,000 hz Coeff. ∂	1,000 hz Absorb. (Sabins)	2,000 hz Coeff. ∂	2,000 hz Absorb. (Sabins)	4,000 hz Coeff. ∂	4,000 hz Absorb. (Sabins)
Terrazzo and stone flooring	1600	0.01	16.0	0.01	16.0	0.02	32.0	0.02	32.0	0.02	32.0	0.02	32.0
Gypsum board ceiling	1600	0.15	240.0	0.10	160.0	0.05	80.0	0.04	64.0	0.07	112.0	0.09	144.0
Gypsum board walls	3200	0.15	480.0	0.10	320.0	0.05	160.0	0.04	128.0	0.07	224.0	0.09	288.0
Room air (1,000 cf)	32	0.00	0.0	0.00	0.0	0.00	0.0	0.90	28.8	2.30	73.6	7.20	230.4
Total absorption (Sabins)			736.0		496.0		272.0		252.8		441.6		694.4
Reverberation time (seconds)			2.1		3.2		5.8		6.2		3.6		2.3

***Figure 6.3.A* Reverberation calculation Example 1: Lobby**

In order to use materials for acoustical absorption, it is necessary to know the acoustical absorption coefficients for those materials. All materials intended for these purposes have been tested and manufacturers publish the test data. (If someone is trying to sell an acoustical product without this data, the product is probably a sham and it should be avoided.) Coefficients range from 0.0 (no absorption at all) to 1.0 (100% absorption, as for an opening). Here are two examples of typical absorption coefficients:

	125 Hz	250 Hz	500 Hz	1,000 Hz	2,000 Hz	4,000 Hz
Heavy drapery	0.14	0.35	0.55	0.72	0.70	0.65
Glued-down carpet	0.02	0.06	0.14	0.37	0.60	0.65

Note that the drapery and carpet perform at the same level, absorbing 65% of the sound, at 4,000 Hz, but that the drapery is seven times more effective than the carpet at 125 Hz. Also note that both materials are at least 4.6 times as effective at 4,000 Hz as at 125 Hz. And note that the carpet is 32.5 times more effective at 4,000 Hz than it is at 125 Hz. This is why performance can vary so widely and why it is necessary to check all six frequencies. When doing these calculations for an actual project, real published data should be used for the products involved, rather than generic data (like that shown previously).

How often does this really matter? More often that one might think, actually. Excessive reverberation can be a problem in many different kinds of spaces: lobbies, large conference rooms, large classrooms, etc. The issue should be considered in any space where there are extensive hard surfaces, little absorption, and relatively high volume. Increasing volume increases reverberation in any space.

Take a public lobby in a high-rise office building, for example. The space is 40 feet square and about 20 feet high; the floor is terrazzo and marble; the walls and ceilings are gypsum board. Figure 6.3.A shows the basic calculations.

As you can see, the range in reverberation times is from 2.1 seconds at 125 Hz to 6.2 seconds at 500 Hz. Is that good? Or bad? Actually, it is pretty bad. There are no rules for acceptable reverberation times, but here are some rough guidelines (for mid-frequencies, 500 Hz and 1,000 Hz):

Pipe organ music	more than 2.0 seconds
Symphonic music	1.7 to 2.3 seconds
Opera	1.5 to 2.8 seconds
Chamber music	1.4 to 1.7 seconds

Rock music	1.0 to 1.2 seconds
Movies	0.8 to 1.2 seconds
Lecture room	0.7 to 1.1 seconds
Classroom	0.6 to 0.8 seconds
Studio	0.45 to 0.55 seconds[3]

The general rule is that reverberation time needs to be relatively low for speech—hence lecture room and classroom near the bottom of the list—and relatively high for music. Reverberation makes music sound richer, particular symphonic and organ music. For singing, something in the middle might be best, depending upon the type of singing that is involved (opera works at higher reverberation than most singing).

In the example, an office building lobby has no particular requirements, but it is good if it is possible to carry on a conversation. The example room would not work well for speech at all, and it would probably be pretty irritating if a large group of people gathered and attempted to talk to one another. The figures above three seconds are truly alarming.

What would happen if we were to replace half of the gypsum board ceiling with a typical 2 × 2 acoustical ceiling? See Figure 6.3.B. As you can see, this has a drastic effect on reverberation time, and the range is now 1.1 seconds at 4,000 Hz to 1.8 seconds at 500 Hz. All of the figures would still be too high for a lecture room, but it is probably not too bad for a lobby. If a spreadsheet is used for these calculations, they are really quite simple and easy to do. And they should be done for spaces that might have issues: conference rooms, lobbies, or other sensitive locations.

As absorption increases in a room, internal sound levels drop because some portion of the sound is absorbed, no longer bouncing around and causing build-up. This also helps to reduce problems for neighbors outside the space because there is less sound to transfer out of the room.

It is possible to have too much absorption, which causes rooms to sound "dead." The ultimate example of this is an anechoic chamber, a room that is designed to absorb *all* sound that hits its surfaces. Such a room has no normal floor because a normal floor cannot be made to be both 100% **acoustically absorptive** and walkable, so these rooms have narrow catwalks out to very small platforms in the center of the room. The experience of being in one of these rooms is very odd, even a little disturbing, mostly because one does not hear oneself talking as usual. Because there are no reflections, little to no sound enters the ears, which is mostly how we hear ourselves talk.

Figure 6.3.B **Reverberation calculation Example 2: Lobby with 50% acoustical ceiling**

Material	Area (sf)	125 hz		250 hz		500 hz		1,000 hz		2,000 hz		4,000 hz	
		Coeff. ∂	Absorb. (Sabins)	Coeff. ∂	Absorb. (Sabins)	Coeff. ∂	Absorb. (Sabins)	Coeff. ∂	Absorb. (Sabins)	Coeff. ∂	Absorb. (Sabins)	Coeff. ∂	Absorb. (Sabins)
Terrazzo and stone flooring	1600	0.01	16.0	0.01	16.0	0.02	32.0	0.02	32.0	0.02	32.0	0.02	32.0
Gypsum board ceiling	800	0.15	120.0	0.10	80.0	0.05	40.0	0.04	32.0	0.07	56.0	0.09	72.0
Acoustical 2 x 2 ceiling	800	0.76	608.0	0.93	744.0	0.83	664.0	0.99	792.0	0.99	792.0	0.94	752.0
Gypsum board walls	3200	0.15	480.0	0.10	320.0	0.05	160.0	0.04	128.0	0.07	224.0	0.09	288.0
Room air (1,000 cf)	32	0.00	0.0	0.00	0.0	0.00	0.0	0.90	28.8	2.30	73.6	7.20	230.4
Total absorption (Sabins)			1224.0		1160.0		896.0		1012.8		1177.6		1374.4
Reverberation time (seconds)			1.3		1.4		1.8		1.5		1.3		1.1

Instead, in an anechoic chamber, we only hear ourselves through the vibrations in our heads—a strange sensation indeed.

The deadest rooms that most people might encounter would be movie theaters (where they want to create the sound "environment" of the room entirely with the sound system) or some recording studios.

6.4 Room acoustics

In most small to medium sized rooms, the shape and arrangement of the walls, floor, and ceiling is unlikely to cause major problems—unless there are very hard and smooth walls (probably glass) that are parallel to each other (which can cause something called a "flutter echo"). Recording studios make up the main exception to this, and such rooms should not have parallel walls or parallel floors and ceilings, rectangular shapes (never square), and never concave curves. Concave curves cause acoustical problems in most spaces because their curvature focuses the sound. Convex curves are good because they disperse sound (see Figure 6.4.A).

When absorptive materials are placed in space—to reduce reverberation or noise or both—they should not be used uniformly, meaning that they are more effective if they are used in "sections." In other words, if 50% of a wall were to be covered with some kind of acoustically absorbent material, it would be more effective to use a checkerboard arrangement than to cover half of the wall contiguously. This is true, at least in part, because the checkerboard pattern would expose lots of edges, thereby increasing the effective area of the material.

What sorts of products are available? Many, but it depends mostly on the application.

For floors, most traditional products (concrete, stone, ceramic tile, terrazzo, etc.) are very highly reflective (with coefficients usually ranging between 0.02 and 0.05). Wood floors can have coefficients up to 0.15 or so. Carpet is somewhat more absorptive, especially if used with a dense but soft pad, but it is always weak at low frequencies.

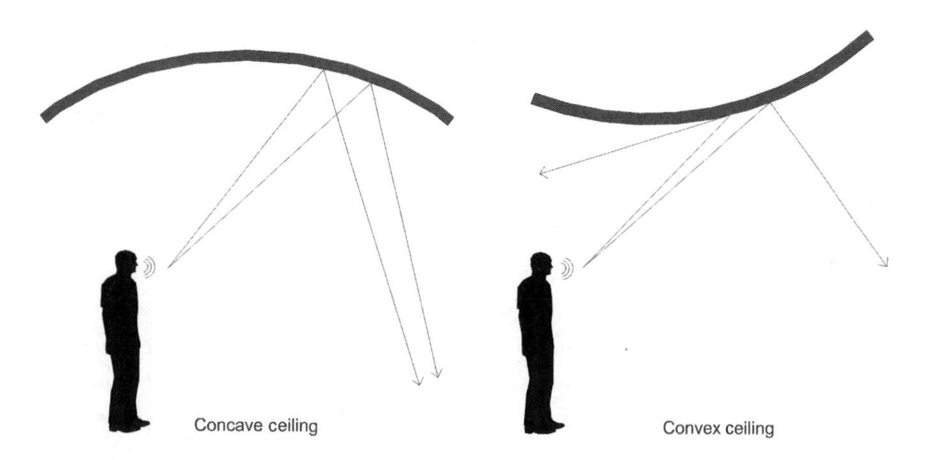

Figure 6.4.A
Concave and convex curves

For walls, gypsum board (the most common material in use today) has coefficients between 0.04 and 0.55, and the mounting method is important. (Mounting methods include rigid attachment to a rigid substrate using screws or nails, rigid attachment to a flexible substrate (such as resilient furring channels), gluing, etc.) The most "acoustical" product for wall applications is the fabric-wrapped panel (Armstrong's Soundsoak is a typical product), which is usually 1″ thick and comes in a wide variety of finishes. These panels can be mounted in two different ways: flat on the wall (Method A) or with an air space behind the panels (Method D), and the coefficients vary from 0.05 to 1.09, depending on the specific product and installation method (see Figure 6.4.B).

The air space behind the panels increases the absorption, so the coefficients are usually a little higher for the D mounting method. Acoustically absorptive concrete masonry units are available too—hollow concrete units, with slotted faces, and acoustically absorbent material inside; this system is very costly, but it is a good option for spaces where fabric-wrapped panels might not be tough enough, e.g. a gymnasium or multi-purpose room at a church or school.

For ceilings, things get more complicated. Gypsum board can be used, with the same coefficients as when it is used on walls (generally speaking); old-style wet plaster can be used, with even lower coefficients; wood can be used with coefficients that are similar to those for wood floors, and various specially designed acoustical products can be used. (Concrete and steel are also seen in exposed structure applications without ceilings.)

Acoustical ceiling products come in four basic categories (see Chapter 5 for more details):

1 Glued-on products (usually 12″ × 12″)
2 Fully concealed framing systems (usually called "concealed spline" and usually done using 12″ × 12″ tiles)
3 Exposed or partially concealed framing systems (usually called "lay-in")
4 Attached panels (such as Soundsoak)

Glued-on 12″ × 12″ tiles were popular in the 1950s and 1960s, before lay-in systems were readily available. They do offer some acoustical absorption, but the effect is usually limited.

Concealed spline systems can be used with a variety of tiles (a wider variety than in the glued-on systems usually), but these systems are costly and they tend to suffer greatly as they age. They are designed to be "accessible" ceilings that can be disassembled to gain access to mechanical and electrical systems (or anything else), but it is difficult to put them back together again exactly as they were originally installed and they tend to look worse and worse each time they are disassembled. They are rarely used these days.

Material	125 hz Coeff. ∂	250 hz Coeff. ∂	500 hz Coeff. ∂	1,000 hz Coeff. ∂	2,000 hz Coeff. ∂	4,000 hz Coeff. ∂
A-mounting, on surface	0.06	0.34	0.84	1.09	1.07	1.02
D-mounting, with 3/4" airspace	0.16	0.52	1.01	1.11	1.02	1.05

Figure 6.4.B
Typical absorption coefficients for wall-mounted fabric-wrapped acoustical panels

Lay-in systems completely dominate the market and they are very cost effective (in their simple forms). They come in 24″ × 24″, 24″ × 48″, 20″ × 20″, and 20″ × 60″ standard grid sizes, although they can be configured in other ways too—12″ × 48″ or 48″ × 48″, for example. The 20″ × 20″ and 20″ × 60″ systems are not often used any more, but they do still exist. (One of the challenges with a 20″ × 60″ ceiling system is that all standard light fixtures used to be 48″ long, and either metal or ceiling-panel filler panels have to be used to fill in at the ends of the light fixtures—an awkward situation at best. This is might be less of an issue with new LED products.) There are also six basic families of grid profiles:

- "Standard" $\frac{15}{16}$″-wide inverted tees, in painted steel, aluminum, and stainless steel (and in gasketed versions for clean room use)
- "Narrow" $\frac{9}{16}$″-wide inverted tees, in painted steel
- "Narrow" $\frac{9}{16}$″-wide inverted slotted (narrow and wide slots) tees, in painted steel
- "Narrow" $\frac{9}{16}$″-wide inverted tees for special slotted panels to conceal the grid
- Extra-wide $1\frac{1}{2}$″-wide inverted tees, in painted steel and aluminum (and in gasketed versions for clean room use)
- Special "extra-wide" inverted tees for hanging gypsum board ceilings (with the grid fully concealed)

Infill panels are available in mineral fiber (the vast majority, usually $\frac{3}{4}$″ thick or so), fiberglass (usually thicker), and even wood. Standard mineral fiber panels are available with three standard edge profiles:

- Square, where the panels simply rest on the upper side of the inverted tees
- Tegular, where the panel edges are notched so that the panel faces will be about $\frac{3}{8}$″ below the face of the grid
- Notched, specifically for slotted tees; in this system, the face of the panel is in exactly the same plane at the face of the grid

Figures 5.6.2.2.A and 5.6.2.2.B illustrate some of the possible combinations. ("Troffer" light fixtures are also available for both flat and slotted grids.)

The acoustical performance of these systems can vary widely because different systems are available for different purposes. Sometimes noise reduction in rooms is the only issue of critical importance, and sound transmission to adjacent spaces is less critical; that is the most common situation, and that is where the most conventional systems are used. If sound transmission is critical, a different panel should be selected that has higher sound isolation performance.

The panels themselves are available in a bewildering array of styles and finishes, from smooth and "omni-fissured" (the oldest, cheapest, and most common option), to smooth and perforated (regular or irregular round holes), to lightly textured, or to heavily textured; in numerous standard colors (or painted in the field—which *does* reduce acoustical absorption), and in many different special materials, from fabrics to wood veneers.

So what can be used in a given situation? That depends mostly on the budget. If there is a desire to use a "non-standard" lay-in ceiling system, costs should be verified before the product is locked into the design. Some of the specialty products are much more costly than the basic products, and there is no advantage to designing something that the owner cannot, or will not, pay for.

On occasion, wall panels and/or acoustical ceilings cannot solve the problem—too little wall surface, difficult placement for ceilings, etc. In these cases it is common to use acoustical baffles, which are two-sided hanging acoustical panels. They can be fabric wrapped (like wall panels) or sealed in mylar (very light-weight plastic—think decorative balloons) and they are available in many different sizes, shapes, and even thicknesses. Baffles tend to be most effective at low frequencies, so they should be considered for use in spaces where low frequency problems dominate.

There are many specialty products out there, so research should always be done when a special need arises during a design project.

6.5 Sound isolation

How do we prevent sound transfer problems—being overheard when we do not want to be, overhearing someone else when we do not want to, being bothered by mechanical equipment, etc.? By controlling sound transmission to improve sound isolation. But how do we go about that?

First, a few basics. Figure 6.5 shows a device that Bob Newman built himself to demonstrate sound isolation to his classes at Harvard and M.I.T. in the 1960s and 1970s. (The author was one of Bob Newman's students at M.I.T. in 1979.) Here is how it worked:

1. The alarm in free space produced about 70 dB at some distance.
2. Add a "box" made of $\frac{3}{4}$"-thick fuzzy acoustically absorptive material (usually called "fuzz" in the industry), which reduces the sound to 67 dB. (The sound inside the box remains at 70 dB because the large amount of acoustically absorbent material prevents sound build-up inside the box.)
3. Remove the fuzz box and add a plywood box with a seal. Now the sound inside the box increases to 78 dB—a clear demonstration of sound build-up. But the sound outside the box drops to 50 dB because the plywood box has better sound isolation properties than the fuzz box.
4. Add the fuzz box inside the plywood box (still with the seal). The sound inside the box is now 72 dB; it went up from the fuzz box alone (at 70 dB) because there is still some sound build-up due to the plywood box. The sound outside the box is now 43 dB due to the added fuzz, the plywood box, and the seal.

So if we want to improve sound isolation between rooms, we should add fuzz, seal all edges, and use relatively dense construction. Do our typical construction methods resemble that? Not at all.

***Figure 6.5* Sound isolation demonstration apparatus**

The most common interior wall construction used in commercial projects consists of a single layer of $\frac{5}{8}$″-thick gypsum board on both sides of $3\frac{5}{8}$″ × $1\frac{1}{2}$″ 22 ga. (sometimes even lighter gauges are used, as are heavier gauges in special applications) metal studs at 16″ on center. This does not include fuzz; the edges are not sealed, but the gypsum board is relatively dense (just not all that thick around a typical room as compared to the plywood box from the demonstration). In addition, these walls do not usually extend up to the underside of the structure above (the floor above or the roof, as the case may be), mostly because it is very costly to build "full-height" walls in most modern buildings. In typical commercial office spaces, the walls actually stop at the underside of the lay-in ceiling in most cases.

Special steps for acoustical isolation (or noise reduction) are not usually taken, simply due to the high costs involved. (Acoustical wall panels like Soundsoak usually cost $15.00/sf, or even more, so it can cost $2,250.00 to cover a single 150 sf wall. That kind of expense is not affordable for most projects.) But if improved performance is required (and it is affordable), here are the steps that can be taken:

1 Seal all edges and openings, using gaskets, tapes, foam, or even wet sealants. This is critical and all other measures will be a waste of time and money if this is not done first.

2 Add fuzz. This would consist of putting insulation inside the stud cavities, either using standard thermal fiberglass batts (cheap) or special acoustical batts (much better and much more costly). Note that the cavities should *not* be filled completely; it is important to leave an air space on at least one side.
3 Add more mass, which can be done by adding more gypsum board, a special product called "mass loaded vinyl," or special products such as Quietrock by Serious Materials, Inc. (www.Quietrock.com). This could be one or more additional layers of materials on either or both sides of the wall.
4 Use an extra-wide sill plate (and head track) and "staggered studs." This has two sets of studs on each side, offset from one another by half of a stud space.

Or

Use two separate sill plates (and head tracks) with a small separation (1″ or so) between them, essentially building two walls, back-to-back (but with no gypsum board or finishes on the inside, of course). In a double stud wall, it is possible to use fuzz twice.
5 Use resilient mounting for the gypsum board (or other material) on either or both sides of the wall. This is done by using resilient furring channels that have flexible joints in them; allowing the wall face to move a little does reduce sound transmission.
6 Add special adhesives by Serious Materials, Inc. ("Quietglue" is the product, which is also available for floors).

As noted previously, cost is a big issue with everything on this list, and each individual item adds significantly to the overall cost. Adding two items adds more. Adding three items adds even more, etc. On the other hand, if it is clear that there is going to be a major problem without special treatment (the CEO's office is directly adjacent to a loud mechanical room), it would be irresponsible not to design appropriate acoustical sound isolation.

For ceiling/floor assemblies, all of these steps still apply (except staggered and double studs). The techniques are a little different. If fuzz is used in a ceiling, it usually just rests on the top of the ceiling, and resiliency is achieved by using resilient blocks (neoprene or rubber usually) or springs in the hangers for the suspended ceiling (see Figure 6.6.B).

There is another new issue for floors/ceilings: impact sound transfer. This is most apparent if someone is walking around in high heels on a concrete slab directly overhead—each "tap" for a step quickly moves through the concrete and enters the space underneath. To minimize impact sound transfer, the steps listed above can be taken, plus the floor itself can be isolated (on resilient mounting pads or even on springs) and the floor surface can be softened—carpet does help considerably to reduce impact sound transfer.

Can sound isolation be calculated? Sure. Here are the two equations that are needed:

$$\text{NR (noise reduction)} = L_1 - L_2,$$

where L_1 = loudness in room 1 (in dB); L_2 = loudness in room 2 (in dB)

$$NR = TL \text{ (transmission loss, in dB)} + 10 \log \text{(Sabins/area)}$$

Here is a specific example. Take two dormitory rooms that are 10′ × 15′ in plan (joining along the 10′ wall) and 9′ high to the ceiling. Each room has direct-glue carpet on the floor; gypsum board on all of the walls and ceiling, and each room has a bed with absorption of 15 Sa. If the occupant of room 1 is playing rock music at a level of 92 dB, how loud will it be in room 2 at 500 Hz if the wall has TL = 35 dB?

Start by determining the absorption in room 2.

Gypsum board walls: (10′ + 10′ + 15′ + 15′) × 9′ × 0.05 (coefficient of absorption at 500 Hz) = 15 Sa
Gypsum board ceiling: 10′ × 15′ × 0.05 = 8 Sa
Bed = 15 Sa (given)
Total absorption: 38 Sa

Now, go to the second equation to determine the actual noise reduction through the wall.

$NR = 35 + 10 \log (38/(10' \times 9'))$ *(10′ × 9′ = wall area in sf)*
$NR = 35 + 10 \log (0.42)$
$NR = 35 + 10 (-0.37)$
$NR = 35 + (-3.7)$
$NR = 35 - 3.7$
$NR = 31.3$ dB or approximately 31 dB

Next, use the first equation:

$NR = L_1 - L_2$
$31 \text{ dB} = 92 \text{ dB} - L_2$
$31 \text{ dB} - 92 \text{ dB} = -L_2$
$-61 \text{ dB} = -L_2$
$61 \text{ dB} = L_2$

Would that be acceptable? Probably not, because 61 dB falls into the "Moderate" category for sound level (and it is close to "Loud"). What would happen if 20 sf of Armstrong Soundsoak 85 were added to the receiving room?

Absorption would be:

Gypsum board walls: [((10′ + 10′ + 15′ + 15′) × 9′) – 20 sf] × 0.05 = 14 Sa
Gypsum board ceiling: 10′ × 15′ × 0.05 = 8 Sa
Bed = 15 Sa (given)
Soundsoak 85 panels: 20 sf × 0.84 (A-mounting) = 17 Sa
Total absorption: 54 Sa

$$NR = 35 + 10 \log (54/10' \times 9')$$
$$NR = 35 + 10 \log (0.57)$$
$$NR = 35 + 10 (-0.25)$$
$$NR = 35 + (-2.5)$$
$$NR = 35 - 2.5$$
$$NR = 32.5\,dB \text{ or approximately } 33\,dB$$

Next, use the first equation:

$$NR = L_1 - L_2$$
$$33\,dB = 92\,dB - L_2$$
$$33\,dB - 92\,dB = -L_2$$
$$-59\,dB = -L_2$$
$$59\,dB = L_2$$

Would such a change matter? Only a little, because we can barely distinguish a change of 3 dB and a 2 dB change would not really be noticeable. If we really wanted to improve this situation, we would want to improve the wall TL from 35 dB to 40 dB or even 45 dB. If it were 45 dB and the Soundsoak were used, the sound level in the room would be 92 dB – 43 dB = 49 dB, which would be a big improvement. Of course, adding even more absorption would help too.

6.6 Mechanical systems sound isolation

What types of mechanical system sounds can be problematic?

- Structure-borne vibration from large equipment (pumps, motors, compressors, large fans, etc.)
- Air-borne sound from noisy equipment (chillers, pumps, motors, compressors, etc.)
- Water noise from piping (heating and cooling water as well as plumbing: domestic water, storm drainage, and sanitary drainage), including water hammer
- Air noise from equipment or ductwork

What can be done about each one of these?

Structure-borne vibration is most problematic for steel-framed buildings, mostly because steel transmits sound at 16,000 fps. Steel buildings are also inherently more flexible than concrete or masonry buildings, which makes the problem even worse. There is really just one solution: use vibration isolation devices for each piece of equipment. For equipment that vibrates a little (like a boiler or a small fan), it is reasonable to use a pad type of isolation device that is made of rubber or neoprene. For equipment that vibrates a lot (pumps, most especially), it is necessary to use spring mount devices. Spring mounts allow the equipment to move quite a lot without affecting the structure.

Figure 6.6.A
Spring isolator, floor

If the equipment itself needs isolation, piping and/or ductwork connected to the equipment needs isolation from the equipment too, which is done with flexible piping couplings and flexible ductwork connectors (usually made out of coated canvas).

To control air-borne sound from noisy equipment, it is necessary to study the sound isolation qualities of the surrounding construction, and make improvements as necessary (as discussed under 6.5 previously).

To control water noise in piping, it is helpful to insulate the piping (which can be costly if it is not required otherwise) and to use resilient (neoprene or rubber pads or springs) hangers and attachments.

Figure 6.6.B
Spring isolator, hanger

To control air noise in ducts, the ducts must be properly designed (poor ductwork designs can easily *increase* air noise), isolated from the equipment (using flexible connectors), and isolated from the structure (or walls) where there is reason to believe that significant vibration transfer can occur.

All of this should be done by the mechanical engineers, but it cannot hurt to ask and to point out areas of concern.

6.7 Speech privacy

Not being overhead can be a serious concern—when talking to one's doctor, perhaps, or when talking to a pharmacist, when having a confidential conversation with anyone, or simply when trying to concentrate. Most typical walls can do a reasonable job of maintaining speech privacy, unless the walls are not sealed (then the leakage paths around or through the wall can be a serious problem). It is common to be able to overhear that someone is talking in the next room with typical gypsum board wall construction, but the speech is often perceived as "mumbling" and it is often difficult to understand words; in other words, speech is often heard but it is not intelligible. So speech privacy is a big issue mostly in open office areas, especially in large open office areas with cubicles.

What can you do to keep from being bothered by the loud guy on the phone three cubicles over?

Increase speech privacy, by reducing speech intelligibility. In Section 6.2 the need for speech intelligibility was discussed in the context of classrooms, lecture rooms, etc.; the open office space is the opposite—it is desirable *not* to be able to understand someone talking at some distance. This is done by using a relatively low, highly absorbent ceiling (definitely a lay-in ceiling, and a high-performance lay-in ceiling would be best) and highly absorbent panels in the cubicles (both to reduce reverberation and general sound spread), and by adding in a "masking noise" system. A masking noise system is a system that adds background sound that is designed to interfere with speech intelligibility. Such systems use sound generators and amplifiers (at some convenient central, or semi-central, location) coupled with a large number of speakers located throughout the space (usually *above* the ceiling, but sometimes in the ceiling). Also, the furniture systems should be planned to keep maximum distances between individuals, although that is very difficult to do. The relatively low ceiling is needed to minimize sound spread by limiting reflection paths. Taller cubicle panels help too. But a large room with a relatively low ceiling (say 10′0″ or less) with high cubicle panels (say 5′0″ or more) is not necessarily a visually appealing work space—everything is about balance. If the room looks open and inviting, it might well be noisy too.

6.8 Electronic sound systems

Sound reproduction began in the late 19th century and numerous developments have occurred since. From the early days of wax cylinders, to acetate platters, to vinyl records, to magnetic tapes, to compact discs (the first digital system), and finally to digital audio files (mp3, among others), systems have achieved better performance with higher reliability and increased fidelity. (There are audiophiles out there who still argue that the analog vinyl record is superior to the digital CD, but that is an esoteric position that few people will accept. Even though vinyl can sound better in some instances, the cracks, pops, and hisses from damage to those records more than offset any improvements. There is little doubt that all-digital sound reproduction is here to stay.)

Despite all of these improvements in recording and play-back technologies, speaker technologies have changed relatively little over the same period. Speakers work by converting the electrical signal into physical motion—in order to start the sound vibration in the air. This is done for low- to mid-range frequencies by simple cone or cone-with-horn speakers (the majority of all speakers) that simply move in and out, under the control of an electromagnet that is, in turn, controlled by the musical signal, and by high frequency speakers that use domes, ribbons, and other exotic devices. Cone speakers start at 2″ diameter (or maybe even smaller) and run up to 18″ diameter (or even bigger); cone speakers have relatively limited "pattern" control. For better pattern control, horns are added to the cones; horns use literal horns to shape and direct the sound.

In most sound systems, the frequency spectrum is divided into ranges: low range, mid-range, and upper range, using different speakers to cover each range. In a very basic system, one speaker might cover all three ranges, but that means that both the very high and very low frequencies will be poorly reproduced (due to the physical limitations of *any* particular speaker). The next step up is a two-way system that uses a woofer (for the low and mid-ranges) and a tweeter (for the upper range); two-way systems start with 4″ woofers and go up from there. The next step is a three-way system that uses a woofer (for the low range), a mid-range, and a tweeter for the high range. Some systems use super-tweeters for the very high range (which most people cannot hear anyway), and many systems use sub-woofers for the very low range.

Systems are available for mono, stereo, and surround-sound. Mono means that there is one channel of sound with no directionality; if there are two speakers in the system, they both play the same thing. Stereo means that there are two channels (to mimic our hearing), which means that we can hear directionality—the whistle blew on the right side, for example. Surround-sound takes stereo to a whole new level, with five channels (front left, front middle, front right, rear left, and rear right), and we can get a complete sensation of sound in front of, next to, and behind us.

Surround-sound systems come in 5:1 and 7:1 configurations. The :1 in both systems is a sub-woofer; a 5 means that there are five speakers, one for each channel; and a 7 means that there are seven speakers, one for each channel plus a middle left and a middle right to enhance the effect even more.

Systems like this can be very small integrated units that sit on a desk, table, or countertop; systems that use floor-mounted speakers in the living room; or they can be elaborate systems used in movie theaters (the ultimate being the sound system in an Imax theater) or at concert venues.

Home theaters form a special sub-classification. The idea of a home theater is to recreate the movie theater experience in someone's home, and there is a big range here too: a 12′ × 18′ room with three seats or a much larger room that seats 25 people could both be considered home theaters. The main issues are:

1 Does it need to be isolated from adjoining areas? (To prevent sounds coming in as well as sounds going out—would you want a master bedroom directly above a home theater with no special isolation?)

2 How good does the video need to be? Is it a 50″ television or a 144″ projection screen? Or something in between?
3 How good does it need to sound?

Sound isolation should be handled as discussed previously under sound isolation, but it should be pointed out that sound isolation is critical in a home theater. Hearing someone shouting, or a telephone ringing, or a door shutting during the movie would tend to ruin the experience.

Video quality is a complicated subject in itself, and, as such, it is far beyond the scope of this book. Suffice to say, the better it is, the more it is going to cost.

Regarding sound quality, the better the room is treated (reverberation control, noise reduction, and sound isolation), the less it becomes necessary to design and install a complicated (and costly) sound system. But, again, the better it sounds, the more it is going to cost (one way or the other). The author works with a home theater specialist who says that a home theater will cost at least $60,000.00 (he works in the high end of the market and he has built $250,000.00 home theaters), but that there are two ways to spend that money. First, you can build a $50,000.00 room and put a $10,000.00 sound system into it, or you can build a $10,000.00 room and put a $50,000.00 sound system into it. Either way, it is going to be expensive.

A few more basics—make sure to carpet the floor between the screen and the first row of seats because that will help to minimize the "first reflection bounce." The first reflection bounce is sound bouncing off the first nearby surface, which can be a problem if the surface is highly reflective to sound. Use absorptive treatments on roughly a third of the walls and ceiling of the room, concentrated toward the front (also to minimize early reflections); use **acoustically diffusive** or absorptive treatments on the back wall (the difference depends on the size and shape of the room). And use a riser for a second row of seats to maximize viewing angles for the screen.

Also, if a client asks for an "open" home theater, the client needs to understand that such a space will not be either a good home theater or isolated from other spaces.

Summary

In this whirlwind tour of architectural acoustics, it is important to remember just a few things:

1 Frequencies really matter. Just because a space works well at 1,000 Hz does not necessarily mean that it will work at all at 125 Hz.
2 It is practical for interior designers to do their own "quick" reverberation calculations to make sure that they are not designing spaces that will cause their clients problems. (Keep in mind that the client's problems tend to revert into the designer's problems.)
3 Sound isolation is key, but high performance is very costly, so it should be used selectively.

4 Projects should be designed to minimize the need for costly sound isolation. The way to avoid having to build an expensive wall between a pump room and the CEO's office is to separate those spaces with a buffer space (a storage room, break room, or other non-sensitive space) or to simply get them farther apart.
5 Always have sound isolation for mechanical equipment where there is a significant risk of problems.
6 Design open office spaces for low speech intelligibility.
7 When using electronic sound systems, try to find a good fit between quality and budget. Get to know an expert who can assist with such work. And do not let clients demand "open" home theaters.

Outcomes

6.2 Understanding basic acoustical theory: frequency, sound pressure, and the decibel.
6.3 Understanding basic sound absorption: absorption coefficients and reverberation.
6.4 Understanding basic room acoustics: room shapes, placement of materials, and types of systems and materials.
6.5 Understanding basic sound isolation: typical wall and ceiling constructions and potential improvements.
6.6 Understanding basic mechanical systems sound isolation: use of isolation devices for equipment, piping, and ductwork.
6.7 Understanding basic speech privacy: relatively low and highly absorptive ceiling with absorptive cubicle panels.
6.8 Understanding basic electronic sound systems and home theaters: sub-woofers, woofers, tweeters, super-tweeters; mono, stereo, and surround-sound.

Notes

1 *Architectural Acoustics*, M. David Egan, page 7
2 Ibid., page 13
3 Ibid., page 64

Chapter 7

What are mechanical systems?

Objectives

7.1 Why do we heat, cool, and/or ventilate?

7.1.1 Heating

It might seem to be obvious why we heat the interiors of buildings—to keep warm—but it is really a little more complicated than that. As a species, humans have learned to adapt to many different climates in virtually all areas of the globe, and we can survive in very hot and in very cold climates without building heating systems. (Has anyone ever camped out when it is 0 °F? Was there heating in the tent?)

Even though we can survive very cold temperatures, mostly by managing our clothing, it is not very practical for large numbers of people to live like that in cold climates for the duration of the winter; it is simply more convenient to spend some, or even most, of our time inside heated buildings. So technologies were developed over thousands of years to provide for warm (well, warmer anyway) interior environments—initially, simple indoor fires; then fireplaces; then stoves (as greatly improved by Benjamin Franklin in the middle of the 18th century); then furnaces and boilers, and then heat pumps.

This might seem facetious, but it is a serious question: why do we want a constant 72 °F all the time?

But the invention and perfection of indoor plumbing in the 19th century provided a new reason for indoor heating—keeping the water pipes from freezing. And this has become the dominant purpose. Even if one wanted to live in one's house in Minneapolis in the winter with the indoor temperature set to 25 °F (still much warmer than outside), that simply would not work because all of the plumbing would freeze.

So there are actually two reasons for indoor heating—comfort and prevention of frozen water piping.

7.1.2 Cooling

Why do we cool the interior of buildings? Because we can and because we want to. Even though all humans managed perfectly well with no mechanical cooling until well into the 20th century, it has become a requirement for most people in the United States, at home, at work, while shopping, going to a movie, play, or sporting event, and just about everywhere else. (That this desire for cooling everywhere requires large expenditures of energy is unfortunate, but even the most dedicated high-performance (i.e. low energy use) building advocates rarely talk about leaving out the cooling system, except in the mildest climates.)

Some people do point out that people who live in cold climates also use a large amount of energy to keep their buildings warm in the winter, and that they might be using more energy for heating than some people in hot places do for cooling. But that analysis ignores the fact that humans survived in hot places for thousands of years without cooling.

As energy costs go up and up, we are likely to start seeing less and less cooling; after all, no one had cooling in the 19th century, right?

Back to climate—in San Francisco, California, many buildings do not have cooling even today because the climate is cool and relatively dry. On the other hand, the climate in the deep south from eastern Texas to Florida has truly oppressive heat and humidity throughout much of the year. Most of the rest of the country is somewhere in between these extremes (central Indiana—home to the author—actually represents one of the most challenging climates in the world: cold to very cold winters (–36 °F record low); hot (115 °F record high) and very humid summers; and large numbers of freeze-thaw cycles), but it gets hot and humid enough in most of the country that we appreciate having the ability to create cool and dry indoor conditions—by using cooling. So we cool just about all of our buildings now, despite the obvious high energy cost.

As more and more people become concerned about potential broad-scope climate change and more specific local energy use, this might begin to change (actually, it already has begun to change). The author and his wife tolerate a wider degree of temperature variation today than they likely would have 20 years ago, partly due to these concerns but also because it is healthier to live in a building that is *not* sealed than one that *is* sealed. What is a sealed building? That takes us right to ventilation.

7.1.3 Ventilation

The air in an enclosed and occupied space will build up carbon dioxide (from breathing, primarily) and carbon monoxide (a poisonous byproduct of combustion, if there is such combustion in the space) if it is not refreshed with uncontaminated air, usually air from outside the building. Refreshing that air, by bringing in some percentage of outside air, is called ventilation, and it can be done naturally, by using open windows, doors, louvers, etc., or by mechanical means, using fans to force air into, and out of, occupied spaces. Modern codes require ventilation, either natural or mechanical, but natural ventilation is quite limited and mechanical ventilation is used in most situations, except single-family dwellings.

Mechanical ventilation, as a process, can be even more energy intensive than cooling, especially when coupled with cooling. But, again, ventilation is necessary for health so the goal is to make ventilation more efficient, and effective, and not to eliminate it altogether.

This is the area where we are unlikely to see decreases in the future; if anything, concerns about healthy environments could drive ventilation rates up, which could put more pressure on reducing the use of cooling.

7.2 How do we heat, cool, and ventilate?

7.2.1 Heating

The process is simply a matter of converting stored energy into heat, which can be done by a number of different methods:

- Burning wood (or other **bio-fuel**) or coal in an open fire.
- Burning wood (or other bio-fuel) or coal in a fireplace.
- Burning wood (or other bio-fuel) or coal in a stove.
- Burning wood (or other bio-fuel), coal, oil, propane, or natural gas in a furnace.
- Burning wood (or other bio-fuel), coal, oil, propane, or natural gas in a boiler.
- Collecting heat directly from the sun.
- From electricity, which is generated through one of the processes listed above, plus nuclear reactors. (Nuclear reactors do not generate electricity directly; instead, they generate steam (from excess heat) which generates electricity using the same technologies that are used in coal-fired and natural gas-fired power generating plants.)

These are all essentially the same process (except solar collection), where fire is used to release the stored heat in wood (or other bio-fuel), coal, oil, propane, or natural gas (or some other material that contains the element carbon). This is done by combining oxygen in the air with carbon released from the fuel, which releases heat. Once that energy is released, the laws of thermodynamics in physics allow for it to be used in only three ways:

1 *Conduction*: best described as heat transfer by touching the heat source. (Think of touching a hot cup of coffee or the handle of a hot pan on the stove.)
2 *Radiation*: best described as the one-directional heat transfer from a hot object to a cooler object through space. (Think of standing next to a bonfire on a cold night—warm front, cold back.)
3 *Convection*: best described as moving heat through air. (Think of a furnace, blowing hot air.)

Of these three heat transfer mechanisms, conduction is the most efficient. But it is not very practical to heat spaces, or even people, this way, so it plays a relatively minor role in actual systems.

Radiation is also highly efficient, though one-directional, and many heating systems use this principle, including: fireplaces (they radiate heat from the hot masonry surrounding the fire as well as from the fire itself); stoves (the hot body of the stove itself); steam radiators (the ones that hiss and clunk); hot water radiators (similar to steam but usually almost silent in normal operation); in-ceiling radiant systems (usually electric heating coils literally buried in the ceiling); and in-floor radiant systems (electric or hot water, with hot water being strongly preferred by most engineers). Until the advent of cooling systems in the early 20th century, most heating was done by radiation.

Convection is far less efficient than radiation because air has a low heat transfer coefficient (calculating heat transfer coefficients is extremely complex; suffice to say, water transfers heat in the general range of 30–50 times more effectively than air), but it is used commonly in HVAC systems because it has advantages (primarily for cooling) despite its weak overall efficiency.

Historically, the only options for heating were fires—in the open, in fireplaces, or in stoves—until the steam engine, or boiler, was perfected early in the 19th century. Even though heat transfer through water is more efficient than through air, heat transfer through steam is much more efficient than heat transfer through water, so it was recognized early on that steam boilers could be used to heat buildings, usually large buildings or groupings of buildings on campuses. Steam boilers are inherently dangerous though, and they were not used in individual buildings until late in the 19th century (after fire-resistant doors and asbestos fire-proofing were invented) when it became practical to protect a building from the potential explosion of a steam boiler. When high-efficiency water boilers were invented in the latter half of the 20th century, new steam systems started to disappear quickly and today virtually no one builds all-new steam heating systems in buildings. Even though the water systems are somewhat less efficient, they are far safer and more appropriate for inhabited buildings.

Building steam and heating water systems is costly and challenging. Both systems require extensive piping, special valves, and radiators, and workmen who build such systems must be highly skilled (especially for steam). So there was incentive for someone to figure out how to distribute heat from a fire without a dangerous boiler or costly piping.

So the gravity furnace was born. In a gravity furnace, there is a large firebox for burning wood or coal (usually) and an air distribution box sitting on top of the firebox. The air distribution box has large individual vertical (mostly—long horizontal runs do not work very well) ducts that terminate in the floors of the rooms above. The natural tendency of warm air to rise allows warm air to fill the building.

The disadvantage is that there was very little control over the system, overheating was quite common, and the products of combustion (waste products—heavy metals and other particulates in addition to carbon monoxide) were sometimes transferred directly through the occupied space. Without adequate ventilation, such a system could suffocate the occupants simply due to excessive concentrations of carbon monoxide. (Some gravity furnaces did use flues to remove the products of combustion.) The furnace also required frequent **stoking** to keep the fire going.

To address the limitations and risks of gravity furnaces, someone came up with the idea of adding a fan to increase air-speed and to make it possible to move air farther and with more control. Separate chimneys or flues were used to remove the products of combustion from the air in occupied spaces. Most furnaces use the design called "blow-through," pushing air through the burner, which keeps the fan out of the highest heat region in the system.

The fan-driven furnace still dominates the HVAC industry today, and many people have some version of it at home. The increased air-speed makes it possible to make ducts smaller, which saves space and money. The disadvantages to the fan-driven furnace are noise (due to the faster moving air), slow response time (due to using convection for heat transfer), and limited controllability (it is necessary to move a large volume of air to achieve a relatively small temperature change).

The last factor to include in deciding whether to use radiation or convection for heating is ventilation.

In the earliest days of mechanical ventilation systems (early in the 20th century), such systems were often separate from the primary steam or hot water heating system, but, over time, the systems often came to be combined so that a single system provided heating and ventilation. The heat source could be a furnace or a steam or hot water boiler that would feed steam or hot water to a coil that would replace the burner in the furnace.

Such a machine is not really called a furnace; instead, it is usually called a fan-coil or an air-handler. This is still an accurate basic description of much of the equipment that is in use today.

7.2.2 Cooling

When Willis Carrier (the founder of the well-known HVAC manufacturer Carrier) perfected practical large-scale cooling using the **refrigeration cycle** early in the 20th century, everything started to change. The systems were first used in large public facilities, especially movie theaters, as a lure for customers, then spread to large department stores, and on to virtually all other types of facilities.

Unfortunately, cooling cannot be done effectively on a large scale using conduction or radiation, so it is necessary to use convection to blow cool air into occupied spaces. (Today, there is work being done in the HVAC industry to perfect radiant cooling systems, but little practical equipment is yet on the market.) The furnace, fan-coil, or air-handler can be used to provide both heating and cooling by adding a separate cooling coil that is connected, via refrigerant piping or chilled water piping, to an outdoor machine that dumps the heat pulled from inside the building into the exterior atmosphere. (Heat cannot be created or destroyed; it can only be moved around.)

In the case of the furnace, this means that the system has no piping for heating and simple piping for cooling. For fan-coils and air-handlers, there is steam or hot water piping and chilled water piping (or refrigerant piping), but only to the equipment and not to each occupied space.

7.2.3 Ventilation

Natural ventilation is accomplished simply by opening windows, doors, or louvers. The International Building Code has strict rules for minimum opening sizes (relative to room sizes) and it limits natural ventilation to spaces directly adjacent to exterior walls (sort of—there are ways to provide natural ventilation to interior rooms too, although tricky to pull off).

Mechanical ventilation is accomplished by pulling in outside air and running it through some kind of system—furnace with or without cooling, fan-coil, air-handler, or **Energy Recovery Ventilator (ERV)**.

The challenge with ventilation in hot and humid climates is that the water vapor load in the outside air tends to overwhelm the capacity of the cooling equipment, so it is necessary to design carefully for humidity control in such situations.

Ventilation is the most widely misunderstood area of HVAC design because so many people seem to believe that it is both optional and non-critical, when it is really neither. Good ventilation is necessary for life—period. And it is useful for health and happiness too.

7.2.4 Systems summary

Here is a brief summary of the various types of systems and how they heat, cool, and/or ventilate occupied spaces:

These descriptions are deliberately generalized so as to cover much territory quickly; obviously, each one can be studied individually in depth.

- *Direct heating from an open fire.* This is effective but very dangerous, especially in combustible (usually wood) buildings, which can easily become more fuel for the fire. It provides no cooling and very limited ventilation. An indoor fire would go out without fuel; some of the fuel for any fire is oxygen, so it is necessary for some air to enter the building to keep the combustion process going. Even at that, there would be a tendency for the combustion products (especially carbon monoxide) to build up, so it remains highly risky to have open fires inside buildings. (Natural gas cooking stoves do contribute to this problem somewhat, but the gas burns more cleanly than wood or coal and it is not dangerous if it does not burn continuously.)

- *Direct heating from a fireplace.* This is actually somewhat less effective than an open fire because much of the heat produced by a fireplace leaves the building through the chimney, but fireplaces are safer than simple open fires (though still dangerous). It provides no cooling and limited ventilation. Some ventilation is necessary in order for a fireplace to "draft," which means that smoke (and much of the heat) from the fire rises through the chimney and exits the building without poisoning or suffocating the occupants.
- *Direct heating from a stove.* Because the stove is completely surrounded by the interior space (as compared to a fireplace which is often located on an exterior wall) and because it burns more efficiently (the enclosed combustion chamber raises the temperature of the fire, which makes the combustion more efficient), it is considerably more effective than a fireplace. But stoves are dangerous hot objects that take up significant useable space. No cooling and limited ventilation are provided. Stoves must also have some source of air for the fire to burn and for the flue to work.
- *Indirect heating (via convection) from a gravity furnace.* Gravity furnaces do work, although they are difficult to control. They provide no cooling and very limited ventilation. If there is no separate chimney or flue, there is a necessity to have air-relief (removal from the building) at the top of the building because the combustion products pass through the occupied space with the heat; not very safe at all.
- *Indirect heating from a fan-driven furnace.* Most furnaces are designed to remove the products of combustion through a chimney or flue that is separate from the air distribution. Conventional furnace burners use air from the space for combustion, so they cannot be installed in very small or fully enclosed spaces. Newer high-efficiency furnaces usually have piped vents to draw outside air directly into the burner, eliminating the need for indoor combustion air. Cooling and/or ventilation can be added to this type of furnace.
- *Indirect heating from a steam or hot water boiler with radiators.* In this case, the system can work extremely well for basic heating, but without cooling or ventilation.

At this point, it must be noted that all of the systems listed so far have a single point of "control," usually some form of wall-mounted thermostat. If there is a single very large air-handler in a large building, there would be only one thermostat for the entire building. But systems that use radiators for heating can have individual controls on each radiator, so there can be room-by-room temperature control. This is a huge improvement that cannot be underestimated.

- *Indirect heating from a steam or hot water boiler with fan-coils and/or air-handlers.* In this case, the systems convert heat from conduction (the steam or hot water in the pipes) to convection before it is delivered to the space, which is safer although less efficient. With fan-coils and/or air-handlers, cooling and/or ventilation can be added.

- *Collecting heat directly from the sun.* In addition to providing light, the sun—a giant burning ball of gases—also provides massive quantities of infrared heat, which is why the earth is not permanently frozen. This heat can be used directly to warm the interior of buildings; in fact, the sun will heat the interior of our buildings whether we want it to or not. In cold climates, thin buildings facing south with large windows can heat themselves by using only the infrared heat from the sun; in fact, it is possible to overheat such buildings. The author designed a building in 1982 that was intended to be passively heated (by using the sun) by collecting heat through a very large (about 6′ 0″ high × 100′ 0″ long) south-facing **clerestory** window; the owner was forced to put in shading film because the window actually overheated the entire two-story (plus basement) 10,600 gsf building.

Of course, actual systems can get far more complicated as buildings grow and owners desire more controllability of temperature and/or relative humidity. Large buildings usually have multiple pieces of equipment of various types in multiple locations—in boiler rooms, chiller rooms, air-handler rooms, above ceilings, in closets, in tunnels, basements, attics, etc. Anything is possible all the way down to having a thermostat in each occupied room; it is just a matter of cost.

7.3 Heat flow in buildings

In order to heat, or cool, a building, it is necessary to maintain a temperature difference between outdoors and indoors. This can be done with virtually any type of construction, but it makes sense to use materials that resist heat flow to make the process more efficient. There are two basic methods that are used:

1. thermal mass; and
2. thermal insulation.

Thermal mass simply means using dense and heavy materials to both absorb and release heat slowly. Keep in mind that the temperature in a cave is roughly 55 °F year-round, no matter how hot or cold it might be on the ground above the cave. This happens because the thermal mass of the material (i.e. the earth) between the top of the cave and the surface isolates the cave interior from rapid external temperature changes. This effect is so pronounced that the earth stays at the nearly constant temperature of 55 °F or so only 5′ 0″ or so below the surface.

A good example of a thermal mass approach would be a typical school, church, or courthouse building from the late 19th century. The exterior walls are probably brick or stone on the outside, brick in the core, and plaster on the inside—solid through and through, and probably at least 12″ and as much as 30″ thick. It takes months for heat to build up enough in so much mass that it affects the interior environment (although the effect is compromised by windows—lots of windows means less effective thermal mass); this is why such buildings usually stay reasonably cool well into the summer and why it usually is not necessary to turn on the heat until well into the fall. (Do not forget that the windows can cause major problems because they might have very little resistance to heat flow.)

The principle of heat radiation is extremely simple: heat flows from warmer to colder. If a person stands next to an ice sculpture, the person's body will radiate heat to the ice sculpture. On the other hand, if a person stands next to a hot Franklin stove, the stove radiates heat to the person's body and every other solid object nearby that is cooler than the stove. This is true for all solid materials at all times, including the walls, floors, and ceilings of buildings.

Heat penetrates solid masonry or concrete at a rate of roughly 1″ per hour. If the exterior wall of the school noted above is exposed to the sun for eight hours, radiated heat from the sun will penetrate about 8″ into the wall. As soon as exterior objects around the wall are cooler than the wall in the afternoon or evening, the wall starts radiating heat outward, reversing the process and stopping heat from penetrating farther into the wall or building. If the wall is more than 8″ thick, the heat might never reach the interior under this sort of exposure. Under protracted hot periods, heat will continue to build up and eventually reach the interior of the building.

So why do we not build buildings this way anymore (and we rarely do)? Mostly because it is labor intensive, which makes it too slow and too costly. It is also space inefficient due to the very thick walls.

And codes do not always help. Even though there are provisions for calculating loads to account for the effects of thermal mass, the codes typically still require insulation, at least partially defeating the purpose in using thermal mass.

For the most part, we now build light-weight buildings and we get little effect from thermal mass. If someone does want to build a passively heated house that takes advantage of thermal mass, the mass is usually put in the floors first (which often are massive to begin with, if concrete slabs-on-grade) and sometimes in the specific walls that are carefully placed for heat collection. In a light-weight building, we use thermal insulation to slow down heat flow. Thermal insulation does not stop heat flow; it simply slows it down, making it easier to balance the amount of heat that we have inside the building in the winter or the amount of heat that we have to remove in the summer if we are cooling.

All buildings fall into one of two categories for thermal characteristics:

1 skin-dominated (or external load dominated); or
2 internal load dominated.

The differences are important. Skin-dominated simply means that more heat is lost through the skin than is generated inside the building, so that it is necessary to add more heat (when it is cold outside, of course). Heat is generated inside a building by people, pets, lights, cooking, laundry, and equipment. Most single-family houses are skin-dominated buildings because there is relatively little internal heat generation—very few people, relatively little lighting, and very little equipment. This means that it is difficult to over-insulate such a building.

For internal load dominated buildings, the reverse is true and more heat is generated inside the building than is lost to the environment, and it is necessary to cool the building interior even when it is cold outside. If a building like this is over-insulated, it acts like a thermos bottle, preventing internal heat from escaping through the walls and roof, which in turn causes the cooling load to go up. Under heating conditions, the reverse would be true and no harm would be caused by over-insulating—unless it caused a condition where it is necessary to cool in the winter.

How is heat flow measured? In British Thermal Units per Hour (BTUH). The basic heat flow equation (which is valid for heating (heat loss) but not entirely accurate for cooling (heat gain) due to sun exposure factors) is:

$$BTUH = A \times \Delta T \times U_o$$

where:

A = area (in square feet)
ΔT = difference in temperature between inside and outside
U_o = heat flow coefficient (this is the inverse of the thermal resistance, or R-value, also expressed as $1 \div R = U_o$)

In all locations, ΔT is a function of the local climate, and the American Society of Heating, Refrigeration, and Air-Conditioning Engineers (ASHRAE) publishes data for all locations in the US. ASHRAE publishes a number of different versions, but codes often require using the 98% figures, meaning that the conditions are exceeded only 2% of the time. In central Indiana, this means that the summer design temperature is 88 °F and the winter design temperature is 3 °F. The corresponding summer relative humidity is 52%. (These conditions do not represent extremes and they are not intended to. The record low temperature in central Indiana is –36 °F from 1994, and the record high temperature is 116 °F from 1936—an overall difference of 152 °F!)

As time goes by, global warming could affect these numbers, making them either higher or lower, as the case might be. It must be noted that global temperature changes that are likely to be problematic are in the range of a few degrees; as one can see from the recent extremes in Indiana alone, actual temperature ranges are already very wide in the temperate zones.

For heating purposes in central Indiana, the temperature difference for design with an indoor temperature of 70 °F (a typical indoor temperature for winter) is calculated like this:

$$\Delta T_{heating} = 70\,°F - 3\,°F = 67\,°F$$

For cooling purposes in central Indiana, the temperature difference for design with an indoor temperature of 75 °F (a typical indoor temperature for summer) is calculated like this:

$$\Delta T_{cooling} = 88\,°F - 75\,°F = 13\,°F$$

As noted above, R-values are published for all insulating materials (always clearly marked on packaging), so it is easy to calculate U_o if that number is not readily available.

The thermos bottle effect is quite real. Several years ago, the author worked on the conversion of an old school building (1915, 1945, and 1955 sections) into low-income apartments. A study was done to determine if it would make sense to insulate new stud walls that lined the solid exterior walls, and the study found that even though heating costs would be somewhat lower, cooling costs would be even higher than the savings for heating. In other words, adding more insulation to the building would make it *more* costly to operate.

The Greek letter D, or delta, is used universally in mathematics to mean "difference".

When calculating overall U_o for a given wall or roof assembly, the interior and exterior air-films add thermal resistance, as do any air spaces within the construction, and as do all solid materials in the construction. See Figures 7.3.A and 7.3.B for the components required to perform the calculations for two typical residential wall assemblies.

Overall U_o for an assembly is simply the sum of all of the U_o numbers for each component in the construction. Here is an example for a simple wall with wood siding:

$$U_{o\ total} = U_{o\ ext\ air\ film} + U_{o\ ext\ siding} + U_{o\ ext\ sheathing} + U_{o\ studs} + U_{o\ insulation} + U_{o\ gypsum\ board} + U_{o\ interior\ air\ film}$$

The same calculation applies to windows for basic heat flow, but additional factors come into play for solar heat gain.

These calculations must also take continuity into account. Another word for continuity in buildings is thermal bridging, meaning that a single solid material bridges between hot and cold. The best example of this is an old-fashioned wood window frame; in the winter, the exterior surface of the board could be 3 °F (or even colder in a colder climate than central Indiana, or just on an extreme day) while the interior surface is 70 °F, or even warmer. Wood, as a material, can tolerate such exposures (due to its

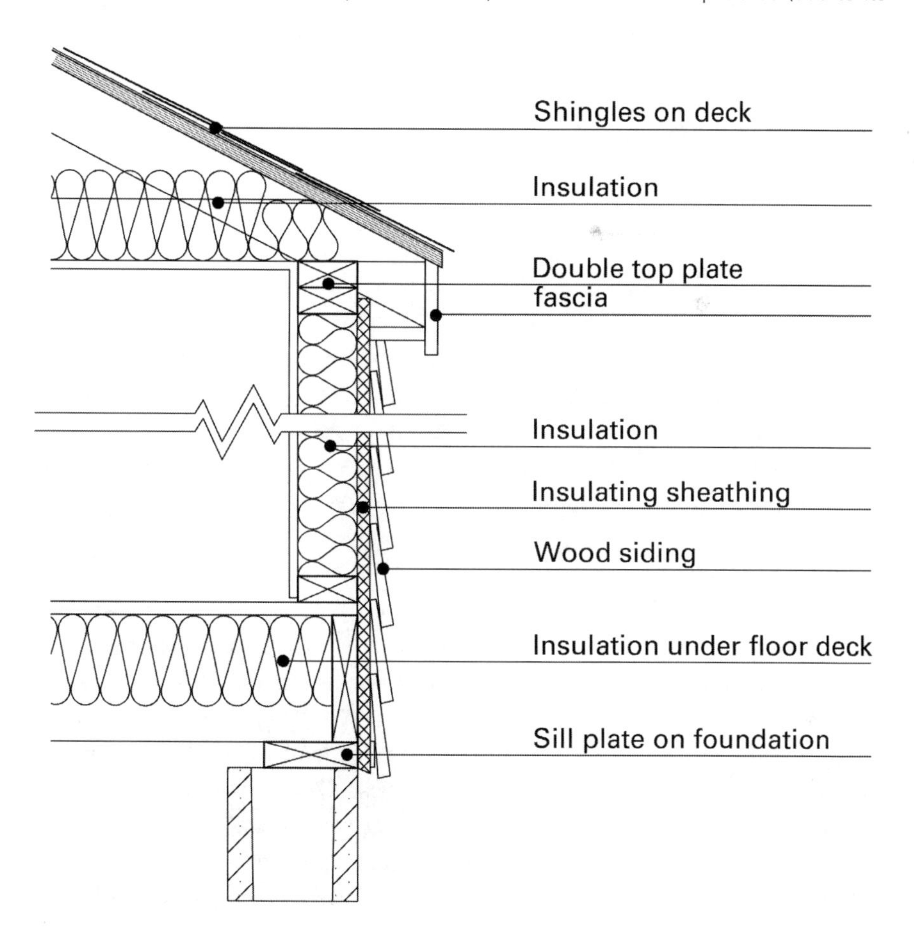

Figure 7.3.A
Section of a typical wood exterior wall

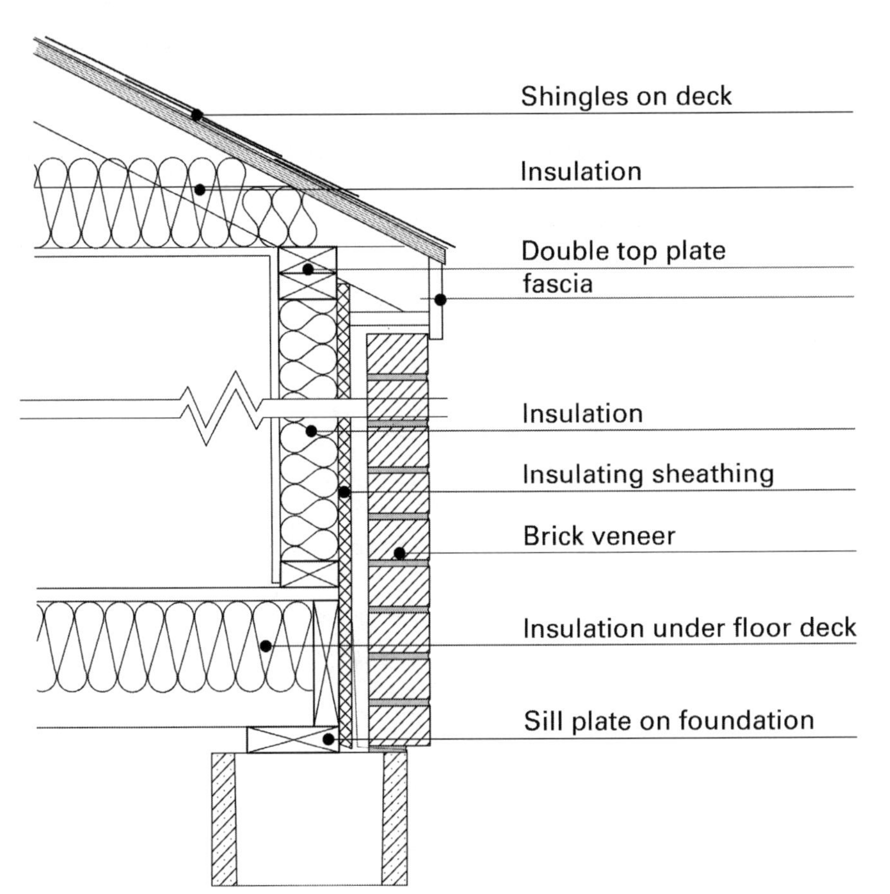

Figure 7.3.B
Section of a typical exterior wall with brick veneer

relatively high thermal resistance) but other materials can create problems, especially metals. The author's 1958 house still has its original aluminum windows and the outer frames are continuous from exterior to interior; in the summer, there is sometimes condensation (liquid, dripping water) on the inside faces because they can be so warm. In the winter, frost is sometimes apparent. This has caused damage to the wood sills under the windows but there is no plan to replace the windows because they are in good condition and changing them would drastically alter the look of the house. (As mentioned under windows in Chapter 6, metal windows often have narrower frame profiles than wood windows, which is the case here. Replacing these windows with modern wood windows would be a major aesthetic change. It would create waste metal and glass, and trees would be cut for new windows, neither of which is sustainable.) Modern windows are designed with "thermal-breaks" in the frames to avoid this kind of problem.

This thermal bridging issue applies to walls too (and roofs, for that matter). In stud framed walls, the studs are continuous from behind the gypsum board (or other material on the inside) to inside the exterior sheathing (on the outside). They are not exposed to the full temperature range but can be close. Modern energy codes typically require continuous insulation outside the studs (in colder climate zones) to minimize thermal bridging and to improve overall performance. Thermal bridging is also a problem at foundations of slab-on-grade buildings because the slab edge can be exposed to the exterior environment; modern energy codes generally require insulation breaks in such situations as well.

Solar heat gain causes HVAC designers much grief because it is powerful and difficult to control (actually, it is really impossible to control). ASHRAE publishes values for solar heat gain for every longitude and latitude, and the numbers are dramatic: 150 BTUH/sf for south-facing glass at 39° North Latitude (central Indiana, again). That means that an unshaded 4′ × 6′ window would bring in 4′ × 6′ × 150 BTUH/sf = 3,600 BTUH, and this is not even a big window. (For reference purposes, the cooling system at the author's 2,600 sf house is a nominal 2.5-ton system. This means that the system produces 2.5 × 12,000 BTUH/ton = 30,000 BTUH of heat removal, or cooling. So this small window would use up about 12% of the capacity. This house has a window in the living room that is 9′ × 6′ 8″ = 60 sf; if that window faced south, the heat gain would be 60 sf × 150 BTUH/sf = 9,000 BTUH, or about 30% of the overall capacity. The window actually faces west, which is even worse from a heat gain perspective.)

All building glass products that are sold in the US are labelled with "SHGC" which stands for Solar Heat Gain Coefficient, with ranges between 0.0 and 1.0. A SHGC of 0.0 would mean that 0% of the heat of the sun is transferred into the building by the glass, and a SHGC of 1.0 would mean that 100% is transferred. Actual values range from 0.23 (very low and very costly) to 0.90 (very high and not so costly). Energy codes usually require a maximum SHGC of 0.70, but that varies according to climate zone. Exterior shading devices can also reduce solar heat gain, but usually only on the south side (because sun angles get too low in the east and west, making shading very difficult at sunrise and sunset). Interior shading devices can also be used but they are less effective because the heat has already entered the building before it can be absorbed or rejected by the shading device.

When cooling and de-humidification are required (in places that can be warm or hot and wet), this is a little more complicated. In order to de-humidify, it is necessary to maintain a difference in **vapor pressure** between the interior and exterior. A temperature difference is required also for cooling. Vapor pressure can be controlled only by using some sort of construction that is highly resistant to the passage of water vapor (not liquid water, which is different and more easily controlled)—a vapor barrier. A solid stone/brick/plaster wall that is 16″ thick usually presents a strong barrier to vapor transmission, but a simple 2 × 4 wood stud wall with interior gypsum board, exterior sheathing, and even exterior brick veneer does not. So the codes require the use of highly impermeable vapor barriers in most buildings (in most climate zones), usually on the warm-side-in-winter (in cold climates), or the inside. The idea of this is to prevent the winter-time interior vapor pressure (which is usually higher than the exterior vapor pressure) from forcing water vapor into the walls from the inside out. In the summer, the reverse problem occurs (in warm wet climates), and water vapor is pulled into the walls from the exterior in; in that situation, the ideal vapor barrier location would be on the outside. In mixed climates, such as is found in central Indiana, where both conditions occur, the vapor barrier will be on the wrong side of the wall at least part of the time. (One might wonder if it would make sense to use vapor barriers on both sides of the wall, but that would not help because it could trap water in the wall, which could cause rot, rust, and even mold—not good.) For buildings in temperate climates, in an ideal world, sophisticated designers (probably architects or professional engineers) could study each building to determine which placement would be the most effective, but the energy codes tend to revert to heating and force the vapor barrier to the inside.

In extremely cold climates (parts of the northern continental US, most of Canada, and parts of Alaska in North America), this is just as important in heating due to the very high vapor pressure variation (exterior vapor pressure can be extremely low in very cold conditions), but in milder climates, many older buildings function quite well—if heated only—without modern vapor barriers, which are either sheets of plastic, plasticized materials, or spray-on elastomeric materials.

The last item to be discussed in wall construction is the air barrier. In addition to heat moving in and out of a building through the walls, roofs, floors, windows, and doors, air can leak through the walls—especially in poorly constructed buildings with wood siding. The air leakage is called **infiltration**. **Ex-filtration** is the reverse process where air leaks *out* of a building. An air barrier is a sheet product that is applied to the outside of the walls, under the finish, to prevent this leakage (the most common trade name is "Tyvek"). However, if a building is well constructed with high-quality materials that are properly installed, this air barrier might offer little benefit. (This is a somewhat controversial topic, and there are many experts who believe that reducing infiltration is the most important way to reduce energy use in buildings. The author does not share that point-of-view.) The marketing materials for most air barrier products note that the products are *not* vapor barriers, largely to prevent the two vapor barrier scenarios mentioned previously.

Cooling and historic buildings—the author's practice has included extensive work in historic buildings, mostly in the form of what is called "adaptive re-use," but also in full restoration. The school to low-income housing project that was mentioned previously is such a project. In projects like this, and even in some "restoration" projects that strive to make an old building look like it did when it was new (sort of), there is usually a desire to adding cooling. This can be dangerous because these buildings almost never have vapor barriers. Even if the walls can hold out water vapor, it still tends to leak in around windows, doors, and other openings, which can cause damage to historic fabric (as original historic materials are called).

Even if a building is made air-tight by using carefully sealed walls, windows, doors, roofs, and air-barriers, air still moves into and out of the building when doors or windows are opened. (From the author's point-of-view, this truth should cast grave doubt on the whole idea of an air-tight building.) Also, if a building is truly air-tight, the occupants would suffocate; ventilation is necessary. The best way to minimize winter-time temperature control problems (commonly seen in grocery stores and similar buildings when doors open frequently) is to make the air pressure inside the building slightly higher than the pressure outside the building (this is known as positive pressurization) because this causes air to move out through the doors when they open, rather than rushing in (if the building were negatively pressurized). Such small pressure variations cannot prevent strong wind gusts from blowing cold air through an open door, but they can minimize issues all the same.

Back to ventilation. Air moves into and out of buildings all the time. If there is a small exhaust fan running in a bathroom at home, air is leaving the building—if there is a source of "make-up" air for the exhaust fan. Fans do not create air; they simply move it around and they cannot move air out if there is no air moving in. (Actually, powerful fans can maintain pressure differences; this is the type of fan that is used in a vacuum cleaner.) In a typical house, the small exhaust fans work because adequate air leaks in through the walls, roofs, doors, and/or windows to make them work. If the building were actually sealed, it would be necessary to create some kind of mechanical air-intake system to allow the exhaust fans to work. This is the arrangement that is preferred by engineers—mechanical ventilation—because it can be designed to be predictable and to do only what is needed. For energy efficiency reasons, it is best to use an ERV for such ventilation, whether in someone's house or in a large commercial or institutional building.

Overall, the idea is to keep everything balanced: heat flow through the envelope (walls, roofs, windows, and doors); infiltration and ex-filtration; and general air-pressure balancing, tending toward slightly positive pressurization.

In truth, most single-family homes are probably negatively pressurized, if exhaust fans are running or if the laundry drier is on. The author also does extensive work in restaurant design, and it is generally easy to tell a restaurant owner why they are having problems with temperature and/or humidity control—negative pressurization.

From an energy efficiency point-of-view, buildings should be carefully designed to relate to their climate and micro-climate, meaning that there should not be large expanses of unshielded glass anywhere (except maybe on the north side in most temperate zones); there should not be unshielded east and west facing glass at all (if feasible); the south-facing glass should be shielded to exclude the summer sun completely; the glass that is used should be at least double glazed with low U_o and low SHGC; the walls should be as heavily insulated as practical (do not forget about the thermos bottle effect though and internally load dominated buildings in general); the roof should be heavily insulated; the building should be no bigger than it really needs to be, and it should use the most efficient heating, cooling, ventilating, lighting (see Chapter 9), and equipment available.

Does it make sense to build a super-efficient 10,000 gsf house for two people? No, because it cannot be made to be efficient enough to offset the high resources used to build it.

Would it make sense for everyone to live in communal arrangements—large apartment buildings, large condominium buildings, etc.? Yes, because such buildings can be made to be more efficient than individual small houses.

We have building technologies available that can offset our desire for more space than we really need, to have larger expanses of glass than should be used in a given climate, or even to use excessively high ventilation rates, but the basic point is that using those technologies has an impact on the environment too. Simpler always seems to be better, and it always seems to make the most sense to build in a way that is most appropriate in a given climate zone. Having railed previously against very glassy buildings, there are exceptions where such buildings do make sense, such as Portland, Oregon. Portland has a mild climate—not too cold in the winter and not too hot in the summer (at least not for extended time periods)—but it is very cloudy much of the year. So it is not crazy to build glassy buildings in Portland, mostly to provide more daylighting, because the heat loss and heat gain penalties are moderate. To build the same building in Indianapolis would be a very different undertaking, with much higher operating expenses (even though it is pretty cloudy in Indianapolis in the winter too).

One last thing—it should be noted that equipment: furnaces, boilers, air-conditioners, etc., is sized according to ASHRAE's design load information. This means that the heating system in a building in central Indiana should run continuously when it is 2 °F outside; if the system does not run continuously, it is over-sized. (This does mean that buildings should be expected to get cooler inside when conditions are colder than the design day.) Similarly, in the summer, if it is 88 °F outside, the air-conditioner or heat pump should run continuously; if it is does not, it is too big. And, as in heating, if the outdoor conditions move above the 2% design day, the cooling system should fall behind. Having said this, there is little risk in over-sizing heating systems (and most such systems are over-sized) because the only downside is that the equipment will cycle However, in cooling, there is a real risk in over-sizing. Typical cooling systems de-humidify only when the compressor is running and the evaporator coil (the coil inside the building) is wet. When the compressor in such a system turns off, humidity will actually go *up* due to reabsorbed water from the wet coil; if such systems cycle on and off frequently, there will be excessive wear-and-tear on the equipment *and* there will little, if any, humidity control.

Here is a detailed example for how to calculate the heat loss (for heating) and the heat gain (for cooling) for a simple small residence:

Heat loss

Total wall area (excluding windows and doors) = 812 sf (given)

Roof area (same as the floor because the insulation is on the ceiling) = 1,000 sf (given)

Slab edge = 130 lf (given)

Windows area = 146 sf (given)

Doors area = 42 sf (given)

$U_{o\ wall}$ = 0.05 (given)

$U_{o\ roof}$ = 0.04 (given)

$U_{o\ window}$ = 0.10 (given)

$U_{o\ door}$ = 0.08 (given)

(Remember: heat loss = Area $\times U_o \times \Delta T$)

Walls heat loss	= 812 sf × 0.05 × 67	= 2,720 BTUH
Roof heat loss	= 1,000 × 0.04 × 67	= 2,680 BTUH
Windows heat loss	= 146 × 0.1 × 67	= 978 BTUH
Doors heat loss	= 42 × 0.08 × 67	= 225 BTUH
Slab edge heat loss	= 130 × 30 BTUH/lf	= 3,900 BTUH
Total heat loss		**= 10,503 BTUH**

Heat gain

Total wall area (excluding windows and doors) = 812 sf (all given again)

Roof area (same as the floor because the insulation is on the ceiling) = 1,000 sf

Slab edge = 130 lf

South windows area = 48 sf

North windows area = 38 sf

East windows area = 30 sf

West windows area = 30 sf

Doors area = 42 sf

$U_{o\ wall}$ = 0.05

$U_{o\ roof}$ = 0.04

$U_{o\ window}$ = 0.10

$U_{o\ door}$ = 0.08

Solar heat gain South = 150 BTUH/sf

SHGC South = 0.50

Solar heat gain North = 30 BTUH/sf

Solar heat gain East/West = 180 BTUH/sf

(Remember: heat gain = area × U_o × ΔT)

Walls heat gain	= 812 sf × 0.05 × 13	= 528 BTUH
Roof heat gain	= 1,000 × 0.04 × 13	= 520 BTUH
Windows heat gain	= 146 × 0.1 × 13	= 190 BTUH
Doors heat gain	= 42 × 0.08 × 13	= 44 BTUH
South windows solar gain	= 48 × 150 × 0.5	= 3,600 BTUH
Total heat gain (w/o ventilation)		**= 4,882 BTUH**

Ventilation heat gain = 50 ÷ 200 × 12,000 BTUH = 3,000 BTUH (for 50 cubic feet per minute of air)
Total heat gain (with ventilation) = 7,882 BTUH

It is evident in this example that this is a skin-dominated building, even without having considered internal loads directly, and that heating loads are higher than cooling loads (10,503 BTUH for heating verses 7,882 BTUH for cooling). The cooling load is quite small, at only 0.66 tons of cooling (7,882 BTUH ÷ 12,000 BTUH/ton); the smallest size available in "standard" equipment is 1.5 tons although some 1.0-ton systems are available at higher cost.

7.4 Space requirements

The main reason for interior designers, and interior design students, to learn about HVAC systems is to understand their impact on space—horizontal space (as in plan) and vertical space (as overhead above ceilings or on roofs). Different types of systems have different space requirements. This section will treat this issue from a somewhat generalized "system type" point-of-view and more detail will be added in the next section when specific types of equipment are described. For now, the basic types of systems need to be identified:

- All-water systems (using radiators with boilers). These are really heating-only systems because adding chilled water (another water system) to this for cooling almost inevitably leads to air-side equipment too (see Combination systems below).
- All-air systems (using ductwork for supply and return for heating and cooling and for exhaust).
- Combination water and air systems (using radiators, fan-coils, air-handlers, etc. with boilers and chillers; see all-water systems above).
- Combination water and air systems (using water-source heat pumps).
- Combination refrigerant and air systems (using packaged terminal air-conditioner (PTAC) units, packaged terminal heat pump (PTHP) units, ductless split-systems or variable refrigerant flow (VRF) or variable refrigerant volume (VRV)).

Generally speaking, water- or refrigerant-based systems take up less space than air-based systems, simply because pipes to move water or refrigerant are much smaller than ducts to move air. But that is a crass generalization. It is also necessary to differentiate between degrees of centralization:

- Fully centralized: in this type of system, all of the equipment (boiler(s), chiller(s), pump(s), even air-handler(s)) would be located in a single room (plus a **cooling tower** on the ground or on the roof). Systems like this are rare, mostly because buildings that use this type of approach are usually too large to have all of the equipment in a single location; however, it is common to group boilers and chillers together no matter how large a building might get. (This is even true for large campuses.)
- Partially centralized: in this type of system, the equipment (boiler(s), chiller(s), pump(s), air-handler(s), air-cooled condensing units (ACCU), cooling tower(s), evaporative cooler(s), dry cooler(s), and/or rooftop units (RTU)) would be located in multiple rooms, on the ground at multiple locations, or on the roof in multiple locations. This category covers the majority of systems, where there are various different pieces of equipment at various locations all around a building.
- Fully distributed: in this type of system, equipment is located all over a building, possibly even in every room. This is also somewhat rare, primarily due to cost.

The more centralized a system is, the larger its components will be. This means that boiler systems that distribute heating hot water around buildings will have larger piping as the system grows (3″ piping is on the small side; 8″ piping is not uncommon in large buildings such as high-rise office buildings or hotels, schools, or hospitals). Heating hot water piping is nearly always insulated (usually $1\frac{1}{2}$″ to 2″ thick) so the overall sizes of the pipes can be 7″ to 12″, or even bigger.

For air-based systems, these sizes get much bigger. Small ducts are 8″ or less in height or diameter; medium ducts range from 10″ to 22″ high; large ducts are 24″ to 36″ high, and very large ducts are even bigger. Widths for medium sized ducts, or larger, can be 10″ to 120″ or even more. Supply ducts (which provide the cooling or heating) are insulated when they are concealed (as in above ceilings or in attics) and that insulation is usually $1\frac{1}{2}$″–2″ thick, making the overall sizes 3″–4″ bigger. (The author had a recent medical office building project where the largest duct is 70″ wide and 50″ high, uninsulated.) Return-air and exhaust air ducts are usually not insulated, but return ducts are larger due to lower air velocities.

This means that buildings with all-air systems tend to be taller than buildings with all-water (or mostly water) systems because headroom has to be provided on each floor for these large ducts. (And do not forget that supply ducts sometimes have to cross return ducts or exhaust ducts, which can double the overall height that is needed.)

The indoor machinery involved here varies in size from a single wall-mounted ductless split indoor unit that is 42″ long × 10″ high × 8″ thick, to a typical basement furnace that is 21″ wide × 29″ deep × 50″ high or long, to air-handlers that are 50′ 0″ long × 14′ 0″ wide × 9′ 0″ high (or even bigger). The outdoor equipment varies in size across a wide range too, but not quite as much as indoor equipment.

The refrigerant compressors that provide the cooling for cooling systems can be air-cooled or water-cooled. Most people are familiar with air-cooling because that is what most people have at home. Air-cooling means that the compressor has a fan (a large fan, relative to the size of the compressor) that pulls a large amount of air across the condenser coil and the compressor, thereby putting the heat that has been removed from inside the building into the atmosphere and cooling the compressor in the process. This is why the typical outdoor "air-conditioner" (actually called an Air-Cooled Condensing Unit, or ACCU) is noisy and blows lots of hot air around. But large commercial systems can use ACCUs too, including chillers, so ACCUs can be as small as the one at someone's small house or 30′ 0″ long, 10′ 0″ wide, and 6′ 0″ tall (or even bigger). A dry cooler is another form of air-cooling, but it is indirect because the cooler only cools water in a loop, which is then used to cool indoor compressors (this is one form of a "geothermal" system; geothermal is used incorrectly here though. Such a system is really an "earth-coupled" system). Air-to-air heat pumps are simply ACCUs that can run backwards to provide heating; unfortunately, typical equipment like this becomes highly inefficient in heating below 30 °F or so, but they are air-cooled compressors all the same. The latest technology is VRF (or VRV) which is a large outdoor air-cooled heat pump (often called a heat-recovery unit) that serves multiple indoor zones.

Compressors can also be water-cooled, which is efficient but often costly. The simplest form of water-cooled compressors is to connect domestic water (drinking water) to the compressor's condenser coil, let it collect heat, and then throw it away. (This is called "pump and dump" and it is unconscionable in this day of water scarcity, but it is often done when other options are too difficult or costly.) The next form of water cooling is the "closed-loop water-source heat pump" system, where the compressors have cooling water running through their condenser coils that is drawn from a centralized loop; the larger loop is then connected to a dry cooler (noted previously) for air-cooled heat removal or, more commonly, to an evaporative cooler, which is a machine that uses evaporating water to cool the loop water more efficiently. Finally, large chiller compressors are often water-cooled by using cooling towers; cooling towers are very similar to evaporative coolers. Both evaporative coolers and cooling towers tend to be tall machines (some 20′ 0″ or more) that are very noisy.

Earth-coupling is very popular in many areas because it can be extremely efficient. In this type of system, excessive heat (from the cooling side of the system) is stored in the ground, put there by running water through a very long loop that is buried in the ground (there are many different ways to pipe these systems), and taken from the ground if heating is needed in the building. When the system is in heating mode, it takes heat from the ground. These systems depend on a degree of balance, which means that they do not work well if they need far more cooling than heating, or vice versa. Earth-coupling is most commonly used for relatively small water-source heat pumps, but it can be used for other types of systems too (especially VRF). Earth-coupling is usually rather costly (and not necessarily feasible in rocky areas), so it is somewhat uncommon.

The other characterization that all HVAC systems share is zoning. Zoning means having more than one point of control—more than one thermostat. Very basic systems, like most home systems, are "single-zone" systems, which means that the furnace burner is on or off and the cooling compressor is on or off; the whole system switches between heating and cooling (manually or automatically) and every room in the house gets heating or cooling at any given time. Everyone knows that this often means that there is a "cold" room in the winter or a "hot" room in summer, and rarely are all rooms the same temperature. Single-zone systems are commonly used in commercial, and even institutional, projects, mostly to save costs (but with the penalty of serious performance compromises).

Zoning can be improved by putting in multiple systems (not uncommon in larger houses) or by putting in more sophisticated systems that incorporate zoning, which are called "multi-zone" systems. Multi-zone system options include:

- *All-air variable air volume (VAV)*: this system uses a single centralized cooling-only air-handling unit (AHU if inside or RTU (rooftop unit) if on the roof) or a few units if it is a large system; the AHU or RTU has a special fan that can speed up or slow down (based on duct pressure) as individual zone terminal devices (air-valves in their simplest form) open and close to control temperatures in multiple zones. Zone sizes can be as small as a single office or as large as a small gymnasium. Typically, electric heaters are added to the zones if heating is needed. Heating is always needed at exterior zones with heat loss but it can be needed for interior spaces too if there is ever a need to raise room temperature quickly. These systems can be very complicated but they are well understood throughout the HVAC industry and they are standard for a number of different project types. It is a struggle to manage ventilation though (the air-valves can shut off, which shuts off ventilation) so they have to be handled very carefully where ventilation is critical (usually in healthcare settings). This can be done by using constant-volume zones, but putting in too many of those would defeat the purpose in having built the system in the first place.
- *Air-and-water variable air volume (VAV)*: this system is the same as all-air VAV except that the zone heaters would use hot water coils and a boiler for heat. This is more costly to build than electric reheat but it is less costly to operate. It also requires more sophisticated maintenance, so it is usually found only in institutional settings with full-time on-site maintenance staff.
- *Packaged terminal air-conditioners (PTAC) and packaged terminal heat pumps (PTHP)*: these are through-the-wall units that are most commonly seen in low-rise (and sometimes mid-rise) motels. The unit is self-contained, with a compressor for cooling (and heating if it is a heat pump), auxiliary electric heat (for low-temperature option), condenser, and indoor fan (with evaporator coil). They are ugly and noisy but they are also the least-cost option to get room-by-room control. The only limitation is that they only work in rooms with exterior walls that are large enough to install the units, which are usually at least 42″ wide and about 16″ high.
- *Water-source heat pumps*: this system can include heat pumps from 0.75 tons up through 10.0 tons (and even larger) in various cabinet configurations: concealed horizontal above ceiling; horizontal exposed; vertical concealed in a closet; vertical exposed; vertical in a finished cabinet; and console (under window) style. The heat pumps are linked together by a water loop, and there is usually a boiler for auxiliary heat (if needed) and an evaporative cooler (to reject excess heat). As noted previously, common variations on this system include using a dry cooler (less costly to install and easier to maintain, but less efficient) and using earth-coupling.

- *Two-pipe or four-pipe hydronic (water-based)*: in this case, fan-coils (FCUs—for heating, cooling, or both) are located all around the building, in types that are similar to water-source heat pumps. In a two-pipe system, the whole system is in either heating or cooling; in a four-pipe system, each FCU has separately piped coils for heating and cooling, and the system can heat one room while cooling another at the same time. These systems usually use boilers for heating and chillers for cooling, but it is possible to use heat-recovery chillers (special chillers that use the heat that is removed in cooling the water for heating in the building) and to earth-couple the systems (which is exotic and highly unusual). These systems can work extraordinarily well and they are standard in hospitals, schools, and some large office buildings. But they are among the most costly systems to build and sophisticated maintenance is usually required (mostly for the boiler and chiller).
- *VRF or VRV*: these systems can resemble water-source heat pump or FCU-based hydronic systems but the individual FCUs are linked together with refrigerant piping, back to the outdoor heat-recovery unit (or possible just a heat pump). These systems can be configured (and usually are configured) to heat and cool simultaneously, similar to a four-pipe system, but they are usually more efficient. The main limitation with these systems is that refrigerant leakage is a concern, because it could dump easily in occupied space; that means that large systems (i.e. greater than 30 tons or so) cannot be done. Instead, systems like this that need to be large are made up of groupings of smaller systems; in other words, if we need 200 tons of VRF cooling, we would simply use eight 25-ton systems.

Of these systems, VAV and PTAC systems automatically incorporate ventilation, but the water-source heat pump, two-pipe, four-pipe, and VRF (or VRV) systems do not. In these cases, the common practice is to use a separate ventilation system, which can be configured to deliver ventilation air to the return side of some, or even all, of the FCUs or heat pumps, or to deliver air directly to spaces. Separate ventilation systems usually utilize special machines called dedicated outside air systems (DOAS), which are designed to heat, cool, and de-humidify incoming outside air under all conditions, and they are very useful in maintaining humidity control. DOAS machines often include ERVs on the return side.

In all of these systems, cooling and heating must be delivered to spaces and occupants. In water-based heating systems, that can be direct via radiation or indirect via air-handlers or FCUs. In air-based systems, there is always supply air and return air (remember: fans cannot create air, they just move it around), making for a looping recirculation effect, with or without ventilation directly tied in. Supply air can be delivered from the floor, walls, or ceilings (or some combination of all three); floor delivery is preferable for heating but ceiling delivery is preferable for cooling. We rarely build separate systems for heating and cooling, so one of them is nearly always compromised; the decision about whether to deliver through the floor or ceiling is often an aesthetic decision. But, in general, if supply air is high, return air should be low, and vice versa; the objective is to achieve good air circulation within every space. Exhaust air can be removed from a space from anywhere—floor, walls, or ceilings, but floor inlets should not be used in wet areas such as kitchens, bathrooms, etc.

The devices used for air movement include grilles, registers, and diffusers (see 7.7 for more detail).

Now it is time to move on to specific pieces of equipment, except for a brief summary of the process of selecting an HVAC system for a particular project. The driving factors are zoning needs/necessities, ventilation, and architectural impact (meaning floor space and overhead space and maybe noise), despite the obvious need for cooling and heating. First, it is *never* a good idea to combine multiple exposures (east and south, say) or different exposures (some glass and no glass) on a single zone, no matter what the owner wants or can afford; this will cause constant and irresolvable problems. Second, it is also never a good idea to combine multiple load profiles (data room and conference room, maybe) on a single zone; in this case, the data room has constant cooling needs (24/7, 365 days per year—no matter how cold it might be outside) and a conference room is highly variable. Also, the data room probably needs minimal ventilation while the conference room might require Demand Control Ventilation.

Few owners want to pay for systems that they do not think they need and some want to buy systems that cannot meet their needs (in the desire to hold down the cost) as a result; designers are owner's advocates and should not relent in the face of pressure to do the wrong thing. A poorly designed HVAC system will be an ongoing headache (perhaps literally) for its lifetime, which is likely to be 15 years or longer. As such, these systems should always be evaluated from a long-term perspective.

And do not forget about noise implications. The author used to be a strong advocate for closed-loop water-source heat pump systems, largely due to their extraordinary efficiency. But there were noise problems in his first project and every water-source heat pump project since. This does not automatically mean that the systems were poor choices or that they were poorly designed; it simply means that having local compressors in every zone makes for a system that is noisier than most alternate systems. This is simply a fact. If an owner wants maximum efficiency, these systems are still viable alternatives. In fact, only water-cooled VRF/VRV technology can come close to the efficiency of a water-source heat pump system.

Finally, always think about headroom. In addition to HVAC ductwork and/or equipment, the overhead space in most buildings also has to accommodate plumbing (sanitary waste and vent), condensate drainage, domestic water (hot and cold) and even natural gas, fire suppression sprinklers, lighting, power, low voltage wiring, etc. It gets crowded quickly so space needs to be planned for at the beginning.

7.5 HVAC equipment

7.5.1 Air-handlers (AHUs): constant volume (CV) and variable volume (VAV)

Air-handlers are the most versatile of all types of HVAC equipment. In an AHU, the fan either pushes (blow-through) or draws (draw-through) air through the heating coil (furnaces are not usually options for conventional AHUs), the cooling coil (DX or chilled water), and the filters (filters can be located before and/or after a fan). These units are usually located indoors, but outdoor options are possible. Large units sometimes have separate return and/or exhaust fans. AHUs come in a wide variety of sizes but in only two types: horizontal and vertical. Small AHUs are usually unitary, meaning that there is a single cabinet that contains the fan, coil(s), and filter(s), but larger units are usually modular, meaning that they are made up of separate sections (often one each for the fan, each coil, and the filters) that are attached together in the field to make a complete unit. The fan module could be the same for a steam heating unit as for a hot water heating unit (or not—the specific fan that would be required would relate to the type of coil used for heating).

Coil-pulling space is a major issue with units like this. Most servicing of this type of unit can be accomplished from a small space along one side of the unit (sometimes both sides though, so be careful), but coil-pulling requires a space that is as wide as the unit so that the coils can be removed for servicing, repair, or complete replacement. This is a realistic requirement because the coils are likely to need servicing and/or replacement long before the overall machine needs to be replaced. (It is OK to use access panels and/or doors to make this space available. In other words, there could be a wall too close to the machine to pull out the coil(s) but that space could be made available by opening an access panel or a door.) If the basic unit cabinet is 5′ 0″ wide, the coil-pulling space should be planned to be at least 6′ 0″ to make sure that there is adequate space to get the coil(s) out of the machine. This issue applies to *all* AHUs, no matter where they are installed but it might be just as easy to take out the whole machine if it is small, rather than trying to dismantle it in place.

AHUs can be either horizontal (long relative to their height) or vertical (tall relative to their length). Horizontal units can be floor-mounted or suspended from structure above (usually hanging on **all-thread**, even for large units. (Structural checking

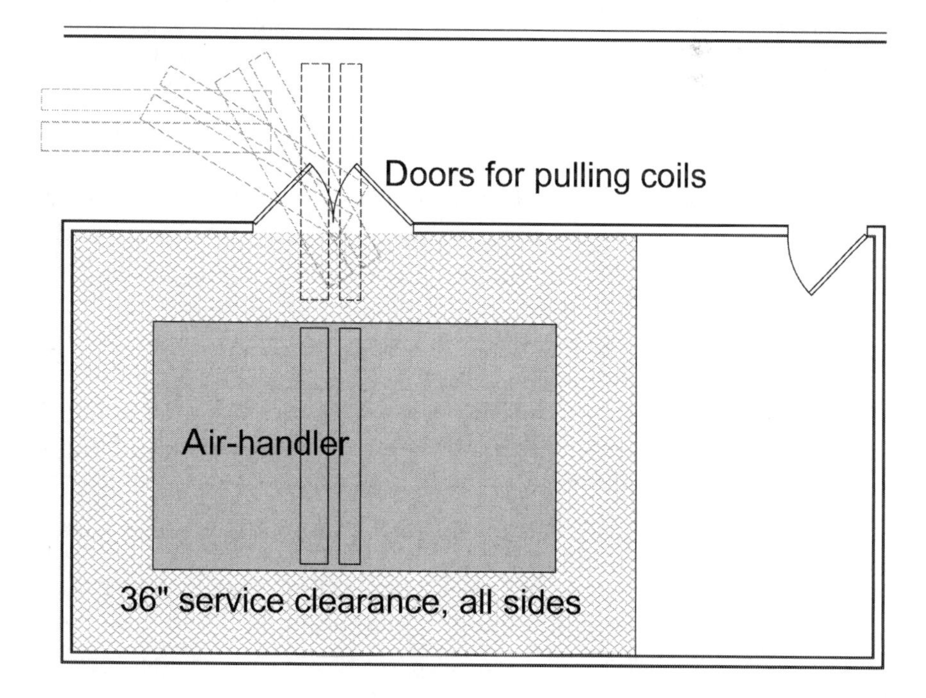

Figure 7.5.1
Coil-pulling space

might be required for hanging large units from existing structural framing if similar equipment was not there before.) Vertical units are most commonly floor-mounted but they could be mounted on raised platforms or rails or even wall-mounted on brackets if they are small. Engineers choose the type of unit that is needed based on performance criteria and available space. Providing more space makes it easier (and less costly) to get the system installed.

AHUs can run at constant volume, which means that the fan speed is constant as is the resulting amount of air movement, or variable volume, which means that the fan speed can vary to cause the resulting amount of air being moved to vary. In the latter case, this is usually done according to duct pressure, which varies according to how many terminals are open or closed. As noted previously, the air terminals include air-valves that close when the zone being served does not want cooling (or the air-valves close partly when the zone wants heating). A CV or VAV system can be configured to run an entire building (some very large old AHUs are actually multiple rooms in a penthouse or elsewhere), a floor, or even part of a floor. It is very common in systems like this to see AHUs on each floor (usually floor-mounted but sometimes hanging, but nearly always in a dedicated mechanical room; such spaces are often quite large and have a major impact on leasable, or merely useable, space), or even multiple AHUs on each floor.

The terminals for VAV come in two basic types: standard and fan-powered and either one can be designed with or without heat (really reheating because the air coming to the terminal has been cooled). Reheat is necessary in a VAV system to prevent over-cooling (remember that the central unit(s) always runs in cooling) or to provide heating. If an occupant wants to raise the temperature of a VAV zone quickly, the air-valve closes to "heating" position (usually about one-third of fully open) and a heat source is activated. This heat source is either heating hot water or an electric coil; the electric coil is less costly to build but more costly to operate and most such systems are built with electric reheat. Some internal zones (without heat loss from walls, roof, or windows) can be done without reheat, but that does mean that the temperature cannot be raised easily. In a standard terminal, the air-valve can shut off almost completely, which means that there is no ventilation in space, which is problematic. There are two techniques for dealing with this: always leave the terminal partially open (with reheating running) and use fan-powered terminals. In fan-powered terminals, the terminal becomes a small AHU that has a return-air connection; if the air-valve closes because the zone does not need cooling, the fan simply recirculates the air in the space. This is not really ventilation either, but the air mixing helps to minimize stuffiness. Ultimately, if ventilation is critical for a zone on a VAV system, the terminal for that zone becomes a CV terminal that will end up using more reheating.

AHUs can be very quiet or very noisy, depending on their size, location, mounting method (see Chapter 6 for more information about sound-reducing equipment mounting), application, and the ductwork (if any) design. But noise-sensitive applications are well suited to AHUs because they can be located far away with ducts configured for slow air movement for quiet operation.

From a maintenance point-of-view, AHUs are good because there are usually relatively few of them, which limits the maintenance to a few locations. Systems that use steam or hot or chilled water require special maintenance for water chemistry, boilers, chillers, cooling towers, pumps, etc.

AHUs are generally mid- to high-quality equipment and good units can last for more than 40 years under heavy use. (The author is familiar with AHUs that were built in the 1920s that are still running, but modern equipment is not quite that tough.)

7.5.2 Fan-coils (FCUs)

Fan-coils are the second most versatile type of HVAC equipment. A typical fan-coil is simply a box with a fan, one or two coils, and a filter(s) inside, which sounds a lot like an AHU. The cooling coils can be DX or chilled water and heating can be provided by hot water, electricity (heat pump), or electricity (electric coil). Cabinet types include the following:

- Exposed horizontal without ductwork: (a finished cabinet intended to be seen in a finished space; often the cabinet is recessed into a ceiling so that only the bottom is visible)
- Exposed horizontal with ductwork (a finished cabinet intended to be seen in a finished space)
- Concealed horizontal with ductwork
- Exposed vertical without ductwork
- Exposed vertical with ductwork
- Concealed vertical with ductwork
- Exposed console on the floor without ductwork, usually under a window
- Exposed on-ceiling without ductwork (this is a VRF-only option or for individual ductless split systems)
- Exposed in-ceiling without ductwork (this is a VRF-only option or for individual ductless split systems)
- Exposed on-wall without ductwork (this is a VRF-only option or for individual ductless split systems)

The main disadvantage to FCU systems is that each unit has a fan and a filter, so there are lots of fans and filters to maintain and there is fan noise in the local zone. Many FCUs are very small though so fan noise might be negligible; VRF fan-coils are usually nearly silent.

The advantages of FCU systems are excellent controllability (limited only by the number of units) and small size so that it is usually easy to locate the units.

FCUs vary in quality, but conventional units (hot and chilled water) usually last more than 30 years while DX and VRF units are likely to last 20 years or longer.

7.5.3 Furnaces

Residential-style furnaces are most widely used in one- and two-family dwellings (the "residential" world in this text), but they are also common in small and low-rise commercial buildings. (The author's building has two constant volume rooftop units (see 7.5.16) for the second floor and it used to have four residential-style furnaces with split-system DX cooling for the first floor. The latter system has been replaced with an 8-zone VRF system.) Furnaces are available in oil-, propane-, and natural gas-burning versions as well as all-electric versions, up to 2,000 cubic feet per minute (cfm: the unit for air-flow in HVAC systems). If larger systems are required, or desired, units can be "twinned" together to cause two units to operate as one. For even larger systems, multiple separate units, or multiple twinned units, must be used. Large-scale commercial furnaces are available too; see 7.5.4.

This type of equipment is low in cost, relatively low in quality, and unlikely to last much longer than 15 years or so under normal use.

In the natural gas-fired furnace sub-group, there are two levels of efficiency: standard and high, where standard is defined as ≥80% and high is defined as ≥90%. There is no reason to use standard efficiency furnaces today, even though they are less costly. It must be remembered that standard efficiency burners require combustion air from the space and making provisions for that can offset some, even all, of the savings for the equipment.

The biggest advantage to most small furnaces is cost, but they are designed to fit into conventional wood buildings with framing at 24" o.c. (which is part of the reason for the size limit) and they can be used as up-flow (return air at the bottom and supply air at the top), down-flow (return air at the top and supply air at the bottom), and even on their sides for attic and crawlspace installations.

Multi-speed furnaces are commonly available and they should be used for systems that both heat and cool. (Cooling requires substantially more air-flow—remember ΔT?—so the unit should run on "high" for cooling and "medium" or even "low" for heating.) Using multi-speed fans can help to make the units quieter too. In general, noise is more likely to be due to ductwork design than to the equipment.

Furnaces are also available in dual-fuel versions, which means that primary heating is provided by an electric heat pump and low-temperature heating is provided by a high-efficiency furnace; in many ways, this would be the best of both worlds in many climate zones.

The biggest disadvantage to small furnaces is that they are not designed to handle substantial amounts of outside air for ventilation, so it can be difficult to use this type of equipment for commercial projects in high-humidity climate zones if ventilation loads are high.

If cooling is desired or required, a separate coil has to be added, on top of unit if it is an up-flow unit, or under the unit if it is a down-flow unit, or next to the unit if it is on its side.

7.5.4 Duct furnaces and duct heaters

Duct furnaces are a special class of furnace that is designed to be used in ductwork. They do not have fans because the fan from an AHU (usually) is used to blow air through the duct furnace. They are generally used in medium- to large-sized systems to provide primary gas-fired heating; this is done instead of having a heating hot water coil in the AHU. AHUs are not usually available with integral furnaces because they are mostly designed to accommodate hot water heating. Duct furnaces are available in a limited range of sizes, so it is not uncommon to see two duct furnaces connected to the outlet of a single large AHU.

These units are generally well made and can be expected to last 30 years or more, unless there is a very high percentage of outside air. Lots of outside air exposes the internal parts of the furnace to challenging conditions and the internal heat exchangers inside the furnaces tend to fail early (this is also part of the reason why conventional small furnaces do not work well with high levels of outside air).

Duct heaters are simply electric versions of duct furnaces. They are electric heating elements that actually slide into a duct through a slot in one side, and they can be used with virtually any size of duct and any air-flow. Electric heaters like this do not necessarily have to be 100% on or off; they can be designed with steps of control ($\frac{1}{3}$, $\frac{2}{3}$, $\frac{3}{3}$, etc.) or for fully modulating (varying) control. The type of control is dictated by the application and the required performance.

These heaters are used for various different purposes (reheat for VAV, local supplemental heat where there is a heating problem, primary heat, etc.) and they are quite flexible. But it must be noted that electric heaters of any kind tend to drive up electric bills, especially commercial bills for large facilities, because the loads are quite high. For large commercial customers, most electric utilities charge for the cost of the electricity used (per kWH, kilowatt-hour) plus various other charges including a demand charge that is based on the peak usage recorded during the billing period. Electric heaters always increase peak demand, especially big electric heaters, so they are often avoided.

Electricity in heating—it is widely, and correctly, understood that electric heaters are 100% efficient in the sense that if you put 2,500 watts of electricity into the heater you get 2,500 watts of heat out. But this should not be compared to 90% efficient furnaces, or even to 80% efficient furnaces, because the efficiency of the furnaces is complete, meaning that fuel is used at 80% or 90% (or even higher). In the case of electricity, to be comparable, the efficiency of the power plant that generated the electricity has to be taken into account; typical power plants are only 25% efficient (roughly speaking), so the actual overall efficiency of electric heating would be roughly 25% and that is the number that should be compared to the furnace's efficiency.

7.5.5 Unit heaters

Unit heaters are free-standing heaters that incorporate a heat source, a fan, and some air distribution device (usually some simple louvers in the front). They can use steam, hot water, natural gas, propane, and electric fuel sources. They are generally very inexpensive, so they are commonly used to heat garages, warehouses, factories, and other large spaces. They are also used above ceilings, under the roof, of some buildings (especially buildings built before the 1990s) to help offset heat loss and to prevent the ceiling plenum from getting too cold (this is required in buildings with uninsulated roofs). This is not commonly seen in new buildings.

Unit heaters can be loud, but they are commonly used in environments where sound is not critical.

7.5.6 Cabinet heaters

Cabinet heaters are often found in vestibules and corridors to offset cold air entering at open doors. They usually consist of steam or hot water coils with a fan, but electric versions are available too. They can be surface-mounted or recessed into the wall; they are 9″–12″ deep or so, so recessing requires careful planning and coordination to assure that there is adequate space for the unit and the piping to serve it.

The advantage of cabinet heaters is that they do an excellent job of keeping spaces warm. The disadvantage is that they are large and relatively unattractive.

7.5.7 Wall heaters

Wall heaters are usually small electric heaters that are surface-mounted or semi-recessed into the walls of vestibules and sometimes restrooms. They are inexpensive but somewhat prone to damage, especially if they have accessible controls (dials, switches, etc.). They can be noisy too.

7.5.8 Draft stop heaters

Draft stop heaters are used where there are large expanses of glass (even high-quality glass) to minimize draftiness and discomfort from being close to large cold surfaces. The most common types use hot water or electricity, and they are available in a wide variety of types and sizes.

They can be recessed into the floor (most commonly under decorative grilles of some kind—a very costly installation) or into a wide window sill, mounted on the floor, raised above the floor on pedestals, or mounted to the wall. They can be used in sections or continuously (including around corners) for a cleaner look. Their cost is modest (or not), but running hundreds of feet of draft stop heaters can easily wreck even a large budget.

7.5.9 Radiant heating

As discussed previously, radiant heating is remarkably effective and most people are happy to have it. Old-style cast iron radiators have fallen out of favor, but numerous other products have taken their place (especially in parts of the world where heating-only is still common). But there are several other families of radiant heating products: electric ceiling panels, electric in-ceiling and in-floor, hot water in-floor, and even overhead linear natural gas-fired. Each will be covered separately as follows.

7.5.9.1 Electric radiant heating

Electric radiant heating takes three forms:

1 On-ceiling (or in lay-in panels)
2 In-ceiling (concealed in a gypsum board ceiling). (The author's first apartment in Eugene, Oregon when he was in graduate school had this system; it is uncommon in cold climates though.)
3 In-floor (buried in a slab or in mats under tiles, etc.)

Figure 7.5.8
Floor-mounted hot water draft stop heater

Of these, the most practical is (1) on-ceiling panels. The reason for this is that electric heating coils break down over time (no matter what manufacturers say), and it is not feasible to replace them if they are embedded in the ceiling or floor. With the on-ceiling panels, they can simply be replaced when they fail. If an in-floor system is small and under ceramic tiles (as in a bathroom), then it might be feasible to replace it in the future.

https://indeeco.com/products/
www.markel-products.com/catalogs.html
www.raywall.com/catalogs.html

7.5.9.2 Hot water radiant heating

Hot water radiant heating comes in two forms:

1 Exposed radiators
2 Tubing embedded in floors

7.5.9.2.1 Radiators

Radiators actually come in many forms, from simple "fin-tube" devices that consist of fins that are attached to pipes (see www.fintube.com) to elaborate cast iron and steel devices that are designed to be features or to fade into the background. The oldest form—the "decorative" cast iron radiator—is the least likely one to be seen today.

Fin-tubes can look like almost anything because they can be hidden behind any kind of heat-resistant screen or enclosure. (Heating coils in AHUs, FCUs, and even cabinet heaters are essentially large fin-tubes.) A pocket can be built into the floor slab so that a fin-tube radiant heater can be installed below floor level, covered with a decorative grille (usually stainless steel; see www.kadeeindustries.com), or pockets can be built into walls, or even ceilings. Enclosures can be built using steel or aluminum (or any other heat-resistant material—not plastics).

Radiators usually use large face surface area to radiate heat, in lieu of large fin area as in fin-tubes. So modern radiators tend to look like flat panels, parallel tubes, or even spirals.

www.aimradiantheating.com
www.runtalnorthamerica.com
www.sterlingheat.com
www.viessman-us.com

Radiators make highly effective (though costly) towel warmers in bathrooms too. Such flat panels can be used as baseboard, along walls, on-ceiling, or recessed into ceilings, in virtually any size and shape imaginable.

7.5.9.2.2 In-floor radiant heating

In-floor radiant heating systems first became popular early in the 20th century when small hot water boilers were first available. The famous architect Frank Lloyd Wright used them in most of his houses. In the early days, steel piping was used and then hard copper piping, but both types of piping require lots of joints that are not easily repaired when the piping is either in, or just below, the floor slab. Copper tubing (without joints) has been used, but it is difficult to bend it to the small radii that are required for a typical piping layout. (Typical piping layouts have piping as close as 4″ o.c.—those are very sharp turns for metal piping of any kind.)

In the latter part of the 20th century (1970s and on), manufacturers began making tubing from cross-linked polyethylene (best known as "PEX"). This type of tubing is very flexible, and it can easily be bent into the small radii that are needed; it is also installed in very long lengths without joints (or if there are joints, they are located so as to be accessible for repairs). This tubing is also used for domestic water piping, for earth-coupled heat pump systems, for in-floor radiant heating (both ground slabs and floors above grade), and for in-grade snow and ice melt systems.

There are two disadvantages to these systems:

1 They do not cool, so a separate cooling system is required if cooling is to be provided.
2 They do not ventilate, so separate provisions have to be made for ventilation (either natural or mechanical).

Otherwise, everyone (who can afford it) would love to have one of these systems. They are silent, they produce a warm floor surface (no more cold feet for those who like to walk around barefoot in the winter), and they are extremely effective in heating. They are particularly good for garages in cold climates because they melt snow and ice off vehicles quickly; as a result, they are found in a number of fire stations.

7.5.9.3 Natural gas-fired radiant heating

Natural gas-fired radiant heaters have been around for a few decades but they have only caught on in a big way since 2000 or so. The most commonly used units consist of natural gas-fired burners with blowers that blow the heat down long tubes; the tubes heat up and radiate heat to the surrounding objects. They are highly efficient and relatively quiet—very quiet, really—and they have become very popular for garages, fire stations, and other facilities with high ceilings or open structures and where people are actively working.

www.infrasave.com
www.reverberray.com
www.tnb.com
www.schwankgroup.com
www.spaceray.com
www.superiorradiant.com

7.5.10 Cooling compressors

Cooling compressors come in three types:

1 Reciprocating
2 Scroll
3 Screw

1 Reciprocating compressors are the oldest type, and they are essentially car engines that are powered by electric motors. (Car engines produce compression too, but the

compression is generated to provide power to move the car. In compressors, it really works the other way around.) They are noisy and inefficient, but they are relatively easy to repair and they tend to have long life-times. Naturally, given the desire to use highly efficient equipment, there are very few reciprocating compressors around now, at least in HVAC equipment. They are more commonly seen in large-scale refrigeration systems like those in grocery stores.

Reciprocating compressors are available in four-, six-, and eight-cylinder versions, and they can run in stages by turning on and off pairs of cylinders. So a four-cylinder compressor can run at either 50% (two cylinders on) or 100% (four cylinders on); a six-cylinder compressor can run at 33% (two cylinders on); 67% (four cylinders on), and 100% (six cylinders on), and an eight-cylinder compressor can run at 25% (two cylinders on), 50% (four cylinders on), 75% (six cylinders on), and 100% (eight cylinders on). This staging, as it is called, is highly desirable so that output can be matched to the load. This is not usually an issue in small systems (as found in most houses) and those compressors usually run only "on" or "off."

2 Scroll compressors operate by driving a continuous spiral, or scroll, impeller to produce compression. They are quieter and more efficient than reciprocating compressors, and they dominate the small compressor market almost completely. Scroll compressors can vary the output by running at variable speed; this is not commonly found in residential equipment (although it is seen in high-end versions of that type of equipment), but it is used extensively in VRF systems.

3 Screw compressors are more-or-less very large scroll compressors, and they are used almost exclusively in chiller and some large-scale refrigeration systems. Screw compressors can run at variable speed too, which is very important in the large facilities where they are usually found.

All compressors are designed to operate at 100% capacity at 100% of the expected load. 100% of the expected load occurs when all heat gains are taken into account on a design day (88 °F at 70% relative humidity (RH), in central Indiana). It is common to size the condensing equipment (whatever it is—air-cooled condensing unit (see 7.5.11 below), evaporative cooler, or cooling tower) at higher design conditions to provide some measure of excess capacity (usually 95 °F at 78% RH in central Indiana). So, during the rare times (2% of the time or so) when design conditions are exceeded, a properly sized cooling system should start to fall behind, allowing indoor temperatures to climb above **set-point**; this is normal and should be expected. If this does not happen, the system is probably over-sized and it would be no surprise to see that there are humidity problems in the building.

7.5.11 Air-cooled condensing units (ACCU)

These are some of the most familiar pieces of HVAC equipment. If your house has "central air-conditioning," you are likely to have an ACCU outside the house somewhere (if you do not have gas heating, this is likely to be a reversible unit that can run backwards to provide heat, and it would be called an "air-to-air heat pump").

ACCUs are available from 1.0 nominal tons (or even a little smaller in special cases) up to hundreds of tons for large systems. They include the following components:

- Compressor
- Condenser coil
- Condenser fan
- Automatic valve (sometimes a thermal expansion valve (TXV) is provided, but that is not necessarily standard)

Each ACCU is linked to an indoor evaporator coil (at the furnace, or in the AHU or FCU) by "**hot gas**" (or suction) and "**liquid**" refrigerant piping. Refrigerant piping is nearly always built using copper pipe (rigid) or tubing (flexible) and the hot gas piping always has to be insulated. Aluminum is starting to be used for refrigerant piping for VRF systems, but it is still relatively rare. For certain types of systems, liquid piping must be insulated too, but that is not always a requirement. That said, it cannot hurt anything to insulate all refrigerant piping.

When the thermostat for the system calls for cooling (because the temperature has exceeded the set-point), the compressor starts and the automatic control valve opens to allow refrigerant to flow through the system. The compressor runs at 100% (in a basic system) until the thermostat turns the device off when the temperature is back to the set-point.

When the compressor is running, the evaporator coil will usually be wet because the water extracted from the air collects on the coil (this might not be true in especially dry climates), so it is necessary to drain water (called "**condensate**") away through a drainage system connected to the plumbing (usually, but not always). When the compressor turns off, the air blowing through the wet coil actually collects some of the water again, which causes the RH to increase. This is why cooling systems that cycle on and off frequently usually do not do a very good job of controlling RH.

Typical residential ACCUs (which vary from 1.0 to 5.0 tons) are usually 3′ 0″ or so square and anywhere from 2′ 6″ to 4′ 0″ high (rough estimates only—sizes and shapes vary considerably from one manufacturer to another). Commercial units are similar, except that they use multiple condenser fans as they get bigger; units with 10, 12, or more fans are not uncommon. Obviously, such units are much larger, especially in length.

For residential ACCUs, see the following manufacturers:

www.bryant.com
www.residential.carrier.com/dealers/index.shtml (Carrier markets residential equipment under the brand names, Bryant, Carrier, and Payne)
www.goodmanmfg.com
www.heil-hvac.com
www.lennox.com/residential
www.trane.com/residential
www.york.com

For commercial ACCU's, see the following manufacturers:

www.aaon.com
www.commercial.carrier.com
www.johnsoncontrols.com
www.lennoxcommercial.com
www.mcquay.com/mcquay/Home/homepage
www.trane.com/residential
www.york.com

7.5.12 Heat pumps

It has already been noted that a heat pump is an ACCU that can run backwards to produce heat. There are two forms of heat pumps:

1 Air-cooled
2 Water-cooled

The air-cooled and water-cooled concepts have been reviewed before, but it should be noted that only air-cooled heat pumps are ACCUs that run backwards (because the ACCU itself is an air-cooled device). For specific equipment for air-cooled heat pumps, see the previous list of manufacturers for ACCUs.

Water-cooled heat pumps are a little different. In water-cooled heat pumps, the heat of the compressor (essentially the same as the heat that the refrigerant takes from the air) is removed through a **heat exchanger** that has refrigerant running on one side and water running on the other side. See the following manufacturers for water-cooled heat pumps:

www.carrier.com/commercial/en/us/products/packaged-indoor/packaged-indoor-wshps/l
www.climatemaster.com/commercial/all-products-index/
www.fhp-mfg.com
www.mcquay.com/McQuay/ProductInformation/WSHP/WSHPpage
www.trane.com/Commercial/Dna/View.aspx?i=1013
www.waterfurnace.com

The water side is looped throughout the facility and excess heat in the overall loop is rejected to the exterior atmosphere by one of five means:

1 Evaporative cooler
2 Dry cooler
3 Earth-coupling
4 Pond-coupling
5 Open well supply and return

1 Evaporative coolers are efficient and are commonly used. They operate by running the loop water (or a sub-loop through a heat exchanger) through a large coil. A large fan draws or blows air through the coil, which is sprayed with water; the evaporation of the water increases the cooling effect. Their advantage is they are usually very small in area—150 sf for up to 500 tons. Their disadvantages include size (they are usually tall—16′ or more), noise (very large fans), and water chemistry (they are very sensitive to water problems, especially if the water has high lime content).

www.baltimoreaircoil.com
www.colmaccoil.com/products/dry-coolers-condensers.aspx
www.evapco.com
http://usacoil.com/fluid-coolers-and-remote-air-cooled-condensers

2 Dry coolers are similar to ACCUs because they rely only on air movement to reject heat from the condenser coil. Their advantages are lower installation and maintenance costs but their disadvantage is lower efficiency (about 15% less than evaporative coolers), which drives up operating costs. If an evaporative cooler is operated without the spray water, it becomes a dry cooler (sort of).

www.colmaccoil.com/products/dry-coolers-condensers.aspx
www.drycoolers.com
http://usacoil.com/fluid-coolers-and-remote-air-cooled-condensers

3 Earth-coupling is more efficient than evaporative coolers but also more costly. With earth-coupling, approximately 400 lf (a very rough rule-of-thumb; the actual number varies according to soil conditions) of piping for each ton of system capacity is buried well below the frost line either in horizontal trenches, vertical boreholes, or some combination of both. The vertical trenches are often 2′ wide by 6′ deep and contain six pipes (three "to" and three "from"); vertical boreholes are probably 6″ diameter and they can be hundreds of feet deep. Systems can combine horizontal trenching with vertical boreholes to conserve space, but vertical boreholes are usually more costly than horizontal trenches. The advantages to this approach are efficiency (the highest possible) and aesthetics—no ugly and noisy outdoor equipment on or above the ground. The disadvantage is cost. Earth-coupling is commonly called "geothermal" but that terminology is incorrect. Geothermal accurately refers to boring holes thousands of feet deep to extract boiling water (due to the heat of the earth's core), which means that heat is drawn directly from the earth. In earth-coupling, the soil (earth) is simply a heat sink that holds excess heat rejected in the summer to be used for heating in the winter.

4 Pond-coupling is very similar to earth-coupling except that the piping is not actually buried—it simply rests on the bottom of a large pond instead. Pond sizing is critical, especially in cold climates, because the systems make the water colder (in winter, as heat is drawn from the heat sink) than it would be otherwise, thereby increasing the likelihood of freezing the pond. (A little ice on the surface of the pond would not

necessarily be a problem but having the entire pond frozen would be.) The advantages and disadvantage are the same as for earth-coupling.

5 Open well supply and return were the first approach to water-cooled heat pumps. In this arrangement, a high-capacity well is drilled and water is taken directly from the earth for cooling the compressors; that water is pumped through the heat exchangers (where it gets cooler or hotter according to the equipment cycle—heating or cooling) and then returned to the earth to a surface stream or deep into the earth through a second well. (The second well method is preferred but it doubles the already high well-drilling cost.) The advantages to this system are efficiency (very high) and aesthetics (no outdoor equipment). The disadvantages are cost and difficult maintenance due to major challenges in water chemistry. (In areas like central Indiana, where water is usually high in both iron, lime and even sulfur, it can be very difficult to keep these systems working. The lime tends to accumulate in the heat exchangers, causing frequent failures.)

7.5.13 Chillers

Chillers are generally used only in mid- to large-sized buildings, mostly because they are not available in very small sizes. (A 50-ton chiller is a very small one, although some **modular** units are built in sizes down to 10 tons or so. A 250-ton chiller is medium sized. A 2,000-ton chiller is a very large one.)

In addition to having either reciprocating or screw compressors, chillers also come in air-cooled and water-cooled varieties. Air-cooled chillers also come in packaged and split-system versions.

Packaged air-cooled chillers are just like large ACCUs—they contain a compressor (or compressors), automatic valves, **chiller barrel** (the heat exchanger), condenser coil(s), and condenser fans in a single cabinet that is installed outdoors near the building. Packaged chillers are noisy (double whammy here—large compressor noise *and* condenser fan noise) and relatively inefficient, but they are easy to install and fairly simple to maintain. Some packaged chillers are available with integral pumping, meaning that the pumps are installed in a separate compartment with the chiller cabinet; see 7.5.19 for more details about pumps.

Split-system air-cooled chillers have the chiller barrel indoors and the rest of the components in an outdoor package (really just a large ACCU). Pumps could be either inside in the mechanical room or outside in the condensing unit cabinet; indoor pumps are far more likely. The main reason for using a split-system air-cooled chiller is to simplify maintenance on the water side of the system by having all of it indoors; this is why the pumps are more likely to be inside.

Water-cooled chillers usually use open cooling towers, which are similar to evaporative coolers but use much more water due to their large size. The main difference is that the cooling water is not channeled through the cooling tower in a coil; instead, it runs through in the open, which increases the cooling effect. Open cooling towers come in a wide variety of types and sizes, some of which look like buildings. Due to their high water flow rates, cooling towers often use vertical turbine pumps in lieu of in-line and base-mounted pumps; see 7.5.19 for more pump information.

www.commercial.carrier.com
www.johnsoncontrols.com
www.trane.com/COMMERCIAL

7.5.14 Chilled beams

Chilled beams are a relatively recent innovation in the HVAC industry and they are used primarily to improve efficiency in cooling systems. In fact, a chilled beam is not a beam at all, nor does it even look like a beam. A chilled beam actually resembles an FCU, but without the fan. A chilled beam relies on the principle of air induction to operate. Air moves in response to temperature variations and air currents, so air will move around a cold coil without a fan; if the coil is positioned correctly in a specially designed cabinet, it can cause the air in a space to circulate without using a fan. The naturally circulating air is also cooled in the process. This process is more efficient because it does not require a local fan—as in an FCU—to circulate air within a space.

www.activechilledbeam.com
www.ltg.de/us/products-services/ltg-comfort-air-technology/air-water-systems/ltg-induction-units-chilled-beams/
www.troxusa.com/products/chilled-beams-4756506a6de845ef

7.5.15 Ductless split systems and VRF (VRV)

In the middle of the 20th century, some equipment manufacturers determined that there would be some advantages to developing cooling and heat pump systems that do not require ductwork. The original products of this sort are called "ductless split systems" or "mini-splits" because they consist of outdoor ACCUs or HPs coupled to indoor FCUs without ductwork. The indoor FCUs for these systems originally came in two basic types:

1 On-wall units (mentioned previously under FCUs)
2 On-ceiling units (mentioned previously under FCUs)

These systems are quite useful for special applications, such as elevator equipment rooms, data server rooms, or rooms that need constant cooling that are located in buildings where the central systems do not operate 24/7, 365 days each year. (If a building has a sophisticated VAV system that gets shut down overnight, or on weekends or for holidays, a separate system might well be needed.) In general, there is no need for ductwork to distribute air in such small spaces.

Over the years, these systems have become more and more sophisticated and now the indoor FCUs are available in all of the types as described under FCUs previously: on-wall, on-ceiling, in-ceiling, exposed on-floor, concealed on-floor, and concealed overhead. Obviously, not all of these systems are necessarily ductless anymore.

Outdoor ACCUs and HPs tend to look a little different for these systems because they tend to be taller and narrower (some can even be wall-mounted).

In addition, for some years now, these systems have been available with multiple indoor FCUs linked to single outdoor ACCUs or HPs, although this is usually limited to three FCUs. The controls for the exposed FCUs usually consist of wireless hand-held remote controls.

www.daikin.com
www.enviromaster.com
www.friedrich.com
www.fujitsugeneral.com
www.lghvac.com
www.mehvac.com
www.panasonic.com

In the 1980s, Daikin Industries, Ltd, a Japanese HVAC manufacturer (listed above for ductless split systems), designed a heat-recovery version of the multiple FCU ductless split system, which they called variable refrigerant volume (VRV). In this system, numerous (40 or so) indoor FCUs can be connected to a single outdoor HP (heat-recovery unit (HRU)) and the system can heat and cool simultaneously. It can do this by using special arrangements in the refrigerant piping system (so that sub-loops can switch between heating and cooling) and by having a very sophisticated control system that tracks the difference between heating and cooling across the loop. If all of the indoor units are in full cooling, the outdoor unit would run in full heat rejection; if all of the indoor units are in heating, the outdoor unit would run in full heating; if 50% of the system is in heating and 50% of the system is in full cooling, the outdoor unit would not need to run at all. So the outdoor unit figures out the difference between heating and cooling and runs only to account for the difference. To make this work, it is necessary for the compressors and condenser fans in the HRUs and for the FCUs to have widely variable speeds; they are not actually continuously variable but they have hundreds of small "steps" of capacity so that they can match virtually any operating condition.

In their pure form, VRV systems are ductless and there is an FCU (or FCUs) in each space. In the US, designers and owners do not always like the look of the exposed FCUs and it has been found that it is less costly to use ducted (maybe concealed) FCUs in VRV systems. A VRV system that uses concealed and ducted FCUs looks just like a conventional HVAC system, with typical diffusers and return-air grilles in ceilings, walls, and floors.

Since Daikin developed this technology, most other large Asian HVAC manufacturers have joined in (all of the Japanese and Korean and some Chinese manufacturers, as of 2017), although all of the other manufacturers call the technology VRF, for variable refrigerant flow. All versions of these are available through the US at this time, and the technology is catching on very quickly. The main reason for this is that this technology offers a lower-cost solution for projects that require multiple zones of control, typically the turf for VAV, water-source heat pump, and four-pipe systems. Earth-coupled water-source heat pumps can be more efficient but they are far more costly to build; also, recently, it has become feasible to do earth-coupled VRF, which is now the gold standard for aesthetics and performance.

www.daikin.com
www.hitachi.com
www.lg.com
www.mitsubishi.com
www.panasonic.com
www.toshiba.com
www.york.com

Mitsubishi even markets a version of this system that is specifically targeted toward the one- and two-family dwelling market.

The outdoor equipment for these systems looks different, because it is taller and more compact than traditional ACCUs and HPs in the US. Each section of an HRU is about 3′ square and about 6′ 3″ tall—about the size of a refrigerator—and it needs to stand on rails about 18″ above the ground or roof. HRUs can consist of one, two, or three modules of this size. (The 10-ton unit at the author's office is about 5′ 0″ × 3′ 0″ × 6′ 3″ high; see Figure 7.5.15 TEC HRU.)

The biggest limitation of a VRV or VRF system is that a single system cannot exceed 40 tons or so, mostly due to the refrigerant dumping concern that was mentioned previously. Refrigerant is odorless and colorless (although the oil in it is visible if it leaks or spills) and it evaporates readily in open air; but if a small occupied space fills with refrigerant, it is possible for some people to suffocate, or least have difficulty breathing until the gas dissipates. For that reason, these systems are kept relatively small to minimize this risk. If larger systems are needed, multiple independent systems are used. (In the author's practice, these systems have been used in a 70,000 sf multi-story building (usually six or seven systems) and, more recently, in a 110,000 sf two-story office application (12 water-cooled systems).)

Here is a comparison of various types of HVAC systems that could be used for a new three-story 50,000 gsf medical office building:

1. Multiple single-zone rooftop units: at least 30 five-ton units, with 80 sf for vertical ducts on the third floor and 50 sf for vertical ducts on the second floor. Space above ceilings for 12″-high ductwork plus 3″ of insulation. (30 control zones and 30 compressors)
2. Multiple residential-style split-systems with separate ventilation via a DOAS: at least 30 five-ton units, with 4 sf for vertical ventilation duct on the third floor and 2 sf for vertical ventilation duct on the second floor. Space above ceilings for 12″-high ductwork plus 3″ of insulation. (30 control zones and 30 compressors)
3. VAV: one 120-ton RTU with one fan and six 20-ton compressors (multiple units are also commonly used, but the system is more efficient with a single unit), at least 30 fan-powered terminals for perimeter zones and at least 30 standard terminals for

Figure 7.5.15
TEC HRU

interior zones. 60 sf used for vertical ducts on the third floor and 30 sf used for vertical ducts on the second floor. Space above the ceilings on the third floor for 30″-high ducts plus 3″ of insulation and space above the first and second floor ceilings for 16″-high ducts plus 3″ of insulation. (60 control zones and six compressors)

4 Water-source heat pumps, closed-loop with separate ventilation via a DOAS: one 120-ton evaporator cooler on the roof, two 500,000 BTUH input high-efficiency boilers, one DOAS, and at least 60 individual heat pumps. For vertical ventilation ductwork, 4 sf on the third floor and 2 sf on the second floor. Space above ceilings on all three floors for 20″-high heat pumps. (60 control zones and 60 compressors)

5 VRV/VRF: five 25-ton systems with two or three HRUs each, one DOAS, and at least 60 FCUs. For vertical ventilation ductwork, 4 sf on the third floor and 2 sf on the second floor. Space above ceilings on all three floors for 14″ FCUs. (60 control zones and 10–15 compressors)

7.5.16 Rooftop units (RTUs)

Rooftop units have already been mentioned, but they should be discussed in more detail. In general, a rooftop unit is a packaged piece of equipment that includes heating, cooling, and ventilation (and possibly RH control and possibly even an ERV) in a single cabinet. Most rooftop units are installed on pre-fabricated curbs that are built into the roof (with ducts connected to the bottoms of the units inside the curbs) but they can be installed on the ground as well, with ducts leaving the sides of the units. This kind of equipment comes in three levels of quality: commodity, commercial, and special:

- *Commodity equipment* is defined to be low cost (very low cost); it is manufactured in sizes from 2.0 to 30.0 tons (roughly, this varies by manufacturer); few options are available, and it is available with short lead times (usually). Units are available for cooling only; or cooling with gas or electric heating; and in CV and VAV (for larger sizes). Several manufacturers sell equipment like this, including Bryant (a brand name of Carrier), Heil, Lennox, Trane, and York.
- *Commercial equipment* is defined to have a wide range of capabilities so as to be able to fulfill nearly all project needs; it is manufactured in sizes from 3.0 tons to 200 tons (and even larger); many options are available, and lead times are long. Units are available in cooling only; or cooling with hot water, gas or electric heating; and in CV and VAV. Several manufacturers sell equipment like this, including Aaon, Carrier, Lennox, McQuay (a brand of Daikin), Trane, and York.
- *Special equipment* is defined as "custom" or "semi-custom" and virtually anything can be done—100% OA, very large sizes, special fans, RH control, ERVs, etc. Naturally, such equipment is much more costly than commodity equipment (and more costly than commercial equipment). In some situations where the budget is extremely challenging, it sometimes makes sense to use one special unit (usually for a DOAS) to handle ventilation so that commodity equipment can be used for everything else.

Commodity RTUs can be used for many applications but not for those where RH control is important without adding special features, such as in grocery stores (refrigerated and frozen cases will not work properly if RH is not controlled), or in heavily occupied spaces.

Commercial and special RTUs can be configured to handle just about anything, including operating rooms in healthcare facilities.

7.5.17 Filters

All-air systems require filters to remove excessive dirt, dust, pollen, particulates, and other contaminants from circulating air. The degree to which such contaminants are removed depends upon the situation. Filters are rated by "arrestance" which is the percentage of particulates that are captured. The MERV rating system is also used because it differentiates between particles of different sizes. Here is how MERV generally relates to arrestance:

Arrestance rating (5 microns)	MERV Rating
>30%	8
>80%	13
>95%	15

More specifically, a MERV 10 filter corresponds to 50–65% arrestance for particles between 1 and 3 microns and 80% arrestance for particles between 3 and 10 microns. A MERV 15 filter corresponds to 85–95% arrestance for particles 0.3–1.0 microns and >90% arrestance for particles 1–10 microns.

This system covers all typical situations, including hospital operating rooms (where >90% arrestance is usually required). Then there are HEPA filters. HEPA stands for "high-efficiency particulate air" and it means (according to the US Department of Energy) that the filter has to remove 99.95% of particles larger than 0.3 microns. There was a time when HEPA filters were routinely used for hospital operating rooms, but that changed some time ago (to the current >90% standard).

HEPA filters are not code-required anywhere (to the author's knowledge) although they are used in clean rooms and various other applications where small amounts of dust can be a problem. The main reason not to use HEPA filters is that they reduce the operating efficiency of the fan due to increased static pressure.

Most MERV-rated filters are available in various types—pleated (various types) and foams, mostly. They are available in 1", 2", and 4" thicknesses, as well as numerous different sizes. Thicker filters tend to be more effective, even at the same MERV rating, simply because there is more material to absorb the particles.

www.airhandlerfilters.com
www.reliablefilter.com
www.filtrete.com
www.yourhome.honeywell.com

With centralized HVAC systems, including furnaces, AHUs and RTUs, the filters are usually located only at the equipment. With FCU, water-source heat pump, and VRV/VRF systems, filters are located at each terminal unit (including fan-powered VAV terminals), which is one of the disadvantages of such systems. Occasionally, multiple filters are used. Most filters need to be replaced (or cleaned in some cases) every three months, but unusually dirty environments could require more frequent replacement while unusually clean environments could allow for less frequent replacement.

Many licensed healthcare facilities require pre-filters (before the fans and coils) and final filters with 90% arrestance and the final filters are required to have gauges before and after them (to notify if there is a problem with the filter).

7.5.18 Boilers

The oldest boilers are the type that are seen on old steam-powered locomotives for trains—the horizontal cylindrical part of the locomotive above the wheels is the boiler vessel—but many other types have been developed since. These boilers are always steam boilers because the steam is used to drive the locomotive wheels. Boilers are really very simple: they add heat to water to boil it, under pressure (which increases the heat that it can absorb) to produce either steam or hot water. Steam boilers must be used for power (electricity generation) applications, such as locomotives, tractors, turbines (as in modern power plants), or other equipment, but hot water boilers can be used if the only purpose is building heating. Steam boilers can be extremely dangerous, mostly because they operate under higher pressure (very high pressure in some cases) than hot water boilers—very hot objects under high pressure have a tendency to explode unless they are carefully manufactured, operated, and maintained. Steam boilers are still used for building heating, but mostly in campus settings where the boilers are located in a remote **power house**, which greatly reduces the risk of accidents. Such boilers can be fueled by coal, oil, natural gas, propane, wood or other bio-fuel, or electricity. Steam piping is extremely dangerous too, mostly because leaks can be deadly, especially if they are explosive leaks.

There are two basic types of steam boiler: fire-tube and water-tube. In a fire-tube boiler, the heat of the burning fuel is directed through tubes that are submerged in water and out through a stack, or chimney. Fire-tube boilers (like old locomotives) do not produce high pressure steam but they can produce a lot of steam. Water-tube boilers have tubes full of flowing water surrounding the fire and are more suitable for higher pressure.

Steam boilers tend to be inefficient, as compared to hot water boilers, so they are not really a good fit for building use: dangerous, hard to maintain, and less efficient—not a recipe for environmental friendliness or appropriateness. When existing steam heating systems are encountered in older buildings, they should be replaced if at all possible.

Hot water boilers were developed early in the 20th century, and several families have been developed since, including:

- *Fire-tube and water-tube (mostly water-tube)*: these are the oldest types and are rarely seen in the 21st century. These boilers are relatively inefficient and they operate best with relatively high water temperatures, 160–180 °F or so, and running such a boiler at lower temperature can cause early failure due to internal corrosion. Such high temperatures are not required to heat buildings in most cases, and using a high temperature anyway can make a system less efficient.

www.bryanboilers.com (the web-site for Bryan water-tube boilers)
www.buderus.com (the web-site for Buderus conventional and condensing boilers)
www.cleaver-brooks.com (the web-site for Cleaver-Brooks fire-tube and water-tube boilers)
www.fulton.com (the web-site for Fulton boilers)
www.hydrotherm.com (the web-site for Hydrotherm boilers)
www.lochinvar.com (the web-site for Lochinvar boilers)
www.peerlessboilers.com (the web-site for Peerless conventional, condensing, and modular boilers)
www.weil-mclain.com (the web-site for Weil-McLain residential and commercial boilers)

- *Condensing boilers*: in condensing boilers, the flue is designed to cool the flue gases quickly so that moisture in the flue gases will condense and run back to the boiler. This return process increases the efficiency of the boiler, and it is very common to see boilers like this operating well above 90% efficiency. Condensing boilers work best, and are the most efficient, at low temperatures (100–120 °F or even lower), which makes them nearly ideal for building heating applications.
- *Modular boilers*: just like modular chillers, modular boilers are available and they can be used in groupings of small sizes to meet almost any conceivable load condition.
- *Miniature boilers*: the new features already mentioned—condensing flues, high efficiency, and modularity—have made it practical to use very small boilers. One good example is the Munchkin boiler (see www.munchkinboiler.net), which is so small that it often hangs on the wall. The smallest size is 24″ high, 12.75″ deep, and 17.5″ wide, weighing only 58 pounds (empty) with inputs from 18,000–50,000 BTUH. The largest unit has a maximum input of 199,000 BTUH. Groupings of small boilers (actually any sort of gas, liquid, or solid fuel burners) to meet larger loads has advantages from a code point-of-view; see 7.9 Code Issues.

7.5.19 Pumps

Whenever water is circulated in a piping system under pressure, it is necessary to have pumps to drive the water movement, which includes hydronic (water) heating and cooling systems; water-source heat pump systems; evaporative coolers and cooling towers; in many (well, nearly all) domestic hot water return systems (see 10.2.14); in some domestic water supply systems (see 10.2.15); and in some fire-protection standpipe and/or sprinkler systems (see Chapter 11). See the following manufacturers for pumps:

www.armstrongpumps.com
www.bell-gossett-pumps.com
www.pacopumps.com
www.taco-hvac.com

For the purposes of these systems, three basic types of pumps are commonly used:

1 *In-line pumps*. These are relatively small pumps (up to a few horsepower) that are literally built into the piping. They can be suspended overhead (in suspended piping), located in vertical piping runs, mounted to a wall, or floor-mounted. Nearly all domestic hot water recirculation pumps (see 10.2) are in-line pumps. Many heating hot water and chilled water pumps, especially for sub-loops, are in-line pumps. Fire pumps for standpipe and sprinkler systems (see 11.4 Fire pumps) can be in-line pumps too. The most important thing to know about in-line pumps is that they take up relatively little space, especially if suspended. (Keep in mind that it would be necessary to allow for easy access for servicing if a pump is located above a ceiling.)

2 *Base-mounted pumps*. These are used for large applications, usually from one to hundreds of horsepower. A base-mounted pump has a base that is usually constructed using small steel channels, with a motor mounted on one end and a pump mounted on the other end. (An in-line pump, and some base-mounted pumps, can be "close-coupled," which means that the motor is attached directly to the pump housing.) The motor is attached to the pump using a shaft and special coupling and alignment is critical for proper operation and durability. Base-mounted pumps are most commonly used in hydronic applications, including heating hot water, chilled water, and water-source heat pump loop water. The most important thing to know about these pumps is that they take up lots of space. If a pump base is 20″ wide and 6′ 0″ long (a medium- to small-sized unit), it takes up that much space plus access aisles (at least 24″ wide but preferred to be 36″ wide) on all four sides. That adds up to about 92 sf for a single pump (if using 36″ aisles)! The aisles can overlap for multiple pump installations, but these things will not fit into small closets or other very small spaces. In most installations, pumps like this are duplexed, which means that there are two pumps running in parallel; this is done to ensure that flow does not stop completely if a pump fails. This does mean that a typical loop has two pumps, and a four-pipe heating and cooling system would usually have four pumps. (Some engineers like to use triplex pumping—three pumps per loop—which takes up even more space.)

3 Vertical turbine pumps are high-capacity pumps that are used mostly for cooling tower spray water. A vertical turbine pump is exactly what it sounds like—long, thin, and arranged vertically. Systems that use these pumps usually have such high flow rates that they use sumps (large pits) in lieu of pipes to provide adequate water intake; a vertical turbine pump body sits on the bottom of the sump and the motor sits on top of the lid; the two are connected by a long shaft (often 10′ or even 20′ long). The pumps themselves take up relatively little space, in plan, but the sumps can be quite large (10′ × 10′ would be considered small).

All pumps can be noisy, especially large pumps with large motors that move large amounts of water, so they should be located carefully to avoid acoustical problems (see Chapter 6).

There is one more space issue for pumps: it is nearly always necessary to have a **service disconnect** (see 9.3.5) next to the pump so that power can be turned off for servicing. For large pumps, this can be a motor starter or even a variable-frequency drive (VFD); either of these pieces of equipment takes up a little more space than a typical service disconnect. Electrical working space rules (see 9.5) apply to all power equipment.

7.6 Ductwork

Ductwork is usually thought of as metal, but there are several types of non-metal ducts too. And metal ducts are available in several different metals, using multiple types of construction.

Even though interior designers typically do not design ductwork, if it is going to be exposed in the space, the interior designer should at least review the proposal layout to avoid design conflicts and simple ugliness.

7.6.1 Metal ducts

Most metal ductwork is made from **galvanized** steel sheet metal (Figure 7.6.1) but ducts are also made of "black steel" (raw steel without galvanizing), stainless steel,

Figure 7.6.1
Galvanized sheet metal ductwork

aluminum, and even copper (very rare). Black steel ducts are required for grease exhaust systems for Type I (grease removing and fire suppressing) kitchen hoods (largely because it is not feasible to weld galvanized ducts); stainless steel is used for an outer covering over exposed grease ducts in kitchens, aluminum is used for ducts that are subjected to lots of moisture (dish machine exhausts in kitchens and sometimes shower exhausts), and copper could be used anywhere aluminum or stainless steel is used. (The author has seen copper ductwork in only one building in 35 years, where it was used for exhaust ductwork for large locker rooms, including showers, in a large university fitness facility.)

Metal ducts are built in relatively short sections that are practical to load onto trucks and move to job-sites (20′ or less, in most cases) and joined together in the field using one of four methods:

1. Welding (used only for grease ducts), in round, square, and rectangular sections
2. Lock-seams (used mostly for small ducts), in square and rectangular sections
3. Flanged joints (used mostly for large ducts), in square and rectangular sections
4. End-to-end joints (used mostly for spiral-wound and low-cost residential ducts), in round and oval sections

7.6.1.1 Welded ducts

Welding ductwork is extremely labor intensive, and therefore costly, so it is avoided to the greatest extent feasible. In recent years, systems have been developed to use special sectional mechanically jointed multiple-wall stainless steel ducts with integral fire insulation to replace welded black steel ductwork, and the market for those products is growing rapidly. These new products are not less costly, but they are easier and faster to install and they can reduce code complications. Welded ductwork for Type I hoods in commercial kitchens is dangerous (very hot and prone to internal fires), so there are stringent limitations on where it can be located in the various types of buildings (especially the types that allow for combustible framing—Types III, IV, and V) and the new mechanically jointed systems are designed to simplify these issues by reducing clearances to combustible materials (especially important in wood structures).

To be more specific, the clearance required between a welded black iron duct and wood framing is 18″, which can be very difficult to do. If that duct is wrapped with temperature-reducing "fire wrap" insulation (also very costly), that clearance can probably be reduced to a few inches or even to zero clearance; by contrast, the mechanically jointed double-wall ducts are usually designed for zero clearance to combustibles.

www.captiveaire.com
www.econair.com
www.firesprayusa.com
www.greaseduct.com

7.6.1.2 Lock-seam ducts

Lock-seam ducts joints are simply flat folded-over sheet metal edges that lock together. Lock-seams are not usually air-tight (important in a duct), so it is necessary to use sealant to make them air-tight.

www.isystemsweb.com
www.larosagroup.com

7.6.1.3 Flanged ducts

Flanged joints are similar to lock-seam joints because they also use sheet metal flaps at the end of each duct section. But in a flanged joint, the flaps (flanges) are perpendicular to the face of the duct and they lock together with separate clips that slide on and hold

Figure 7.6.1.3
A flanged joint

the joints together. Flanged joints usually have internal gasketing (sealing), so they are air-tight when they are properly installed. They are also tougher and able to resist much more air pressure than lock-seam joints, so they are used almost exclusively in large ducts and ducts with high air-speed. There are two disadvantages to flanged joints: (1) they are more costly than lock-seam joints (even when factoring in the cost of the sealant for the lock-seam joints); and (2) they take up more space because they make the ducts bigger by about 1.5″ on each side. Of course, if the duct is insulated, the insulation will be as thick, or thicker, than the 1.5″ for the flanges so the flanges are no longer an issue.

7.6.1.4 End-to-end joints

In residential and some low-end commercial duct systems, sheet metal ducts are used that arrive on the job-site curved but not in tubes; the longitudinal joints of such ducts are formed with lock-seams, where the plain metal on one side slides into a formed slot on the other side, making a tube. The end joints of such ducts are formed by a pinched end on one section that simply slides into the next section, secured by tape (infamous duct tape) or screws or both.

For spiral-wound round ducts (which are made by winding a continuous narrow strip (3–6″ wide) into a tube), the end joints are similar to the end joints described above.

The overall look of spiral ductwork is usually considered to be more attractive than lock-seam and flanged ducts, so spiral duct is usually used in applications where it will be exposed in finished spaces. (There is another sub-class here, which is "flat-oval"

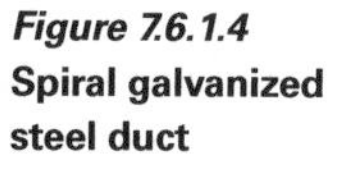

Figure 7.6.1.4
Spiral galvanized steel duct

ductwork, which is essentially a flattened spiral duct with flanged joints. This is used to reduce headroom required, as compared to round ducts, but it is very costly.)

www.spida.org
www.spiralmfg.com
www.zenindustries.com/

7.6.1.5 Duct sealants

Duct sealants come in two basic types: tape and liquid. Tape sealants are simply special very sticky air-tight tapes that secure and seal joints. (But this tape is not the common "duct tape".) Liquids are brushed on by hand to seal up all gaps, whether those gaps are in lock-seam joints, flanged joints without integral gaskets, or simply holes in the ducts.

www.hardcast.com
www.rcdmastics.com
www.spiralmfg.com

7.6.1.6 Duct insulation

Some ducts are insulated for two reasons: to save energy and prevent condensation (sweating). Exhaust and return ducts are rarely insulated because there is little to be gained in energy savings and because there is little risk of condensation. Condensation occurs mostly on cold air (cooling) supply ducts because the surfaces of those ducts can be 55 °F (or even colder), which can be well below the dew-point even inside a building. Insulation is even more important for outside air-intake ducts, which can be extremely cold in the winter in cold climates (as this is being written, it is 5 °F outside in central Indiana; that does means that incoming outside air is also 5 °F).

It is possible to insulate ducts internally, by using various board duct-liner products (including fiberglass), but such liners must be selected carefully to make sure that hazardous particles (such as small bits of fiberglass) cannot get into the air stream. In the author's practice, duct-lining is avoided if at all possible. Duct liners can be required in acoustically sensitive environments; one just has to be careful.

The ultimate in insulated ductwork is dual-wall ductwork where the insulation is in between the two ducts; needless to say, that is so costly that it is rarely done.

When metal ducts are insulated externally, one of two basic means is used: rigid boards or blanket wrap.

7.6.1.6.1 Rigid boards

Rigid board insulation is the right way to do it because it is more durable and more attractive than simply wrapping the ducts with soft fiberglass (see 7.6.1.6.2). The external boards can have paintable outer jackets, so the ducts can look smooth and clean (and can be painted easily). This only works with square and rectangular duct sections though and it is very costly so it is usually only seen in government or other institutional projects with substantial budgets.

7.6.1.6.2 Blanket wrap

Most insulated ducts are wrapped with soft fiberglass blankets with various facings, the most common of which is aluminum foil. The joints are taped with matching foil tape and the whole thing looks like shiny puffy ductwork when it is finished. Blanket wrap is fragile and easily damaged, but it is inexpensive and used almost everywhere (see Figures 7.6.1 and 7.6.1.3).

7.6.2 Non-metal ducts

Non-metal ducts include reinforced foam, fiberglass ductboard, concrete pipe, fiberglass pipe, plastic pipe, and flexible ducts (which include, in turn, fabric and spiral-wound).

7.6.2.1 Reinforced foam

Reinforced foam ducts are constructed from special rigid foam (usually phenolic materials) boards that are covered with aluminum foil on both sides. These ducts are very strong, cost effective (higher material cost but lower installation cost), and they do not pose a risk to health because there is no chance of hazardous fibers getting into the air stream.

www.igloothermal.com

7.6.2.2 Ductboard

Ductboard ducts are constructed from rigid fiberglass boards that are coated on the inside and covered with aluminum foil on the outside. This product is weak, very inexpensive (although also very low quality), and there is a risk of putting hazardous fibers into the air stream if the internal coating wears away. This material should be avoided if at all possible.

www.certainteed.com/products/insulation
www.knaufusa.com/products/commercial_industrial/air_handling_insulation/duct_board_m_with_ecose

7.6.2.3 Concrete pipe

Concrete pipe is used occasionally for underground ductwork; therefore, it is not particularly relevant to interior design, except to note that it is possible to use underground ductwork. Modern energy codes require underground ducts to be insulated (in some climate zones), so this is no longer feasible. A modern replacement called "Blue Duct" is a form of reinforced foam duct that is designed to be water-tight and to be used underground; this product can be used successfully, but it is costly.

7.6.2.4 *Fiberglass pipe*

Fiberglass pipe used to be used for underground installations, but the same insulation problem noted above applies here too.

7.6.2.5 *Plastic pipe*

This is where "Blue Duct" comes into play, as a proprietary plastic duct system for underground applications. It can be very effective but it is costly so in-building options should be pursued first.

7.6.2.6 *Flexible Ducts*

Flexible ducts include inflatable plastic and fabric ducts and spiral-wire-reinforced insulated and uninsulated ducts.

7.6.2.6.1 Inflatable ducts

Inflatable ducts, using plastic or fabric, have been around for decades, but they have caught on in the architectural world only since 2000 or so. Prior to that, they were used (usually in plastic form) mostly in industrial settings to provide inexpensive and simple air distribution for relatively crude ventilation systems. In architectural (i.e. aesthetically sensitive) applications, solid vinyl and shiny fabrics do not perform all that well (and they look terrible), so new fabrics have been developed to improve the appearance and performance.

If a fabric duct is made using impervious vinyl, then it will be shiny, probably wrinkly, and it will not clean itself. In other words, dust will collect on top on it as with all other solid ducts. But fabric ducts can be made from porous fabrics that let a little air leak through the entire surface of the duct, which makes the duct somewhat self-cleaning—literally blowing the dust off the top (and bottom and sides) of the duct whenever the fan is running. Primary air distribution is done with larger openings in the duct that are engineered specifically for the application at hand. The openings can vary from open weave meshes that run continuously like zippers to individual round holes, from 0.25″ to 1.5″ diameter.

Fabric ducts usually hang from stretched stainless steel cables (one or two cables, depending on the design) and they do deflate and "sag" if the fan is turned off. If high-quality fabrics are used, the ducts do not wrinkle and they look at least as good as spiral sheet metal ducts. Fabric ducts usually distribute air better than spiral sheet metal ducts, and unattractive branch ducts can often be eliminated when using fabric.

The downside to fabric ducts is that they do make a loud popping noise whenever the fan is turned on (due to the inflation of the duct); this can be avoided by leaving the fan running during occupied hours (as is often recommended by consulting engineers). Some manufacturers also offer structures that can be installed inside the ducts to hold their shape when they are not inflated—obviously a costly add-on.

Fabric ducts: in the author's practice, fabric ducts have become commonplace since 2000 or so and they have been used in school gymnasiums, restaurants, offices, and even in a pediatric dentist's office where color-changing LED lights were installed inside the duct! We have heard that the kids who visit the facility love the color-changing ductwork.

www.adctubes.com
www.airmax.net
www.ductsox.com
www.durkeesox.com
www.kefibertec.com
www.qsox.com
www.zip-a-duct.com

7.6.2.6.2 Spiral-wire-reinforced insulated and uninsulated ducts

This material is usually called "flex duct" and it is used extensively on low-budget projects. It is available in materials that are useable in a plenum (see Chapter 3), both in insulated (for supply air) and uninsulated (for exhaust and return air) forms. Short lengths of flex duct are often allowed for commercial and even institutional projects to make it easier and faster to connect ductwork to supply air diffusers, exhaust grilles, and return-air grilles (where applicable). But long lengths should be avoided because it increases static pressure in the system, which has the effect of either increasing fan horsepower (and making the system less efficient) or of making some, or much, of the air apparently disappear. It should never be used for elbows and it must be installed (supported) in such a way that it is not pinched. In a perfect system, there would be no flex duct at all.

www.atcoflex.com
www.hartandcooley.com
www.quietflex.com
www.thermaflex.com

7.7 Grilles, registers, and diffusers

This is yet another vast topic in itself. Recall the descriptions in Section 7.5.

Grilles are simply frames with grids or bars (single or double bars) that are intended to keep large objects out of the ducts and to trim out openings. They are available in painted steel (inexpensive), aluminum (costly), and stainless steel (very costly), in virtually any size and shape imaginable.

These air distribution devices are all-important from a design point-of-view because they are the most visible elements of a mechanical system. And interior designers should be actively involved in their selection and placement.

Registers are simply crude devices for air supply and return (mostly common seen in residential projects). They look like single- or double-bar grilles but they have dampers behind the bars so that air volume can be controlled (well, partially controlled); they can be duct, wall, or ceiling mounted, and they are available primarily in painted steel. (Air volume is controlled for grilles by dampers in the ducts, which are not visible.)

Diffusers are designed for specific types of supply air distribution—low ceilings, high ceilings, sidewalls, etc. They are available in painted steel and aluminum, and they can be hidden behind stainless steel grilles. They can be duct, wall, or ceiling mounted. The most common diffuser in the industry is the painted stamped-steel multi-cone (usually three or four cones) 2 × 2 for mounting in a lay-in ceiling, and it is not particularly attractive (see Figure 7.7.A).

Several improved options are available:

- *Metal plate or plaque*: this is essentially the same as a multi-cone diffuser but with only one cone. It is slightly more attractive and a bit more costly, but it is still just white painted folded steel.

Figure 7.7.A
Multi-cone diffuser

Figure 7.7.B
Metal plate or plaque diffuser

- *Ceiling plate*: this is a metal plate diffuser where the central metal plate is replaced with a frame that can be filled in with a ceiling panel. Much more attractive; much more costly.
- *Perforated diffuser*: this is really a register with a perforated face. While they look better when they are new, they tend to collect dirt more rapidly than the other types, making them dirty and ugly sooner rather than later. They are more costly too.
- *Linear diffuser*: available in a wide variety of styles and sizes, from one slot to eight slots, and at almost any imaginable length. They can be ceiling- or wall-mounted, but special wall framing (usually double walls) is required for horizontal applications.

Figure 7.7.C
Linear diffuser

These are far more attractive, especially in contemporary design, but much more costly. When using linear diffusers, it is possible to use blank sections (only the front), return sections, and/or exhaust sections, so it all looks the same and so that it can look continuous.

Several duct-mounted options are available too: round multi-cone, square or rectangular register, and linear. These are used primarily with exposed spiral sheet metal ductwork, although they could be used with lock-seam or flanged ductwork too.

How does a designer know which type of device to use in which situation?

First, most commercial office buildings (and many other buildings) have "building standard" grilles and diffusers for ceiling mounting, so projects in those buildings are generally required to use those devices. (Anything can be negotiated away in a lease though, so it is possible that the standards might not apply.) If it is an existing building, the most common practice is to move around existing grilles and diffusers, adding new ones only when truly necessary. (If the devices are more than 20 years old or of unusual types, it might not be feasible to get the same devices any longer. If that happens, new devices would be used in isolated areas, especially if the look is substantially different.)

Second, if there is no building standard (or if it does not apply), what is the budget? All (well, most) diffusers and grilles are relatively inexpensive compared to an overall project budget, but they are also devices that are purchased in the dozens or hundreds (or more for really large projects) so a small increment can add up to quite a lot. In other words, if the premium is $20.00/diffuser, but there are 500 of them, that is still a $10,000.00 budget item; $10,000.00 can get someone's attention on almost any project. If the budget is very low, it is probably necessary to use standard painted steel multi-cone 2 × 2 diffusers, plastic egg crate return-air grilles, and very basic painted steel exhaust grilles. If the budget is moderate, it might be feasible to upgrade the diffusers (even up to linear diffusers in many cases) and grille types (painted steel or aluminum return grilles). If the budget is high, then just about anything goes and aesthetics can play a much larger part.

Here are two brief case studies to illustrate what can happen:

High-end accounting firm

In this project on the 47th floor of the tallest building in Indiana, there was a beautiful conference room separated from the reception area by a glass wall framed in solid figured anigre wood mullions and with full-custom anigre-veneered five-light (three lights frosted) doors; custom millwork inside the room included a large oval conference table and a large AV cabinet. The room needed several supply air diffusers, at least two return-air grilles, and an exhaust grille for a locally switched exhaust fan (to remove food odors, etc. should the need arise). Even though the author was acting as the interior design project manager on the project, he also happened to be designing the HVAC system (a situation that came about because the firm's engineers were too busy at the time), so he decided to use a single run of linear diffusers/grilles along the inner wall of the conference room, incorporating supply, return, and exhaust in a way that is visually indistinguishable. In other words, there are ten 2′ 0″-long individual slots in the lay-in ceiling for an overall length of 20′; most of the slots are for supply air, two are for return air (this is a performance compromise but a reasonable one in a situation like this), and one is for the exhaust. And it looked fantastic.

Inn ballroom

In a large ballroom at a state-owned inn (in a state park), the architect wanted exposed wood trusses—real old-fashioned large-scale structural wood trusses. The owner paid an extra $40,000.00 (in 2001) just to have the exposed trusses, so it was unacceptable to fill up the space with ductwork, diffusers, light fixtures, sprinklers, speakers, etc. In order to conceal the systems, it was decided to build a 20″-high cavity above the ceiling at the tops of the trusses and to run all of the ductwork, piping, and wiring in spaces that are 20″ high (perpendicular to the steeply sloping ceiling) and 24″ wide, running parallel the length of the room. To minimize the visual impact even more, it was decided to use continuous supply air slot diffusers from one end of the room to the other, only about a third of which is active. Most people walking into the room would never notice that there is any HVAC distribution at all, let alone find it objectionable.

The difference between painted steel and aluminum products is significant. Painted steel diffusers and grilles are manufactured by bending steel sheet metal. When sheet metal is bent (any sheet metal), it is not possible to achieve sharp corners, so the corners in such products are always slightly rounded. It is not feasible to bend aluminum in this way (it is too weak and soft), so aluminum is used primarily in the form of extrusions. To make extrusions, molten aluminum is forced though a die into some pre-determined shape—with very sharp corners and details. So if the project screams out for a really sharp and tight look, it would be a good idea to look at aluminum diffusers and grilles, even though they are substantially more costly.

Another issue with return-air grilles in particular is durability when they are mounted in walls, or just within reach somewhere. Standard grilles are not particularly robust (in painted steel or in aluminum), so they should not be exposed to direct contact. What should a designer do if a grille has to be very close to the floor—which the HVAC designers will want if the supply air is high in the space—or if there is reasonable likelihood of abuse (as in a gymnasium)? Specify special and unusually tough products, probably in steel. Such products might be called abuse-resistant, institutional (meaning penal), or correctional (meaning dangerous penal), but they should be used wherever a lot of contact is expected, even if it is just from kids leaning on the grilles. Interior designers will not often write such specifications themselves, but they should ask their engineers to make sure that such issues are addressed.

7.8 Controls

As in so many other cases, the subject of HVAC controls is vast. The simplest controls are wall-mounted thermostats that use mercury switches (no longer used in new projects but many old ones are still around) but modern thermostats are usually all-electronic and can be built into whole-building (even whole-house) control systems or even energy management systems.

Interior designers should review proposed control locations to avoid design conflicts—for example, a thermostat in the middle of an art wall.

A mercury-switch thermostat is a simple device that consists of a metallic coil attached to a small glass capsule that contains a small amount of mercury; it is called an analog device because it directly senses changes and responds simultaneously (analog is the inverse of digital). Mercury is liquid at typical room temperature, so it will move from one end of the capsule to the other if the angle of the capsule is changed. The angle of the capsule is changed by the metallic coil, which moves in response to room temperature changes. If the thermostat is operating heating mode, the coil shrinks as the room gets colder; when it has moved far enough to cause the mercury to move to the opposite end of the capsule, electrical contact is made, sending an "on" signal to the furnace. When the temperature rises as the furnace runs, the reverse process happens: the coil grows and the capsule moves in the opposite direction until the mercury moves back to the other end, breaking the electrical contact and stopping the "on" signal. The oldest type of thermostat operates at line voltage (usually 120 volts in the US—see Chapter 9), which means that the power to operate the furnace and the fan actually flows directly through the thermostat. With cooling systems that usually require two-pole power (usually 208 volts or 240 volts in the US—see Chapter 9), this is too dangerous, so low-voltage controls are used. In low-voltage controls, small transformers are used to reduce the voltage at the thermostat to 24 volts or less, which makes systems safer. This type of thermostat is available for heating-only, cooling-only, or heating and cooling with manual changeover. Manual changeover means that there is an accessible switch on the thermostat that has to be moved to "heating" or "cooling" by hand. Most thermostats of this type control the furnace or heat pump fan too, and there is usually a separate switch for this with "on," "auto," and "off" positions. Auto means that the fan turns on automatically if either heating or cooling turns on. Automatic changeover thermostats are also available with dual settings.

Electronic thermostats mimic this behavior by using electronic components, and such devices are usually more accurate than old mercury-switch devices. Electronic thermostats are inherently low-voltage, and they offer the user the ability to program different settings for different schedules. In many homes, there is no reason to keep the house warm to 72°F from 12 p.m. to 5a.m. if everyone is asleep, and an electronic programmable thermostat allows the users to program the temperature to drop to 65°F (or whatever) during those hours, automatically coming back up to 72°F at 5 a.m. (or whenever). This is called "night setback," and "night setup" can be used for cooling. The most sophisticated thermostats can actually learn typical occupancy and activity patterns, automating some of the programming.

In simple systems, there is a single control device for a single furnace, heat pump, RTU, etc. If a building has multiple furnaces, heat pumps, or RTUs, each one would have its own thermostat. When there is a single control device for a single system, all of the spaces served by that system would be under a single control. This is why some rooms in a house (or apartment or other facility) are consistently warmer or colder than others; the thermostat controls all of the rooms from its location, even though the loads vary between rooms.

Simple systems can have one type of added level of control, which is called the "zone **damper.**" A zone damper is an automatic (i.e. power driven) damper that will open and close according to a signal from a secondary thermostat. If there is a room that is consistently overheated (or over-cooled) and the room is served by a single duct, that room would be a good application for a zone damper. The damper would be installed in the duct before any supply air diffusers or registers, and the secondary thermostat would close off that duct when the temperature is satisfied in that room. If there are multiple zone dampers, it is possible to close off too much air-flow for the fan to work properly, so these approaches are limited. This approach can be used to prevent localized over-heating or over-cooling but they cannot be used to solve under-heating or under-cooling, nor to provide simultaneous heating and cooling in different spaces. To do the latter, some kind of control system is needed.

A control system is used to integrate control of a complex system that has multiple zones, pieces of equipment, types of equipment, etc. In such a system, there are components that sense temperature, water flow, air-flow, etc. as well as monitoring (and operating) each piece of equipment: fan, ACCU, heat pump, burner, RTU, chiller, boiler, cooling tower, FCU, radiator, PTAC, etc.—whatever the equipment might be. Each local zone might have a thermostat but it is more likely that each zone would have a temperature sensor (and sometimes a humidity sensor) with or without a control device. Sensors with "warmer" to "cooler" slide switches are common; these will prevent over-heating or over-cooling a space because the central system controls the range of temperatures between "warmer" and "cooler." Systems like this are usually operated centrally (meaning on a computer connected to a network or remotely via the web on a computer, tablet, or even smart phone) by building management. If there is no building management, someone in the space usually has access to the system.

Pneumatic (air-driven) versions of control systems have been around since early in the 20th century and many such systems still exist (often in schools). One can identify a pneumatic system by the look of the devices, but the dead giveaway is if a device is hissing due to an air leak. New systems are all-electronic and they are called direct digital control (DDC) systems. DDC systems are inherently low-voltage and many of them are simply software running on networks these days.

If the control system includes lighting controls, the system becomes a building or energy management system (BMS or EMS) or a building automation system (BAS). If additional systems are included, such as security, automatic window shades, smart glass, etc., it becomes a smart building system.

www.carrier.com/commercial/en/us/products/controls/
www.buildingsolutions.honeywell.com
www.johnsoncontrols.com
www.lennoxcommercial.com
www.buildingtechnologies.siemens.com
www.trane.com/commercial

7.9 Code issues

The code issues that most directly affect interior designers are requirements for access (to service, test, repair, and/or replace equipment, dampers, etc.), ventilation, fire protection for large fuel-burning equipment, and automatic shut-down of large fans in the case of fire.

If a building has permanent HVAC equipment installed on the roof, the International Mechanical Code will require a permanent means of access to the roof, which can be a permanent ladder on the side of the building, a roof hatch with a permanent ladder inside the building, or a stairway inside the building.

All operating equipment, fire dampers, smoke dampers, etc. must be accessible for service and operation at all times. If such equipment is located above a lay-in (i.e. removable) ceiling, the lay-in ceiling is usually considered to be a sufficient means of access. If the ceiling does not have panels that are removable without special tools (including, but not limited to, plaster, gypsum board, metal, wood, glass, etc.), then access panels are required for all items. The access panel must be large enough to replace the item in question and a workman must be able to reach anything that could need to be reached from such panels (it is acceptable to use a single panel to reach multiple items). For example, if an FCU that is 50″ wide × 36″ deep × 15″ tall is located above a ceiling, an access panel must be provided to allow the unit to be replaced without destroying, or even modifying, the ceiling; in this case, such a panel could be 54″ long × 18″ or 40″ long by 18″ wide if, and only if, there is adequate headroom to turn the unit on its side to get it out of the ceiling (and to get a new unit back into the ceiling). That access panel, or additional access panels, must also be large enough and located so that all elements can be accessed. If there are operating parts on both sides of the unit, the access panel might need to be 98″ wide! (In such a case, it would be better to provide two panels, one on each side.) From a design point-of-view, the issue with access panels is that they disturb the visual integrity, continuity really, of the ceiling. Even if they are carefully detailed to blend into the ceiling, it is very difficult to hide them completely.

Ventilation is key because it can drive the necessary space above a ceiling (to conceal ductwork) even in a VRF system that has no ducts otherwise. Especially when dealing with assembly spaces (see Chapter 3), meeting code requirements for ventilation can be challenging both from a headroom point-of-view and also from a systems point-of-view. Residential-style split-systems, commodity RTUs, water-source heat pumps, and even VRF systems cannot manage high levels of ventilation, so if such ventilation is required, more sophisticated systems have to be used or independent ventilation systems have to be added.

For fuel-burning equipment, large burners represent a substantially higher fire risk than smaller burners, so most codes (certainly the International Building Code) require fire protection, in the form of fire-rated enclosures, for rooms containing single burners larger than 399,000 BTUH input. This is not an aggregate though, so there could be twenty 199,000 BTUH input burners in a room without having to have rated construction. It is all about the largest single burner.

Fans are designed to move air around building, and they also easily move smoke around too. So circulating fans (with supply and return) that are larger than 2,000 cfm must shut off automatically in case of fire. In a building without a fire alarm system, this takes the form of a local smoke detector in the return-air duct, which is wired to shut off the fan when the detector senses smoke or when the fire alarm system is in alarm mode. In a building with an automatic fire alarm system, the unit can be shut off by the local detector, a nearby smoke detector, or an alarm in the system. (There is confusion about this requirement for ventilation systems because such systems sometimes have no return ducts. If this is the case, this requirement does *not* apply and a smoke detector is *not* required for the unit, no matter how large it is.)

7.10 HVAC design documentation

HVAC drawings and specifications can be complex and difficult to read because they must include a large amount of information: equipment sizes and locations as well as duct and pipe sizes and locations. Systems often include supply, return, outside air, ventilation, exhaust, grease exhaust, contaminated exhaust, and lab exhaust ductwork; it is difficult, if not impossible, to show all of those systems on a single drawing, so multiple drawings are often used. The same applies to piping: heating hot water, chilled water, dual-temperature water, heat pump loop water, earth-coupling loops, condensate drainage, refrigerant suction, and refrigerant hot gas.

Equipment (and air inlets and outlets) is usually described on detailed schedules, and numerous details are used to explain how mountings and connections are required to be done.

Summary

This very brief, and likely confusing, introduction to HVAC in buildings only serves to lay out the most basic elements of a very complex subject. But most interior designers will never actually design one of these systems, so there is no reason to go into the specific technical aspects of how to size ductwork, how to lay out ductwork and piping, how to locate controls, how to plan for variable ventilation in a complex system, etc.

Key elements have been covered, including vocabulary, equipment types, basic systems, space requirements, and even basic code requirements. Each of these areas could become its own subject, with extensive research, in an effort to achieve deeper understanding.

Even though these technologies have been used—more likely abused—over the past half century or so to enable architects and other building professionals to ignore the local climate, the recently renewed emphasis on energy conservation and green design is turning us back to the past. Back to a time when buildings were designed to function without cooling systems; back to when natural ventilation really worked, and back to when we did not maintain a constant 72°F indoor temperature year-round—and that is all to the good.

Outcomes

7.1 Understanding why we heat, cool, and/or ventilate.
7.2 Understanding how we heat, cool, and ventilate.
7.3 Understanding basic heat flow in buildings, including the BTUH and construction parameters: U_o, SHGC, etc.
7.4 Understanding the space implications of HVAC equipment, and basic system arrangements.

7.5 Understanding the basic types, configurations, and sizes of HVAC equipment, from residential furnaces to giant boilers and chillers.
7.6 Understanding basic types, configurations, and uses of ductwork, by understanding that there is both metal and non-metal ductwork; that there is rigid and flexible ductwork; and that ductwork sizes vary according to system type and size.
7.7 Understanding the basic types and configurations of air grilles, registers, and diffusers, by understanding aluminum and steel products in various types, sizes, and configurations.
7.8 Understanding the basic types, configurations, and uses of controls and control systems, and the basics of thermostats, HVAC control systems, energy management systems, and even smart building systems.
7.9 Understanding that basic HVAC code requirements affect interior design, especially access requirements and fire-protection requirements for large fuel burners.

Chapter 8

What are architectural lighting systems?

Objectives

8.1 To understand basic indoor architectural lighting and daylighting
8.2 To understand human vision
8.3 To understand the history of artificial lighting
8.4 To understand quantity of lighting
8.5 To understand general optics
8.6 To understand quality of lighting
8.7 To understand sources
8.8 To understand energy conservation and sustainable design in lighting
8.9 To understand lighting design by layers
8.10 To understand daylighting
8.11 To understand luminaires, including controls
8.12 To understand basic lighting calculations
8.13 To understand lighting applications
8.14 To understand egress and emergency egress lighting and exit signs
8.15 To understand lighting design documentation

8.1 Introduction

Of all the building systems, lighting—both daylighting and artificial lighting—is probably the most important from a design point-of-view, simply because excellent lighting can elevate a mediocre design while poor lighting can ruin the finest of designs.

Because lighting is such an important facet of both architecture and interior design and because IDs often design interior lighting, it will be covered in greater depth than some of the other topics. Broadly speaking, there are highly artistic aspects, highly technical aspects, and everything in between. Hand calculations are rarely done these days, but they will be explained in detail because seeing hand calculations does help to understand the output from computer calculations and modeling. The treatment will include basic human vision, history of lighting technology, light quantity, general **optics**, quality of light (including **color temperature** and **color rendering**), light sources (including **incandescent**, halogen incandescent, fluorescent (linear, round, and compact), high-intensity discharge (mercury vapor, metal halide, ceramic metal halide, low-pressure sodium, and high-pressure sodium), induction, cold cathode and neon, plasma, and solid-state (light emitting diodes (LEDs) and organic light emitting diodes (OLEDs)), designing using layers (ambient, task, accent, and decorative), luminaire types (recessed, surface-mounted, suspended, etc.), lighting calculations (zonal cavity and accent lighting), lighting controls, lighting for various types of projects (including residential, retail, restaurants, healthcare facilities, educational facilities, recreational facilities, offices, industrial facilities, and religious facilities, egress and emergency egress lighting, and lighting design documentation (drawings and specifications).

All building technologies change over time, some a little and others quite a lot, sometimes very slowly and sometimes very quickly. Since Thomas Edison perfected the incandescent lamp in the 1880s, electric lighting has been the fastest changing of all building systems. In fact, this accelerated in the 1930s with the invention of what can be called "low intensity discharge" lamps—commonly called "fluorescent," which was rapidly followed by high-intensity discharge (with several sub-families), plasma, induction, and solid-state (i.e. LEDs) over just a few decades.

But today, after only ten years or so of intensive development of solid-state lighting, *all* of the previous sources and technologies are now called "legacy" sources because they are no longer relevant to new work. In other words, to all intents and purposes, all lighting specifications going forward will be for solid-state sources only—no incandescent, no fluorescent, no metal halide, etc. (There will always be exceptions but they will become rarer as time goes by.) Yet it is still necessary to talk about the legacy sources because all of them still exist in buildings and working IDs will encounter them in renovation projects. This might mean nothing more than replacement with new LED technology, but it is still good to understand what the existing systems are and how they work.

Another general issue to cover before diving into the vast sea of information to come is the structure of the lighting design business. In the United States, there are three parts to the system, although the details vary between the commercial and residential worlds:

1 Manufacturing
2 Installation
3 Design

Manufacturing is exactly what it sounds like: thousands of companies worldwide (many hundreds just in the US) who design and manufacture electric lighting hardware—called luminaires or fixtures—for sale on the vast US market for a wildly diverse group of applications. Commercial manufacturers usually go to market in one of two ways: direct sales to owners or through what is called the "specifier" community. Specifiers include lighting designers, engineers, architects, IDs, or anyone else who selects and specifies luminaires. Residential manufacturers usually go to market through retail channels—web-sites, on-line retail, or brick-and-mortar retail.

Installation means on-site work to get hardware into place and operating. This is generally done by electrical contractors, from single individuals to large highly organized national organizations. But this also includes purchasing, which brings manufacturers and their marketing back into the picture. In the commercial world, each order is priced individually (that can be a little hard to believe but it is true); even though all luminaires have nominal list prices, nearly every order is discounted to some degree (e.g. small order = small discount, big order = bigger discount, giant order = even bigger discount). This can make it difficult to determine what hardware actually might cost for a given project. Also, most manufacturers use independent sales agents (often called "lighting reps") to handle contacts with contractors and specifiers. And, there is yet another level called "distributor" between the agents and the purchasers. In most cases, lighting reps do not actually sell products (selling is done by distributors) but the reps in some markets control all of the pricing (Indiana is such a market). In other markets, the reps are weaker and the distributors drive pricing. This matters because it can be a challenge to get exactly what is desired in a rep-dominated market. A large rep might handle 80 or more companies, so they have a strong incentive to package the project to sell all hardware on a given project for a single lump sum number (thereby concealing actual luminaire pricing). But if the specifier wants a luminaire from a manufacturer that is not available to a particular rep, that rep will attempt to substitute another product (during design, during bidding, even after bidding, with or without the knowledge of the specifier), which might or might not be acceptable. (In the worst case, this is not discovered until just before hardware is ordered; needless to say, it can be very difficult to resolve disputes at that stage of the game).

On the residential side, things are simpler in a sense because everything is purchased essentially at retail pricing.

Design includes everything that it takes to get from an idea to detailed specifications for the manufacturers and installers. Those on the design side of this are usually called "specifiers" whether they are lighting designers, IDs, architects, or engineers. What does a specifier do? Well, that varies considerably from one project to another but, basically, someone who designs and specifies lighting takes input from an architect or ID (or an owner or an owner's agent), puts together design concepts (up-lighting, downlighting, or side-lighting; light or dark; soft or harsh, etc.), and then develops detailed luminaire layouts and selections to achieve the design concepts. But this is done amid constantly changing products from hundreds of manufacturers. It is difficult to keep up.

8.2 How we see: basic human vision

Human eyes function in a way that is similar to cameras, especially digital cameras. Light passes through the outer protective covering (the cornea), then through the iris and pupil (equivalent to a shutter), then through the lens (for focusing), and onto the **retina** at the back of the eye. The retina is covered by the photoreceptors called **rods**, except in a small area called the **fovea**, although not at a constant density. The second type of photoreceptors called **cones** are concentrated in the fovea, but are present throughout the retina in much lower density. The cones provide color vision because there are three sub-types of cones, each of which has unique color sensitivity. The optic nerve is located just below the fovea. Just as in a digital camera, the photoreceptors convert the light and color information into electrical pulses, which are sent to the brain via the optic nerves. The brain receives the information from the optic nerves and converts it back into what we perceive as images of the world around us.

The basic measurements for lighting include the lumen, which is a quantity of light (think gallon of water); the candela or candlepower, which is a quantity of intensity (think of water pressure), and footcandle, which is a quantity of density (think of water flow). Lumens divided by area gives footcandles; candela divided by distance squared gives footcandles. The range of our vision is much narrower, mathematically, than the range of our hearing (remember 10^{-12} w/cm^2 from Chapter 6?); our eyes are sensitive from just above 0 fc to just above 10,000 fc (outdoor brightness on a bright sunny day), which is a range of 10^5. Indoor lighting tends to range from 0.1 fc to 100 fc, but most of the time that range narrows to 1 fc to 30 fc or so.

The cones perceive what is called **photopic** light (above 2.0 fc) and provide color vision, and the rods perceive what is called **scotopic** light (below 0.2 fc). The rods and cones operate together, but in a manner that varies according to environmental lighting conditions. Under low light (0.2–2.0 fc), our eyes operate in the **mesopic** range, meaning that both the rods and cones contribute, but with limited color vision. The cones (photopic) provide the detailed, and full colored, vision that we are accustomed to, but only under moderate to high (> 2.0 fc) lighting conditions. Under the low mesopic range, we see better due to higher proportions of scotopic light (the blue, or short wavelength, end of the visible spectrum), but the effect is more pronounced under true scotopic vision, which occurs under very low light conditions (< 0.2 fc). This effect can be dramatic, leading to perceived doubling or tripling (or maybe even more) of the photopic level. We have full color vision only under photopic lighting conditions.

The differentiation between scotopic, mesopic, and photopic lighting has become far more important in recent years due to the development of solid-state lighting—LEDs and organic LEDs. There will be more discussion of this later, but, for now, it can be said that most white LEDs produce white light by altering ultraviolet or blue light. Using either method results in a higher "blue" component in the resulting white light than one sees (no pun intended) from other sources (e.g. incandescent, fluorescent, etc.). Human vision is more sensitive to blue light than to other colors (despite red stop signs and yellow-green fire trucks), so we perceive more light in any given situation if that light is blue, or has a high blue component. As a result, we often perceive more light from LED sources than we do from other sources, even if a photopic meter shows no difference in level.

In 1924, the photopic and scotopic lumens were defined by the Commission Internationale de L'Éclairage (CIE) in France, and it was believed at that time that scotopic lumens are insignificant under normal lighting conditions, especially because most sources (lamps) produced relatively few scotopic (blue) lumens. As noted above, all of this has changed due to LEDs and researchers are working hard to quantify the "blue shift" effect. To date, only the bravest (or most foolish) researchers have published any recommendations for the magnitude of this effect but the early indications are that it could be as much as 100%, or a perceived doubling of the light level.

8.3 History

Humans began using artificial light sources thousands of years ago in the form of fire—wood fires, animal fat fires, and oil fires, among others. Such fires could be in the open, on the ends of torches, on candles, or in lamps. The fuel for these fires changed over the years, adding coal, coal oil, charcoal, kerosene, wicks (fibers and fabrics), and eventually natural gas. If it can be burned to generate heat and/or light, someone found a way to burn it. This did not change until the late 19th century when Thomas Edison perfected the incandescent lamp, often called the "light bulb." Edison did not actually invent electric lighting. Electricity was known and somewhat understood in the 18th century (remember Benjamin Franklin's work with keys and kites?) and it had certainly occurred to other scientists that there must be a way to make light from electricity. Edison was simply the first person to find a practical way to generate electric light that was useable on a large scale. The device that he developed consists of a tightly wound filament made from carbon, enclosed in a vacuum inside a glass bulb—the "light bulb" as it came to be known. (The correct technical term is "lamp" and not "light bulb," and the term lamp will be used from here on.) This device operates by passing electricity through the filament until it gets so hot that it glows "white hot," really more yellow/orange for carbon filaments. The filament does not break and fail instantaneously because there is no air inside the bulb. An Edison lamp is literally nothing more than a very hot piece of carbon in a vacuum.

Since then there have been many developments, but the Edison lamp is still with us in 2017. The US Federal Government (via the Energy Policy Act) has been regulating electric lamps since 1995 and more and more old-style inefficient lamps drop out of the market with every increase in minimum **efficacy**. The original technology is still around because it is very low in cost and familiar to everyone, even though it is low in efficacy. In the early days (until the 1920s or so), these lamps were mostly used in "bare bulb" applications, where the lamps hung from cords from the ceiling or in sockets buried into the ceiling or soffit. This is a relatively efficient way to use an incandescent lamp, but there is no control over where the light goes—it simply splashes around in all directions. By the 1920s, some inventors had come up with shades and reflectors, both of which were designed to contain and even aim the light from these lamps. Both types—shades and reflectors—still exist and are widely used today.

Not long after Edison perfected his lamp, it was discovered that filaments made from tungsten metal are more durable than carbon and the tungsten filament incandescent lamp became the standard of the industry (even now, tungsten filament lamps are commonplace). The next development after that was to add halogens (active chemical elements) to the vacuum inside the bulb, which made the light output whiter and extended the life of the filament. Standard tungsten filaments typically last only 800 hours while typical tungsten halogen lamps last about 2,000 hours; standard tungsten lamps produce light that is roughly 2,700 K (yellowish) while tungsten halogen lamps produce light that is roughly 2,900 K (yellowish white). The lumen output of these lamps tends to be 10–15 lumens/watt.

In the 1930s, the fluorescent lamp was invented. This is different because the filament (the hot wire) was replaced by a discharge system. In this case, there is an anode (negative electrode) at one end of a glass blub, a cathode (positive electrode) at the other end, and a small amount of mercury (in its pure form) in the vacuum inside the bulb. When electricity is applied to the cathode, the mercury is vaporized and the electricity flows (discharges) through the mercury vapor. The mercury atoms give off ultraviolet light under these conditions, which is converted to visible white light by the **phosphor** coatings on the inside of the glass bulb. The earliest fluorescent lamps were roughly four times more efficacious than incandescent lamps (at 60 lumens/watt or so) and modern fluorescent lamps have nearly doubled that efficacy again, reaching 100 lumens/watt by the mid-1990s. This large increase in output allowed for increasingly higher level interior lighting than had ever been seen before. Some of the first applications for fluorescent lamps were in department stores, where the stores were able to reduce the amount of window area (and increase the amount of wall display area) and to remain open after (or before) dark. By the 1970s, fluorescent lamps dominated the entire industry, in their various shapes and sizes. The most common shape is designated "T" for tubular; the first "standard" fluorescent lamp was a 40 watt T12 that was 48″ long and 1.5″ diameter; the next standard of lamp (starting in the late 1980s) was a 32 watt T8 that was 48″ long and 1.0″ diameter; the current state-of-the-art fluorescent lamp is a 28 watt T5 that is 46.5″ long and 0.625″ diameter. But fluorescent lamps are available in other lengths (2′, 5′ and 8′), in circles, and in U-shapes.

The latest development in fluorescent lamps is the T4 compact lamp. The earliest ones are called long compact lamps; they are roughly 22″ long and they are very narrow U-shapes (about 1.5″ wide). They are usually called biax lamps and they are some of the highest output options for fluorescent sources (40–55 watts). Later, the small, or short, compact fluorescent lamp was developed, in sizes including 7, 9, 13, 26, 32,

43, 57, and 70 watts. Later still, the "twist" self-ballasted compact fluorescent lamp was developed, which many people have at home.

Research did not stop with the development of fluorescent lamps. Such lamps are far more efficacious than incandescent lamps, but they are unwieldy and they do not lend themselves to certain applications (mood lighting, artwork lighting, stage lighting, even some display lighting). They also do not lend themselves to optical control with shades or reflectors, due to their large sizes and unwieldy shapes.

There was a need for smaller highly efficacious sources, and the first one came along in the 1950s—mercury vapor. In a fluorescent lamp, there might have been 10 milligrams of mercury in a 48″ long by 1.5″ diameter tube, bombarded by 40 watts of electricity. What would happen if the amount of mercury went up, the envelope got much smaller and stronger (now called an "arc tube" and not just a glass bulb, maybe 3″ long by 0.5″ diameter or even less), and the mercury were to be bombarded by a few hundred watts of electricity? One would get a "high-intensity discharge" lamp, usually called HID where there is a powerful arc (or spark) generated directly inside the arc tube. HID lamps are not necessarily more efficacious than fluorescent lamps (some are and some are not) but the arc tubes are much smaller than fluorescent lamps, so shades and reflectors are back into play. In later years, it was discovered that adding metal halide compounds to the arc tube would improve color and efficacy, leading to the metal halide HID (usually noted MH) lamp (and later to the ceramic metal halide—usually noted CMH—lamp where the arc tube is constructed from clear ceramic material instead of glass). Today, metal halide lamps dominate sports lighting, arena lighting, and outdoor lighting.

At about the same time, someone discovered that sodium could be used in place of the mercury to form an arc, and the "low-pressure sodium" (or LPS) lamp was born. This used to be the most highly efficacious of all artificial light sources (at roughly 140 lumens/watt—LEDs have surpassed that now) but it produces ghastly yellow light that cannot be used indoors and which is horrid to use outdoors. The color performance is so bad that colors of cars cannot be distinguished on a roadway illuminated with LPS lamps. (It was used for highway lighting for a while in the 1970s and 1980s but little of it remains today.) In order to address the horrible color of the light, the pressure was increased inside the arc tube, which resulted in the "high-pressure sodium" lamp (HPS). HPS lamps produce yellowish white light and they are still commonly used for roadway, parking, pathway, and other outdoor applications. They are rarely used indoors. (If outdoor lighting is "yellowish," it is probably HPS; if it is "whitish" or "bluish-white," it is probably metal halide, but LEDs are available in a wide range of white colors so LED lighting can look like anything.)

By the 1980s, development was well under way (by Philips and Osram-Sylvania—two of the largest manufacturers of lamps in the world, with General Electric being the third) for "electrode-less" lamps, also called "induction" lamps. Fluorescent lamps fail when the anode or cathode fails; HID lamps fail when the arc tube explodes (often). What would happen if a discharge lamp could be made without electrical parts inside the bulb? It would last much longer. This is done by using a radio frequency generator to create a strong electrical field around a bulb that contains vapor; the electrical field excites the atoms in the vapor, which cause the phosphors that line the bulb to glow. This is a little bit more like fluorescent than HID but it is neither. The one, and only, advantage of induction lamps is lifetime. In a world where incandescent lamps last 1,000 to 4,000 hours; fluorescent lamps last 10,000 to 80,000 hours, HID lamps last 10,000 to 30,000 hours, and LEDs last 25,000 to 100,000 hours (maybe), induction lamps have a proven life of 100,000 hours. That equates to 11.4 years, if burning 24 hours per day; 22.8 years at 12 hours per day; or 45.7 years at six hours per day. This is not forever, but it is getting close, practically speaking. If a lamp can out-last the building where it is installed, that is essentially forever. Not surprisingly, they are very costly, so it is not affordable to use them on a large scale on most projects.

In the 1960s, the LED was invented. An LED is simply an electronic component that emits visible light (or ultraviolet light) when electricity is passed through it. The earliest LEDs were red, and then green, then yellow, and then blue. Once blue LEDs existed, it became feasible to generate white light by blending red, green, and blue light together. Adding light colors, unlike adding paint colors, is a subtractive process, and the combination of red, green, and blue is white, not black. Also, blue LEDs can be filtered, by using phosphors (like in fluorescent, HID, induction, and even incandescent lamps) to generate high-quality white light. LEDs are now the most efficacious sources available, with products on the market that produce 180 lumens/watt. They are also very small so they can fit into very small packages and their small size makes high performance optics—shades and reflectors—far more effective. Optically, LEDs can do things that no other source can do, especially to produce very narrow bands of light.

Over the course of the last 100 years or so, we have also seen the development of neon, cold cathode, plasma, and organic LED technologies, each of which will be covered separately. The early 21st century might be the most exciting possible time to get involved in architectural lighting, not because there is a wide range of options but because LEDs have replaced nearly all of those options with even more flexibility. Here are the basic parameters:

- *Wattage*: 1,650 watt MH sports-lighting lamps are still available but LEDs, from 0.1 watts to 2.0 watts each, can do virtually anything that the other sources can do.
- *Lumen output*: 15 lumens for a 0.1 watt LED upward to virtually no limit, in steps as small as 50 lumens for LEDs (steps for legacy sources were usually several hundred or more)
- *Centerbeam candlepower*: 100 to 150,000 or so was the range for legacy sources; LEDs can duplicate the full range now
- *Lifetime*: 1,000 hours (incandescent) to 100,000 hours (induction and LED, maybe even longer for LED)
- *Color temperature*: 2,700 K, 3,000 K, 3,500 K, 4,000 K, 5,000 K, 5,700 K, 6,500 K (all LED)
- *Color rendering*: 70 to 100
- *Hardware types*: A nearly limitless range of sizes, shapes, colors, designs, etc.

Given all of this, virtually anything is possible.

8.4 Quantity of light

Determining how much light is needed in a given situation is one of the most challenging aspects of interior (and exterior) lighting design. But first, how do we measure light quantity? And what are the units?

The most basic unit of light quantity is the lumen, which is equivalent to a pound of water—just a quantity. But we do not perceive light quantity (lumens). Instead, we perceive light density, so we need another unit, which is the footcandle. A footcandle is defined as the density of light on the inside surface of a 1 foot diameter globe surrounding a candle flame (sort of like water pressure). (In the metric system, the unit is lux. One footcandle is 10.764 lux.) The other unit used for lighting is the "flow rate," so to speak, which is defined as candlepower or candela; more on that under accent lighting later.

How many footcandles do we need? It is very hard to say, but typical levels in commercial buildings tend to vary from 5 fc to 100 fc or so. Office lighting tends to vary from 20 fc to 50 fc or so. Home lighting tends to vary from 1 fc to 50 fc or so. Outdoor lighting should vary from 0.1 fc to 5 fc or so, but it is common to see much higher levels in some parking lots (especially at car dealerships) and at gas station canopies.

The middle-aged state of the author is relevant here. Everyone's eyes degrade with age. There are two major effects: presbyoculosis and the thickening and darkening of the cornea. Presbyoculosis is the gradual stiffening of the eye, which makes focusing more difficult and which is the main reason why most people need reading classes starting in their 40s. The thickening and darkening of the cornea is the more important issue related to quantity of light. Obviously, as the cornea darkens, it transmits less light to the photoreceptors in the eye, which means that one simply requires more light as one ages. It has been shown that the level of light that reaches the retina in average 65-year-old eyes is roughly $\frac{1}{3}$ of the amount that reaches the retina in average 20-year-old eyes. Clearly, such a substantial difference matters.

Much has changed in regard to recommended lighting levels over the years. While there are no laws that govern such things, the Illuminating Engineering Society (IES) has been the most authoritative source for many years. But even here there has been much inconsistency. Here are the IES's recommendations for office lighting (vaguely defined) since 1913:

1913	2 fc
1936	10 fc
1949	30 fc
1966	45 fc
1972	70 fc
1981	40 fc
1990	32.5 fc
2000	30–50 fc, depending on tasks
2011	14–186 fc, depending on age and tasks

How can one reconcile such widely divergent recommendations—2–186 fc—from a single organization? It is not easy. In the earliest days, certainly before fluorescent lighting, it was not really feasible to achieve indoor lighting levels much higher than 10 fc due to the low efficacy of incandescent lamps, so it makes sense that the recommendations were low. As it became feasible, even easy, to achieve higher levels with fluorescent and HID technologies, the recommendations went up. After World War II, when the country was in a "use more energy" mode, everyone was encouraged to use more power for everything, including electric lighting, so the recommendations went up again, again, and again. The oil crisis of 1973 (when OPEC, the Organization of Petroleum Exporting Countries, cut off oil supplies to the US for an extended period) led to greatly increased concern about energy usage for everything (especially cars, but lighting, heating, equipment, and everything else too) and there was incentive to bring levels down in 1981, again in 1990, and again in 2000. The expectation was that levels would be reduced again in the 10th edition (2011) of *The Lighting Handbook* (the official IES publication) but that did not really happen.

In the 7th edition of the handbook (1981), age was added as a consideration for determining light levels. The basic task categories were kept more or less the same, but it was recommended that the level be increased one category for the elderly or decreased one category for the young. This system was abandoned in the 8th edition of the handbook (1990) and it did not reappear in the 9th edition (2000).

But it is back in the 10th edition (2011), along with a far more complex set of categories. Prior to the 10th edition, there were seven categories[1]:

A	Public spaces	3 fc
B	Simple orientation for short visits	5 fc
C	Working spaces where simple visual tasks are performed	10 fc
D	Performance of visual tasks of high contrast and large size	30 fc
E	Performance of visual tasks of high contrast and small size, or low contrast and large size	50 fc
F	Performance of visual tasks of low contrast and small size	100 fc
G	Performance of tasks near visual threshold	300–1,000 fc

The author commonly used category D for most office lighting and has been using 30 fc as the target average level since the 1990s.

Here are the recommendations from the 10th edition of *The Lighting Handbook* (the entire table has been changed over to lux; the conversions here have been done by the author)[2]:

	< 25 years	**25 years < × < 65 years**	**> 65 years**
A	0.05 fc	0.1 fc	0.2 fc
B	0.1 fc	0.2 fc	0.4 fc
C	0.2 fc	0.4 fc	0.8 fc

	< 25 years	**25 years < × < 65 years**	**> 65 years**
D	0.3 fc	0.6 fc	1.2 fc
E	0.4 fc	0.8 fc	1.6 fc
F	0.5 fc	1.0 fc	2.0 fc
G	0.7 fc	1.4 fc	2.8 fc
H	0.9 fc	1.8 fc	3.6 fc
I	1.4 fc	2.8 fc	5.6 fc
J	1.9 fc	3.8 fc	7.6 fc
K	2.3 fc	4.6 fc	9.2 fc
L	3.5 fc	7.0 fc	14.0 fc
M	4.6 fc	9.2 fc	18.4 fc
N	7.0 fc	14.0 fc	28.0 fc
O	9.3 fc	18.6 fc	37.2 fc
P	14.0 fc	28.0 fc	56.0 fc
Q	18.6 fc	37.2 fc	74.4 fc
R	23.2 fc	46.4 fc	92.8 fc
S	34.8 fc	69.6 fc	139.2 fc
U	69.6 fc	139.2 fc	278.4 fc
V	92.8 fc	185.6 fc	371.2 fc
W	139.2 fc	278.4 fc	556.8 fc
X	232.3 fc	464.6 fc	929.2 fc
Y	464.6 fc	929.2 fc	1,858.4 fc

The patterns are clear: the mid-range numbers are usually double the low-range numbers and half of the high-range numbers, which also means that the high-range numbers are four times the low-range numbers. Also, the large number of categories makes this even more difficult.

In theory, the way to use this system is to buy the 10th edition of *The Lighting Handbook* ($595.00/each or $350.00/each for IES members) and find the relevant task, or tasks, in the extensive listings. These listings provide recommendations for vertical and horizontal categories for thousands of different tasks. But even if the task is found on the list, what happens if there is a wide range of ages in a facility—a school classroom where the students could be six years old and the teacher could be 65 years old, or a church sanctuary where there could people of all ages? The only safe thing to do is to use the highest recommended level; however, doing so could make it very difficult, or even impossible, to meet the requirements of the energy code (which are usually based on the use of the space and not on the age of the occupants).

To show how much light is actually needed for basic reading tasks, the author has developed the following footcandle demonstration for students in his lighting design classes:

Footcandle demonstration

Hardware: a blacked-out upside-down recessed downlight mounted to a board with a 120 watt R120 incandescent lamp (roughly 1,500 lumens) and a dimmer
Room: any completely dark room with a light-colored (preferably white) ceiling
Text samples: several samples of text, from 8-point type to 14-point type in black and gray on white, gray, and green backgrounds
Procedure:

- Turn off all lights.
- Turn over the text samples (right to left or left to right, to keep them right-side-up).
- Turn on the lamp at the lowest level possible (usually below 0.01 fc).
- Slowly raise the level until students can read the text samples (take readings at the desks with the light meter).
- Raise the level to the maximum (usually between 1.5 and 3 fc, depending on the room).

Every time this demonstration has been done, students (usually youngish adults) have been able to read the black-on-white text samples (down to 8-point type size) at 0.1 fc or even lower. All of the text samples are readable at 0.5 fc or lower, including black-on-gray and black-on-green, and including the middle-aged (58 in 2017) instructor. So why does the IES say that we need 15 fc (or much more) for reading tasks? That is not entirely clear. The author works in an office with roughly 10 fc on his desk and has for more than 25 years, but he has noticed that he needs a little more light to read very small or low contrast text in recent years.

The 10th edition the IES *Lighting Handbook* is highly authoritative in the sense that much of it was written by some of the most highly educated people in lighting. But that does not necessarily mean that there is solid science between all, most, or even some of the content. There are two main reasons for this: first, scientific research in this area has been very limited historically, and, second, the IES operates on a consensus decision making basis, which means essentially that most of the people on the committee agree. So the footcandle recommendations are really nothing more than an agreement among experts with little science behind it.

It all boils down to what is acceptable to a client. Despite the federal government's stress on energy conservation, the General Services Administration (GSA) still requires 50 fc on desks in offices, which the author has considered excessive for over 20 years. Also, it is important to remember that the apparent brightness is actually more important than the actual brightness. The actual amount of light in a space is truly less important than the perceived amount of light and there can be a big difference between what we think we see and what is really there when it comes to lighting.

This perception issue mostly comes down to the difference between horizontal and vertical light. Horizontal light is light on a horizontal surface (or light that is moving vertically) and vertical light is light on a vertical surface (or light that is moving horizontally). Given that our eyes are in the vertical plane (when we are standing or sitting) most of the time, it should come as no surprise that we see vertical footcandles better than we see horizontal footcandles. The author has a demonstration for this too.

Horizontal footcandle demonstration

Hardware: a powerful flashlight and a light meter

Room: any darkened room, preferably with a white ceiling at roughly 9′ 0″ aff and a dark floor

Procedure:

- Use the flashlight to create a very bright spot on the dark floor (usually more than 600 fc from about 3′ 0″ high).
- Ask students to look around the very dark room. (Are colors visible? Usually not. Faces? Sort of.)
- Flip up the flashlight to put light on the ceiling from approximately 6′ 0″.
- Given that light intensity (candlepower or candela) falls off by the distance squared, the bright spot on the ceiling should be about $\frac{1}{4}$ the intensity on the floor (or about 150 fc).
- Ask students to look around again (the room will appear to be much brighter; faces are visible, as are colors).
- Aim the flashlight at a light-colored side wall.
- Ask students to look around again, noting improved facial definition due to increased vertical footcandles.

What does this demonstration show?

1. That dark colors really do absorb lots of light. The total amount of light in the room is the same whether the flashlight is pointed at the dark floor or the light ceiling, but the appearance of the room changes drastically due to the much higher level of light from the ceiling.
2. That increasing vertical footcandles does increase apparent brightness (and visibility).
3. That a room with a 600 fc spot on the floor can look almost completely dark while the same room with a larger 150 fc spot on the ceiling looks much less dark. (The room does not look bright with the flashlight bounced off the ceiling—after all, it is just a flashlight—but it certainly looks much brighter.)

The point is that a room with high vertical footcandles might be functional even though the horizontal footcandle readings are low, according to the IES recommendations. The opposite is possible too: a room with high horizontal footcandles can appear to be dark, largely due to a very low level of vertical footcandles. (The author has seen this effect in the field very clearly. In a conference room in a community center building, the exterior wall is outward-leaning blue-tinted glass that is 14′ 0′ high, and 300 fc has been recorded on the surface of the conference table. But the occupants claimed that they could not use the "too dark" room. The exposed structure at the top is painted a medium silvery-gray (maybe 60% reflective), the floor is very dark (almost black), the table top is dark (dark walnut finish), and the walls are painted dark brown and dark red. This room would look dark at any fc level, and it does. The situation was resolved by increasing the level on the table even more to create a "bright" spot in the not dark room.)

Back to the client. If the client wants 50 fc average on the desks, give it to them—if it can be done within the limits of the energy code. This might require more efficient (more costly) hardware (see 8.8) but the client will simply have to accept that (and pay for it) if they want to insist on 50 fc. If the client does not know, or does not care, the author would recommend bringing levels down to 30 fc, or less, even when older people are around. (The author recently took fc readings in mock-up rooms at a nursing home in central Indiana. The rooms are widely perceived to be "very bright," bordering on too bright, and the readings vary from 12 fc to 25 fc. It is all in the perception, right?) In lobbies, walkways, even corridors, it might be acceptable to drop down to 5 fc.

Also, human vision is not very sensitive to light intensity, so much so that we can barely perceive a range of 3:1. In other words, if it is 10 fc on the floor here and 30 fc on the floor 5′ away, we can just barely see the difference. For us to see a clear difference, the levels have to be at least 5:1 and 10:1 is much better. Dramatic lighting contrasts are 20:1 or even more. This is critical to accent lighting, but it affects more general lighting too.

And, finally, watch out for contrasts. If the range of levels in a facility is 20 fc to 50 fc and there is a small dark area at 5 fc, the darkness would be noticeable but probably not striking. If the same dark area occurs when there is a 90 fc area nearby, it will look much darker by comparison. So do not forget about context.

It might be necessary for a novice designer to get a light meter and carry it around for a while to get a sense of what brightness levels really are like. There is no substitute for experience. Also, facilities that have absolutely even illumination are boring and dull and do not enhance the experience, whether that experience is work, play, entertainment, etc. But many facilities are evenly illuminated, usually because it is inexpensive and flexible.

The specific recommendations (from the 10th edition of *The Lighting Handbook*) for a few office tasks are as follows:

Reading handwritten pencil: Category P (28 fc, not young or old, horizontal) and Category L (7 fc, not young or old, vertical)[3]

VDT screen: Category N (14 fc, not young or old, horizontal) and Category K (4.6 fc, not young or old, vertical)[4]

Copy room Category M (9.2 fc, not young or old, horizontal) and Category I (2.8 fc, not young or old, vertical)[5]

Clearly, this supports the idea that vertical illumination levels can be lower than horizontal levels.

8.5 General optics

The use of electric lighting is all about how we control the output of the luminaires that have been selected, and this involves the application of the science of optics (a branch of physics). Clearly, this begins with the lamp or source, or lamps or sources, inside (or even outside) the luminaire.

Every lamp—or light source—that has ever been invented has a number of important characteristics:

Shape
Base type
Wattage
Voltage
Filament type (for incandescent and halogen only)
Arc tube size and shape (for high-intensity discharge only)
Physical size (diameter, length, etc.)
Rated life
Initial lumen (photopic) output
Mean lumen (photopic) output
Initial color temperature
Mean color temperature
Color Rendering Index (CRI) rating
Centerbeam candlepower or centerbeam candela
Ballast type
Driver type
Transformer type

Some of these are critical to the optical performance of the lamp: shape, size, filament type, lumen output, and centerbeam candela, and the others are not. In all cases, the only way to transfer 100% of the lumen output of a lamp to a space is to hang a bare lamp in that space, unencumbered by any other hardware. For some lamp types, that is reasonably close to feasible (for example, that bare "light bulb" in a socket on the end of a cord hanging from the ceiling) but it is very difficult for other lamp types (for example, a 4′ long 1″ diameter T8 fluorescent lamp—which would require two sockets and something to hold the sockets in the correct position). Even where it is practical, it is not usually a very good idea because it tends to result in excessive glare (brightness) and inadequate spread.

So innumerable optical systems of shades, reflectors, lenses, and refractors (a specialized type of lens) have been developed by luminaire (fixture) manufacturers for virtually every source type to make it applicable to a broad range of situations.

Let us start at the beginning: the "A-lamp," which is the modern equivalent of Edison's original light bulb. This lamp has a relatively small filament, which is the light source, and a relatively large envelope (the bulb). Sizes are available from $\frac{1}{2}''$ diameter up to $4\frac{3}{8}''$ diameter in wattages from 0.5 to 1,500. The filament emits light in a nearly spherical pattern, but the base of the lamp blocks a large segment, resulting in a truncated spherical distribution pattern from the lamp assembly. If the intention is to use this lamp in a controlled way—say to light a limited area of a wall—then it is necessary to use optics (a reflector around the lamp in this case) to redirect the truncated spherical output into a "cone of light" that runs in a single direction with a limited boundary. Over many years, thousands of such reflectors have been designed, and it is now possible to achieve "cones of light" or beams that range from 6 degrees to over 100 degrees. LEDs make many more options possible than the legacy sources could ever do. If it is not required to have a beam of light from a lamp like this, but there is a desire to conceal the very bright lamp itself, then refractors or lenses or simply shrouds (or shades) can be designed to result in more uniform distribution of light output—or not, at the designer's discretion.

High-intensity discharge lamps (mercury vapor, low-pressure sodium, high-pressure sodium, metal halide, and ceramic metal halide) are similar to the A-lamp in the sense that the arc in the arc tube is relatively small and the lamp envelope is relatively large. This is especially true for high wattage lamps (175 watts to 1,500 watts) which can be quite large. All of the same optical techniques for incandescent lamps apply here too—just on a larger scale for larger scale lamps.

Linear fluorescent lamps brought a new challenge to optical designers. While they are relatively small in diameter (1.5″ or less usually), they are usually long (24″ or more), and the result is a very large lamp. A lamp designer at GE once said that it would be no problem to design an A-lamp-style reflector for a 48″-long T12 fluorescent lamp, but that it would be the size of a medium sized room and there would be little market for such a giant optical assembly. He was quite right about that. So no one has really attempted to design conventional reflector type optics to achieve "cones of light" from linear fluorescent lamps. Instead, much work has been done on optics to achieve linear patterns of light from these linear sources, and these linear optics get better and better as the lamp diameter gets smaller and smaller. Compact fluorescent lamps (which are often less than 6″ long) are more amenable to optical design and good reflectors can be designed around such lamps; they are not necessarily efficient but they can be optically effective.

Optics for induction lamps vary according to which technology is being used: the Philips QL system or the Osram-Sylvania Icetron system. The former resembles a large sized A-lamp and it can use similar optics. The latter resembles a large rectangular donut and it is very difficult to design effective optics for it (à la linear fluorescent lamps). As a result, the Icetron system is mostly used in applications that are not optically sensitive: outdoor area lighting, some parking garage lighting, etc.

Neon and cold cathode (the higher output big brother to neon) generally do not use optics, except maybe to shape the back of a light cove.

Plasma lamps are another very small source, so they can be used with highly effective optics. These are high-output, very small sources, so they are used primarily for outdoor lighting: roadways, parking lots, parking garages, etc.

And then there is solid-state lighting—LEDs and OLEDs. An individual LED is much smaller than any previous light source, including very small incandescent lamps. In fact, a single LED comes very close to a true point source, except it emits light in only one direction—really a half sphere. This very small size allows for optical opportunities never before seen in the lighting industry. For several years now, there have been linear LED products on the market that can achieve 8–10 degree linear beams of light, while the "tightest" comparable fluorescent beam would be 30 degrees or so. This new kind of optical control is certain to change virtually everything we know about electric lighting; it is only a question of how long it takes. Field changeable optics are now available for relatively low-cost LED recessed downlights, which would make it feasible to change a spot to a flood, or vice versa, for $10 or so; that was unimaginable only two years ago!

8.6 Quality of light

Another critical factor is light "quality," meaning clarity, brightness (and glare), and color.

Non-point-source lamps, such as linear fluorescent lamps, produce light without generating strong shadows. If such lamps are used in fixtures for up-lighting (bouncing light off a ceiling or exposed structure), for example, it is possible to produce a nearly shadow-free environment. But shadow-free environments can be odd, and even a little unnerving. On the other hand, spaces illuminated only with point-source type lamps tend to have excessive shadowing. In many situations, there is a need for balancing the degree of shadowing by using a combination of approaches. In other cases, that is not desirable. But in all cases, it is an important consideration that should not be ignored.

In addition to shadowing, different sources produce different qualities of light. The standard frosted A-lamp is frosted because it is intended to produce broad diffuse light. A-lamps are also available without the frosting, and those clear lamps are intended to produce broad light with strong shadows—even sparkle at low wattages when the filaments are intended to be seen directly. Quartz, halogen, and quartz-halogen lamps tend to produce "crisper" white light that is less diffuse than the light produced by typical incandescent lamps.

It does not really matter very much how much light we have if we cannot see to perform the task at hand. Many of the tasks that we perform involve color differentiation and even color matching. Unfortunately, the various light sources (lamps) are not at all alike in terms of the color of the light that they produce.

Sunlight is a continuous spectrum of color from the ultraviolet (short wavelength and invisible to the eye) all the way through the infrared (long wavelength and invisible to the eye—but easily perceived by the body as heat). The reader is encouraged to look up "light color spectrum" on-line to see color representations of various sources. The spectral distribution is a curve that varies continuously throughout the day (and according to the viewing angle to the sun), but it remains a continuous curve. At dawn and at dusk (looking east or west, respectively) that curve would peak in the orange range, with some red and lots of yellow. At noon, facing south, the curve would peak near white, with some noticeable yellow but still substantial amounts of green, blue, etc. At mid-afternoon, facing north, the curve would peak in the blue-white but still with red, orange, green, etc.

Incandescent lamps, which are really nothing more than glowing hot metal (usually tungsten), also have continuous spectral distributions, though more like singular curves with the upper end of the curve in the yellow-white (standard) or white (halogen). But there is always some green, some blue, some orange, etc.

Fluorescent, HID, and LED lamps do not have continuous spectral distributions. Instead, they have highly "spikey" distributions, with three or four distinct spikes at particular wavelengths with a much lower level bumpy curve across the other wavelengths. It is important to note that, even when these sources appear to be very white, certain colors will "pop" out if they happen to coincide with one of the spikes in the distribution spectrum.

For LEDs, there is one more color measurement that is critical and that is called the "McAdam Ellipse." A chromaticity diagram shows all possible colors of visible light but it does not show where the limits are. In other words, just like the 3:1 ratio for the limit of human vision to distinguish light intensity (fc), human vision has a limit for color differentiation too. On the chromaticity diagram, all of the points around a single color that appear to be the same color to most people can be defined by an ellipse (oval). This curve is called a McAdam Ellipse. It is not practical to select LEDs that actually fit within this basic ellipse (because it is too difficult and therefore too costly), so manufacturers use a series of nested ellipses (one within the other) to define the color accuracy of the LEDs that they are using. In today's market, the best color performance available is 2 McAdam Ellipses, which is truly excellent; 3 McAdam Ellipses, which is good, is more common. Products with lesser color control than 3 McAdam Ellipses probably will not mention this measurement at all. If color consistency and accuracy matter, it is important to get LEDs with no worse than 3 McAdam Ellipse performance, if not 2. The earlier version of this measurement is called "binning" where the color spectrum is divided into consistent color regions—literal bins in an LED factory; the smaller the bin, the higher the cost. McAdam Ellipses simply add another level of nuance to the binning system.

In order to deal with variations in color spectrum between various sources, the CRI was developed by the CIE, and it has been revised and updated recently. Essentially, this system establishes nine standard colors and compares the performance of a given lamp to "perfection" for each of the colors. The final CRI rating is a blended average of sorts across all of the colors. While this is a useful system, it is highly imperfect, and it is possible to see sources with the same CRI and significantly different color performance in the field. For the most part, most incandescent lamps (including halogen) are rated CRI = 100, and all other sources have lower ratings (largely due to the spikiness in their distributions). This is unfortunate because there are significant differences in color performance among incandescent lamps themselves (the author believes that halogen incandescent lamps perform substantially better than non-halogen lamps, for example). Generally speaking, a CRI >70 would be considered "fair," a CRI >80 would be considered "good," and a CRI >90 would be considered "excellent." There are fluorescent, ceramic metal halide, and LED sources available at CRI >80, some even at CRI >90.

Color temperature is another important consideration, which also interacts with CRI. Due to the constant variability of sunlight and the wide variations between electric lamps, it is necessary to define the "color" of the light. This is done by using the Kelvin scale, which is the absolute temperature scale developed by the scientist Lord Kelvin.

(It is considered the absolute temperature scale because absolute zero is zero on this scale. If memory serves, absolute zero is –273 degrees Celsius.) So when we refer to a lamp as "3,000 K" that means that its light output closely resembles—in color only—the light output of a theoretical black body that has been heated to 3,000 Kelvin. (In reality, tungsten performs very much like the theoretical black body.)

The full range is roughly 1,500–10,000 K, which represent "red" and "north sky blue," respectively.

The most neutral-appearing white is usually accepted to be somewhere close to 3,500 K, and virtually all fluorescent and compact fluorescent lamps and many LED sources are available in this temperature. (Some people claim that neutral white is as high as 4,000 K, but the author considers 4,000 K to be noticeably blue.) Other standard temperatures include 3,000 K, which appears slightly yellowish, and 4,100 K (for fluorescent; the standard is 4,000 K for LEDs), which appears slightly bluish. HID lamps (except low- and- high-pressure sodium) are usually available at 4,000 K and sometimes 3,000 K (especially for ceramic metal halide lamps). LEDs are available from 2,700 K up to 6,500 K.

The importance of color temperature to interior design cannot be overstated. If the light sources do not produce neutral white light, the color scheme will be affected, and often not in an attractive way. (In the late 1980s, the author was brought into an unfortunate situation where the ID had specified "deluxe warm white" fluorescent lamps (fortunately no longer on the market—they had a distinctive and unattractive peachy-pinkish light output) for a project with a peach/coral color scheme. The result resembled being underwater—if the water were peach colored. The designer thought the lamp selection would complement the color scheme, but in truth the lamps accentuated the colors in entirely the wrong way.) If 4,100 K lamps are selected because they produce light that looks "clean" (and they do) but the color scheme is composed of mostly warm tones (browns, beiges, reds, etc.), they will tend to make the colors look gray. On the other hand, 3,000 K lamps will tend to make cool color schemes look gray. That is why the author has been a believer in 3,500 K lamps since they became available in the early 1990s; 3,500 K lamps are the least likely to distort a color scheme, whether it is cool or warm. Also, it might be inadvisable to match cool lighting to a cool color scheme or warm lighting to a warm color scheme because color schemes change over time—more often than lamps change.

There is much recent interest in "tunable white" LED luminaires, where the exact white can be selected as a static color or where the color of the white can be programmed to change automatically throughout the day. This is related to the desire to make artificial lighting mimic natural lighting (especially important in Alzheimer's disease treatment), but it must be noted that the color scheme should probably be highly neutral and balanced if the lighting could change from 2,700 K to 5,000 K and back again during a single day

Partly due to the contributions of scotopic lumens, even at photopic light levels, our eyes tend to perceive more light at higher color temperatures. As a result, many people advocate using 4,100 K (or even cooler) lamps to make lighting installations more efficient and "brighter" looking. But the author considers this far from resolved and cautions against automatically jumping into the use of 5,000 K or 6,000 K lamps in normal environments. In fact, the author believes that high color temperatures are inappropriate at most typical photopic levels (5–50 fc or so), mostly due to our natural understanding that low light levels (dawn and dusk) equal "warm" light, while high light levels (noon) equal "cool" light. This is an untested hypothesis at this point, and this issue affects our vision differently during the day-time than at night. Again, the efficiency difference for most sources (except LEDs) is almost insignificant between 3,000 K and 4,100 K, so the choice should probably have more to do with general appearance (and possibly the color scheme) than with maximizing efficiency.

Glare is nothing more than excessive unwanted light. It can occur directly—excessive light right in one's eyes (as from on-coming headlights when driving at night), or by reflection—the reflection of excessive light from somewhere else. The total absence of glare usually leads to a perception of inadequate light level—no matter what that level might be. Glare can be defined as "disability glare" where the glare is so excessive that one simply cannot see at all; "discomfort glare," and "reflected glare." Small quantities of glare in a visual field are usually called "sparkle" and they can enliven a scene, especially if a festive atmosphere is desired. But too much sparkle can easily become glare.

Overall, spaces can appear to be anywhere between completely dark and excessively bright; they can be crisp or murky; and they can be comforting or discomforting. The goal of the lighting designer is to control the way the space looks and feels so as to support the overall design goals of the project. This is done through the selection of source types, specific lamps, and specific luminaire types. The simplest way to achieve shadow-free lighting (usually considered appropriate in work environments) is to use up-lighting, bouncing all the light off a light-colored ceiling or structure. This results in highly diffused light that is nearly shadow free, but often with little character or clear sense of space. Also, even if an indirect luminaire is highly efficient (say 90%), bouncing the light off the ceiling causes the loss of at least 15% and more likely 20% of the light, which reduces the net efficiency dramatically. Early in the 21st century, efficiency is a key element in lighting design, which is largely driven by various energy codes that have been adopted around the country. This does not mean that indirect lighting cannot be used, but it does mean that it should be used judiciously and that all other potential approaches should be considered first.

8.7 Sources

8.7.1 Bases

There must be some means of connecting power to every source, or lamp. In many cases, this consists of a combination of a standard **base** and a standard **socket**. Most of this information relates only to legacy sources, but there are some LED retro-fit lamps for a few of the bases, so it is not entirely irrelevant. The base is part of the lamp and the socket is part of the luminaire or fixture. The oldest and most common base is the "medium screw base" but there are a number of others as well, as follows (this list is based on Osram-Sylvania; others manufacturers might vary):

4-pin for circline T5 (circular T5 fluorescent lamps)
4-pin for circline T8 (circular T8 fluorescent lamps)
4-pin for circline T12 (circular T12 fluorescent lamps)

Axial T2 subminiature	A 2-pin base for T2 fluorescent lamps
BA15d DC bayonet	A pin type of base (INC)
BY22d Pin base	A pin type of base (HID)
E10 Miniature screw	The smallest screw base (INC)
E11 Mini candelabra	The second smallest screw base (INC)
E12 Candlebra	The third smallest screw base (INC possibly LED)
E17 Intermediate	An intermediate size screw base (INC)
E26 Medium base	The most common base (INC, HID, CFL, and LED)
E26D Three contact medium	A specialized type of medium base (INC) for 3-way
E26/50x39 Medium skirted	A medium base with a special skirt (INC)
E38 Screw base	Screw base (HID)
E39 Screw base	Screw base (HID)
EX39 Screw base	Screw base (HID)
G4	A 2-pin base with the pins 4 mm on center (INC and possibly LED)
2G 7	A 4-pin base for compact fluorescent lamps
2GX 7	A 4-pin base for compact fluorescent lamps
G9	A 2-pin base with the pins 9 mm on center (INC and possibly LED)
G23	A 2-pin base for compact fluorescent lamps
GX23	A 2-pin base for compact fluorescent lamps
G23-2	A 2-pin base for compact fluorescent lamps
GX23-2	A 2-pin base for compact fluorescent lamps
G24 d2	A 2-pin base for compact fluorescent lamps
G24 d3	A 2-pin base for compact fluorescent lamps
G24 q1	A 4-pin base for compact fluorescent lamps
G24 q2	A 4-pin base for compact fluorescent lamps
G24 q3	A 4-pin base for compact fluorescent lamps
GX24	A 4-pin base for compact fluorescent lamps
GX24 d2	A 2-pin base for compact fluorescent lamps
GX24 d3	A 2-pin base for compact fluorescent lamps
GX24 q1	A 4-pin base for compact fluorescent lamps
GX24 q2	A 4-pin base for compact fluorescent lamps
GX24 q3	A 4-pin base for compact fluorescent lamps
GX24 q4	A 4-pin base for compact fluorescent lamps
GX24 q5	A 4-pin base for compact fluorescent lamps
2G-11 4-pin	A 4-pin base for compact fluorescent lamps
G53 2-pin	A 2-pin base with the pins 53 mm on center (INC)
GU4 2-pin	A 2-pin base with the pins 4 mm on center (INC)

GU5.3 2-pin	A 2-pin base with the pins 5.3 mm on center (INC)
GU10 2-pin	A 2-pin base with the pins 10 mm on center (INC)
GY6.35 2-pin	A 2-pin base with the pins 6.35 mm on center (INC)
Medium side prong	A base with contacts on the sides (INC)
Medium bi-pin for T5	A 2-pin base for T5 fluorescent lamps
Medium bi-pin for T8	A 2-pin base for T8 fluorescent lamps
Medium bi-pin for T12	A 2-pin base for T12 fluorescent lamps
Mogul base	A larger version of the medium base (INC)
3 contact mogul	A specialized type of mogul base (INC)
Mogul end prong	A mogul base with prongs (INC)
Recessed single contact	(INC)
Recessed double contact T8	For T8 fluorescent lamps
Recessed double contact T12	For T12 fluorescent lamps
Screw terminal	(INC)
S14s	(INC)
Single pin for T6	A single-pin base for T6 slimline fluorescent lamps
Single pin for T8	A single-pin base for T8 slimline fluorescent lamps
Single pin for T12	A single-pin base for T12 fluorescent lamps

Most of these bases are useable without tools—pin base friction fit and lamp bases are simply pushed into sockets (sometimes sockets are spring-loaded, but the concept is the same) and screw bases simply screw in. Screw terminal bases require screwdrivers to attach wires to the terminals; such bases are most commonly used for AR111 and PAR36 lamps.

8.7.2 Voltage

Incandescent lamps are available to operate at a number of different alternating current voltages: 6, 12, 24, 30, 34, 75 (direct current only), 120, 125, 130, 145, 230, 250, and 277. Of these, by far the most common are 12 volts for low-voltage lamps and 120V and 130V for line-voltage lamps. To extend lamp life 130V is used. Incandescent lamps see a 100% increase in lamp life when operated at 90% voltage, so 130V rated lamps operated at 120V realize that increase. Incandescent lamps operated at 50% see a 1,000% increase in lamp life, but it is not usually recommended to run these lamps at 50% due to the typical significant color shift to the yellow-orange. Generally speaking, all incandescent lamps should be operated at 90% (or less) to realize maximum lamp life. Typically, these lamp life increases do not apply to LED lamps in the same way, if at all.

Fluorescent, compact fluorescent, HID, neon and cold cathode, induction, plasma, and many solid-state sources will usually operate at single-phase 120V, 208V (2-pole), 277V, or 480V (2-pole). The voltage is determined by the ballast, transformer, or driver and not by the lamp. The same lamps that run at 120V also run at 277V, for example. Higher voltages are often used to account for voltage drop over long distances, so they are especially common for outdoor lighting where long distances (several hundred feet, or more) are commonly seen.

8.7.3 Lamp life

For conventional lamps (everything except LEDs), lamp life is defined as the point at which 50% (of a very large quantity) of the lamps have failed. Even with this definition, there is a wide range. A few lamps fail much earlier than "lamp life" would imply and a few lamps fail much later, but half of them will have failed at the rated "lamp life."

Incandescent lamps have very constant output over their lives, but the output of all other sources starts to drop as soon as they are turned on the first time. The term for this is "lumen maintenance," which means what percentage of output is maintained at the given lamp life rating. For T5 linear fluorescent lamps, lumen maintenance is roughly 90% at rated end-of-life; for metal halide lamps, it is roughly 50%, and 70% is most commonly used for LEDs.

There is no official standard for LED (or organic LED) lamp life, but the most common rating is L-70, which means how long it takes for the output to degrade (drop) to 70% of the initial output. The most common life at L-70 is 50,000 hours, but ratings vary from 25,000 hours to 70,000 hours and even higher. L-80 and L-90 ratings are sometimes published too.

In general, lamp life ranges between 750 hours and 100,000 hours. The life of the old Edison-style A-lamp is 750 hours (or sometimes 1,000 hours or even longer) and that of induction lamps and LEDs is about 100,000 hours. All other sources fall in between these extremes. Halogen incandescent lamps usually lie in the 2,000–3,000 hour range (but 6,000 is possible); compact fluorescents are usually in 8,000–12,000 range; linear fluorescents are in the 20,000–36,000 range; HID lamps range from 10,000 to 24,000 or so; and neon and cold cathode last 30,000 hours or more (sometimes a lot more). Now that all of the legacy sources are fading away (sort of literally), everything about lighting is moving toward much longer source life. Obviously, that is a good thing but it also means that it will be more difficult to convince clients to change lighting hardware in the future, no matter how out-dated it might seem.

8.7.4 Incandescent lamps (not halogen or quartz-halogen)

- *A-series*: This is the first standardized incandescent lamp, hence the "A" designation. It comes in the following sizes and wattages, all with medium bases (mostly frosted, but some clear). Efficacy varies from 10 to 18 lumens/watt (l/W). "Standard" lamp life is usually about 800 hours, but some lamps last considerably longer (especially if 130V rated). (The US Federal Government began regulating minimum efficacy in lighting in 1995, and the minimum threshold has been raised a few times since then. Most recently, the minimum has been set at 18 l/W, which has effectively made it illegal to manufacture a number of old-style incandescent lamps. New versions are already on the market though, just at much higher cost.)

  The nomenclature for lamps is consistent worldwide and is as follows (note this does *not* apply to most LED sources):

- First letter(s) = type of lamp (fluorescent, metal halide, etc.—this is usually blank for incandescent lamps)
- First number(s) = wattage
- Second letter(s) = shape
- Second numbers = largest diameter of the outer bulb, in $\frac{1}{8}''$ increments

For example, 60A19 means that it is a 60 watt A-shaped lamp that has a $\frac{19}{8}''$ (or $2\frac{3}{8}''$) diameter at the largest part of the outer bulb. Lengths vary and are indicated in lamp catalogs as MOL (which stands for "maximum overall length"). F32T8 denotes a fluorescent lamp of 32 watts in a tubular shape that has $\frac{8}{8}''$ (1″) maximum diameter.

Here is a list of all the standard incandescent lamps, from the 2011 Osram-Sylvania catalog:

A15 15, 30, 40, and 60 watts. This is the smallest "A" type of lamp.
A17 38, 57, 71, and 95 watts. This is the second smallest "A" type of lamp.
A19 25, 32, 34, 40, 50, 52, 58, 60, 67, 75, 90, 100, and 135 watts. This is the most common size of "A" lamp.
A21 50, 75, 100, 135, 150, and 200 watts. This is a slightly larger A-lamp.
A21 15/135/150, 30/70/100, 45/90/140, and 50/100/150 watt 3-ways
A23 100, 105, 150, and 200 watts. A large A-lamp.

- *PS series*: The PS series consists of larger and higher wattage versions of the A-series lamps, including PS25, PS30, and PS35, from 200 to 500 watts. PS25 and PS30 lamps use medium bases, and PS35 lamps use mogul bases. Efficacy varies from 18 to 20 l/W.
- *B-, C-, and S-series*: These are decorative lamps in both blunt-tip and candelabra shapes, with various bases and in a wide variety of wattages (both clear and frosted and some textured). Efficacy varies from 4 to 15 l/W or so.
- *R, BR, & ER series*: These are "reflector" lamps that incorporate a very crude internal reflector and a crude lens to achieve very rough "beam control." Some of these lamps were the first ones to go under the 1995 EPAct law, and more of them have been banned since. Efficacy for the remaining lamps varies from 6 to 10 l/W. Lamp life is usually increased to 2,000 hours or so.
- *T-series*: These are tubular incandescent lamps, in sizes including T6, T6.5, T7, T8, and T10, with various bases. Efficacy varies from 5 to 11 l/W.
- *PAR-series*: These lamps are available in PAR38, PAR46, PAR56, and PAR64 sizes, in wattages from 65 to 500 watts with efficacies from 11 to 15 l/W. Lamp life is usually 2,000 hours or so.

8.7.5 Incandescent halogen and incandescent quartz-halogen lamps

Adding halogens (non-metallic elements, including fluorine, chlorine, iodine, and astatine) to the vacuum inside the bulb of an incandescent lamp increases lamp life and light output, while also raising the color temperature of the lamp to the high 2,000s (i.e. 2,900–3,000 K). Lamp life is increased because the presence of the halogens causes the atoms of tungsten that boil off the filament under normal operation to re-deposit right back onto the filament. Using quartz to enclose the filament (instead of plain glass) in combination with halogens increases lamp life and color temperature, while allowing for smaller lamps. Some of the most common quartz lamps are T4 double-ended lamps (in 150, 300, and 500 watts), which are often found in residential pendants and torchieres.

Tungsten halogen lamps are available in some of the same shapes and sizes as traditional (non-tungsten halogen) lamps, but they are also available in shapes and sizes that are not available for traditional lamps. Typical types are as follows:

- *A-series*: 42, 50, 52, 60, 72, 75, 100, and 150 watt A19s at 14–20 l/W and 3,000-hour lifetime.
- *AR-series*: AR70 and AR111 sizes (in millimeter, not $\frac{1}{8}$" increments) are available in 20 to 100 watts at 3,000-hour lifetime. These are unique lamps in the industry because they are indirect incandescent lamps, meaning that the filament is completely concealed by a metal capsule. They also have the tightest beam control available from any lamp. They operate only at 12V.
- *B-series*: 25 and 40 watt B11s at 7.2 and 9.5 l/W and 3,000-hour lifetime.
- *BT-series*: 60, 75, 100, and 150 watt BT15s at 14–16 l/W and 3,000-hour lifetime.
- *F-series*: 60 watt F17 with 16 l/W and 3,000-hour lifetime.
- *G-series*: 40 and 60 watt G25s at 9.5 and 10.5 l/W and 3,000-hour lifetime.
- *MR-series*: MR8, MR11, and MR16 sizes from 20 to 75 watts at 2,000–6,000-hour lifetimes. They usually operate at 12V, but there are special exceptions.
- *PAR-series*: PAR14, PAR16, PAR20, PAR30, PAR30LN, PAR38, PAR56, and PAR64, from 35 to 1,000 watts at 9–20 l/W and 2,500–4,500-hour lifetimes. These are usually 120V or 130V lamps, but there are some 6V and 12V lamps as well. PAR30s come in standard and "long-neck," styles. When using PAR30s, it must be verified if a particular luminaire will work with regular or long-neck lamps.
- *T-series*: Single- and double-ended lamps, from 5 to 1,500 watts at 10–22 l/W and 2,000–4,000-hour lifetimes.

8.7.6 Linear (and round) fluorescent lamps

Fluorescent lamps started out in large sizes and they have been getting smaller ever since. The "industry standard" lamp for many years was the F40T12/CW, meaning 40 nominal watts, $\frac{12}{8}$" diameter (1.5"), cool white (and 48" long). The industry standard lamp today is the F32T8/735, meaning 32 nominal watts, $\frac{8}{8}$" diameter (1.0"), 3,500 K color temperature, and CRI = 78 (and 48" long). Newer and very common lamps are the F28T5/835 and the F54T5/835, which are, respectively, 28 watts, $\frac{5}{8}$" diameter, 3,500 K, and CRI = 85 (and 45.8" long), and 54 watts, $\frac{5}{8}$" diameter, 3,500 K, and CRI = 85 (and 45.8" long).

Most T8 lamps are available at CRI = 85 as well, which is usually designated as "835" in lieu of "735."

Few T12 lamps are still available on the market (including the round versions), and those will be ignored. None of these fluorescent lamps should be used in new projects.

The T5 lamps are 45.8″ long because they were developed in Europe and they are metric sizes. This was originally thought to be a nearly insurmountable problem in the United States due to our very limited use of the metric system, but the concerns have proven to be over-blown and these lamps are now used extensively throughout the US.

All fluorescent lamps require ballasts, which are available in a dizzying array of types and sizes. Ballasts affect the actual power draw and the actual light output of the lamp (both of which can be higher or lower than the nominal ratings), and are available in "Rapid Start," "Instant Start," "Programmed Start," and dimming. Dimming ballasts alone are available in several different types for different control systems and requirements.

All fluorescent lamps are sensitive to environmental conditions, meaning that light output falls in situations both above and below the optimum operating temperature. This effect can be minimized by using special lamps and ballasts and through luminaire design, but it should not be forgotten. For this reason, it is difficult to imagine a successful fluorescent lighting scheme for a large walk-in freezer or for an unusually hot industrial environment. This is true for all sizes and configurations of fluorescent lamps, including compact fluorescent lamps.

Operating temperature can be a major issue, especially for recessed luminaires with high wattage compact fluorescent lamps. This is also related to the photometric testing procedures used by luminaire manufacturers.

8.7.7 Compact fluorescent lamps

After the development of the narrow leg T8 U-lamp, someone (in Europe) had the brilliant idea of bending smaller fluorescent tubes to get more light out of a shorter lamp. This began with T4 tubes bent 180 degrees into a very tight U-shape; these lamps are "long compact fluorescent lamps" and are usually called biax lamps. They come in 18, 24, 36, 39, 40, 50, and 55 watt sizes, with 12,000–20,000-hour lifetimes, and CRI from 80 to 90 or so. They are dimmable, and the high wattage versions represent the highest lumen package (per unit of length) that can be achieved in an "instant-on" lamp.

Then someone started making much smaller versions—called compact fluorescent lamps. The earliest lamps are "single twin-tube" lamps, meaning that they consist of a single bent tube, and they are available in 5, 7, 9, and 13 watt sizes with 10,000 hour lifetimes, and in 827, 830, 835, 841, and 850 color temperatures. These lamps have efficacies in the range of 50 l/W, so they are much more efficient than incandescent lamps. But a 13 watt lamp still generates only 688 maintained lumens, which is too small for extensive commercial use (where the dominant lamp is a 2,900 lumen F32T8).

The early compact fluorescent lamps all used 2-pin bases, which made dimming impossible. The advent of the 4-pin base (and some internal modifications and special ballasts) made dimming practical and effective. Today, nearly all 4-pin lamps are dimmable.

The next step was to put two twin tubes on a single base: the double twin tube. These are available in 9, 13, 18, and 26 watts. A single 26 watt lamp is useable in a recessed downlight for general purpose lighting.

The next innovation was to put three twin tubes on a single base: the triple twin tube, or triple tube. These are available in 18, 26, 32, 42, 57, and 70 watts.

In parallel with the developments above, there was a desire to have compact fluorescent lamps that could be used in conventional medium base sockets, so lamps were developed with integral ballasts (all of the lamps noted above require separate ballasts, as do all conventional fluorescent lamps). These now come in many different sizes and shapes for use in many different situations. Most of these lamps are not effectively dimmable, so care should be taken when trying to use them with existing dimmers.

8.7.8 High-intensity discharge (HID) lamps

In an effort to find a more compact highly efficient source than a 48"-long T12 lamp, it was discovered that higher light output (as well as higher efficacy) could be achieved by using high pressure inside the lamp envelope as compared to the very low pressure inside a fluorescent lamp. All HID lamps share slow starting (usually 5–15 minutes), slow re-starting (usually 3–10 minutes), and limited to no ability to be dimmed.

The first such design was the "mercury vapor" lamp. In this lamp, a mixture of gases with elemental mercury is placed into a very strong glass tube which, in turn, is placed inside an outer glass bulb. The atmosphere inside the arc tube is subjected to a high-intensity electrical field, which generates an electrical arc inside the tube. This arc directly produces visible light, which then emanates from the arc tube in a manner similar to the emanation from an incandescent filament except that the arc is physically larger than most filaments. Mercury vapor lamps last virtually forever; their output simply gets lower and lower and lower until the lamp appears to be off while it is actually still burning. This makes the lamps highly inefficient anywhere near true end-of-life (because the same amount of power is being used to generate a small amount of light). These lamps produce light that is very strong in the blue/green part of the spectrum and they have very poor color rendering (15 to 50, typically). Efficacy is roughly 75 l/W, substantially higher than early fluorescent lamps. The only good application for this technology is to light exterior greenery—trees and shrubs.

The first improvement was to add compounds called "metal halides" inside the arc tube. This was done primarily to improve the color of the light and lumen maintenance. Early metal halide lamps used "probe start" technology for starting (i.e. very slow) and had mediocre to good color rendering characteristics (usually about 65 to 70). The light output is usually described as "white" but it is actually substantially shifted to the blue end of the spectrum. Early lamps were relatively large, running from 175 watts to 1,500 watts.

Over the past few decades metal halide lamps have been greatly developed. The first big improvement was "pulse start" technology, which starts the lamps more efficiently, reducing the start and re-start times (and the overall efficiency of the system). The second big improvement was the development of a transparent ceramic material to construct the arc tube, which greatly improved color performance. Today, some ceramic metal halide lamps have CRI ratings of 94. The third development was to reduce the sizes of lamps, and ceramic metal halide lamps are now available down to 20 watts. They are available in T, PAR, and MR shapes, which made them applicable in a broader range of situations. Even the newest ceramic metal halide MR lamps are still slow to start and re-start though, and they are not dimmable.

While these developments were taking place in the metal halide arena, there was a parallel track in the development of sodium lamps. The first lamps were "low-pressure sodium" lamps, which achieve the highest efficacy of all lamps available more than five years ago: 140 l/W. However, their light output is very yellow and color rendering is very bad, usually rated at CRI = 0.

The first improvement to low-pressure sodium was to change to high-pressure sodium, which lowered the efficacy but greatly improved the color rendering and usability of the lamps. The output of high-pressure sodium lamps appears to be yellow-white, and these lamps are used extensively for outdoor lighting. (HPS is appropriate when lighting buildings having reddish or red-brown colors due to the graying effect of MH on such colors.) HPS was rarely used indoors, except for a special lamp called "White Sun" that was made by Philips.

8.7.9 Induction lamps

The induction lamp is either a cathode-less fluorescent lamp (according to Osram-Sylvania) or a cathode-less HID lamp (according to Philips), but the commonality is that there are no electrical components inside the lamp envelope. The lamps operate by using a special generator to develop an electrical field around, and inside, the lamp envelope, which excites the gases inside the envelope, which causes the phosphorescent coatings inside the lamps to glow (similar to a fluorescent lamp).

There are two basic designs:

1. Osram-Sylvania developed the Icetron lamp (now manufactured mostly in China), which is available in 70, 100, and 150 watt sizes, at 835, 841, and 850 color temperatures (all CRI = 80), at 65 maintained lumens/watt or higher, and all at 100,000-hour lifetime. These lamps are quite large and rectangular in shape and are used in relatively few luminaire designs.
2. Philips have the QL lamp, which is available in 55, 85, and 165 walls, at 2,700, 3,000, and 4,000 K color temperatures (all CRI = 80), all at about 50 maintained lumens/watt, and all at 100,000-hour lifetime (and also manufactured in China). These lamps are shaped like A-lamps, although they are larger, and a number of fixtures are readily available to use them.

8.7.10 Neon and cold cathode sources

Generally speaking, neon can be used only for lighting effects because it has inadequate output for general room lighting. There are some cases where cove lighting can be done using neon (usually in white), but only if it is for an effect and not for general lighting. There are new LED products on the market that can readily substitute for neon.

If curving lamps are required (as can be done for neon) but high output is required, the best option used to be cold cathode (but now it is LED). Essentially, this amounts to custom-made fluorescent lamps. Unlike neon, which is all hard-wired, cold cathode lamps have bases and are mounted into sockets, and are therefore more easily replaced than neon (although replacement lamps would have to be custom made). Cold cathode can have output that approaches that of typical fluorescent and can be used to provide general illumination. Cold cathode can be dimmed, but only with special power supplies. Cold cathode is available in various different white colors and many neon-like colors as well. Lamp life is good at 30,000 hours or so—nearly seven years at 12 hours/day, 7 days/week.

8.7.11 Plasma lamps

Plasma lamps are essentially high-intensity induction lamps that generate light directly from a plasma inside a ceramic vessel; the plasma is bombarded with induction power from a remote radio frequency field. High efficiency plasma lamps generate light at a rate of at least 90 lumens per watt, making them only moderately efficient while being high in cost. In addition, plasma lamps are physically very small, which allows for high performance optics, similar to LEDs.

8.7.12 Solid-state sources

The LED—an electronic device—was invented in the 1960s, but the first version did not provide useable visible light. The first visible light LED was red, then green, and then blue. Given that light coloration is subtractive, and that adding red, green, and blue light in balanced quantities results in white light, white light from LEDs became feasible as soon as the blue LED was invented. But white LEDs now exist as well, using several different technologies.

First, LEDs are so small that three individual LEDs can be grouped on a single chip (one red, one green, and one blue), and generate what appears to be white light.

Second, phosphorescent coatings can be used on lenses over LEDs to emit white light from LEDs that actually emit ultraviolet (UV) light. (This is the process that occurs in fluorescent lamps.) These coatings can be applied to a lens that is directly attached to the diode or on a remote lens. Remote phosphor, as it is called, has become very popular and it has proven to provide more consistent color and output than other technologies.

Third, phosphorescent coatings can be used on lenses over blue LEDs to emit white light. These coatings can be close or remote as well.

The third technique is the most common in the industry.

Luminaires that use LEDs fall into five categories:

1 Red, green, and blue sources for color changing.
2 Red, green, and blue sources for white (oftentimes adjustable, or tunable, white).
3 Red, green, blue, and white sources for color changing, white, or adjustable white.
4 White sources, static or tunable.
5 Single color sources (or RGB locked to a single color).

The development of LEDs for general lighting in buildings is moving at an extremely rapid pace. The first general-use fixtures came onto the market in 2003 or so, and were limited to specialty applications such as wall grazing or floodlighting. Claims that LEDs do not generate heat have been made, but that is completely false. LEDs actually generate *more* heat than some other types of lamps, but they do it in a completely different manner. In an old-style incandescent lamp, 75% of the energy put into the lamp exits the lamp as heat, in parallel with the useable light; in other words, the lamp projects extensive IR (infrared) energy along with the light. LEDs emit no IR energy at all, especially not along with the light. So there is no perceived heat in the light beam emanating from an LED. But LEDs generate a large amount of heat at the junction where the LED is attached to its circuit board (on the back of the LED), and it is vital to remove heat from that point to allow the LED to develop high efficacy and long life. There is so much heat involved in this junction that it is necessary to use completely different heat management strategies in LED luminaires than is required for other sources. This often takes the form of large heat sinks that might or might not be incorporated into the luminaire's housing (outer enclosure).

When the junction temperature is well managed, LEDs can achieve extremely high efficacy (in theory up to 300 l/W but still about 180 l/W in practice, as of early 2017) and long life (most commonly about 50,000 hours at 70% light output, but beginning to creep up), making them attractive sources for virtually all applications.

The very small physical size of an LED creates optical opportunities never before seen in the lighting industry. Optical systems have been very well developed over the past 100 years for incandescent, fluorescent, and HID lamps, but those sources are many times larger than LEDs. This means that the existing optical systems (lenses, refractors, louvers, and reflectors) are too large to use with LEDs and that it is possible to develop optical systems for LEDs that can achieve effects never before possible with the larger sources. This means that we are beginning to see fixtures with many more beam options for accent lighting luminaires and many more optical patterns for linear and troffer luminaires.

8.7.13 Transformers, ballasts, and drivers

Low-voltage incandescent and halogen incandescent lamps require transformers to reduce voltage, usually from 120V to 12V, but sometimes from 277V to 12V or from 120V to 6V, etc. Transformers are required for neon to increase the voltage from 120V to 600V.

Transformers cannot be 100% efficient—they usually use 3–5% of the power that runs through them (mostly in the form of heat)—so all systems using transformers use more power than the nominal rating of the lamps.

All fluorescent and HID lamps require ballasts to operate. Such ballasts provide starting currents and stable power to operate lamps after they start, and the ability to dim for some fluorescents.

All LEDs require drivers to provide stable DC power (some AC LEDs are coming to the market but that just means that the driver components are on the same circuit board as the LEDs themselves) and to enable dimming.

When dimming incandescent lamps, special devices are required to vary the voltage in order to change the light output of the lamp.

All of these devices—transformers, ballasts, drivers, and dimmers—use some power on their own. The amount of power they use is shown in the power factor (for ballasts and drivers) or simple efficiency (for dimmers).

A power factor of 0.88 (the standard back in the ballast days) for a ballast/lamp combination means that the ballast uses 12% (1.0 – 0.88) of the power for itself. So a standard ballast would use 1.12 × 32 = 35.8 watts to drive an F32T8 lamp at standard output. Ballast factors vary from 0.70 to 0.95 or so, so there is a wide range of performance available. Naturally, high power factor ballasts are more costly.

Drivers are similar but there are no standard ratings, so it is necessary to verify the power factor on a case by case basis. And dimmers are similar to drivers in this sense.

8.8 Energy conservation and sustainable design

Several factors affect energy efficiency and sustainability in lighting design:

1 *Illumination levels*: The easiest way to save energy in lighting is to use less light when the lights are on. This means pushing footcandle levels down to, or even below, the current official IES recommendations, supplementing with highly efficient task lighting where necessary. No more 100 fc, 75 fc, or even 50 fc workspaces and no more 30 fc or even 15 fc corridors and lobbies. (These recommendations are the author's opinions, and other designers and lighting experts could well have different recommendations.) This is trickier than it sounds due to the perception issues involved. Here is an example.

   In 1992, the author designed lighting for a large cardiology office (40,000 sf) that had a waiting room that could seat over 100 people. The ambient lighting system consisted of recessed very low glare radial-louvered compact fluorescent downlights and the target level was 10 fc or a little less. The downlights were carefully arranged to light the fabric-covered perimeter walls as well. The reception desk was illuminated to roughly 30 fc by a concealed double ring of cold cathode at the base of a shallow dome. In order to keep the waiting room, with its relatively low and dark (meaning un-illuminated—it was white) ceiling and dark blue carpet, from appearing to be a dark hole, the author told the IDs that they had to use table lamps scattered throughout the room. He told them that they could use any table lamps that they liked as long as they had luminous shades. The reason for this was to provide a minimal amount of glare—and vertical illumination—to prevent a sense of darkness in the room. The result was highly satisfactory, and no one ever commented about the relatively low light level in the waiting room—which is especially notable because cardiac patients are usually middle-aged and older where low light levels might be unusually troublesome.

2. *Lamp efficacy*: Here, it is best to use the most highly efficacious sources that are practical for the project—meaning the best LEDs that are affordable. From a sustainability point-of-view, many early LED boards used lead solder, which was very bad, but this has been eliminated almost entirely from the market. But LEDs with lower output and lower color performance are less costly, so one does need to be careful.
3. *Luminaire efficiency*: This used to be a major point in luminaire selection because there was a wide range of performance, but this has all changed with LED products. There is still a range (noted under (2) above) but that range is smaller and more controllable than ever before.
4. *Lighting control*: This is where the real pay-back can be found. Even if levels are appropriate, sources are highly efficacious, and luminaires are highly efficient, it remains highly wasteful to leave the lights on when they are not needed. Even today, it is not uncommon to see lights turned on at the start of the business day and left on all day. In some instances, they are left on until late at night, after cleaning crews have done their work. In other instances, the lights are turned off at the end of the business day. But even during business hours, many rooms are not actually in use throughout the day, and the lights could be turned off. This requires automatic lighting controls of some type: local occupancy sensors, local daylight sensors, or centralized computerized control systems. Occupancy sensors turn the lights off when no one is in the room and then back on automatically whenever someone enters the room. Vacancy sensors turn the lights off when a room has become unoccupied but they do not turn the lights on automatically if someone enters the room (this is required in California and a few other places with advanced energy codes). Daylight sensors can be used to dim or to turn off (some or all) electric lights when daylight levels are high. Centralized control systems can combine local sensors and controls into a single, trackable system. Centralized systems can also use scheduling to turn on and off lights that do not have occupancy sensors. Centralized systems can be operated from computers, wall panels, the internet, wireless hand-held remote controls, and even PDAs and smart phones, depending upon the specific vendor, the budget, and the options that are desirable.

Most modern energy codes (including California's Title 24, the International Code Council's 2003, 2006, 2009, 2012, and 2015 International Energy Conservation Codes, and the American Society of Heating, Refrigeration, and Air-conditioning Engineers' (ASHRAE) 90.1-2004, 90.1-2007, 90.1-2010, 90.1-2013, and 90.1-2016) require limited automatic lighting control, in most facilities. ASHRAE 90.1-2004 and 90.1-2007 require only automatic on-off for buildings larger than 5,000 sf, but the 2009 IECC (and ASHRAE 90.1-2010) also require multiple level control. It is necessary to be very careful to determine exactly which code applies in each case to make sure that only applicable provisions are covered. Of course, any given owner might decide to do more than what is required by the applicable code, but that should be the owner's decision and not the designer's.

Finally, remember that indirect lighting is probably less efficient (in most cases) than state-of-the-art direct technology. The reason for this is that the light has to be bounced off a ceiling, which can never be a highly efficient reflector. Most typical ceilings have a reflectivity of about 80%. So if a 100% efficient LED (that sounds impossible but it is true given the way that such luminaires are tested) indirect luminaire is installed under such a ceiling, the net efficiency of the system (luminaire and ceiling reflector) would be $1.00 \times 0.80 = 0.80$ or 80%. Eighty percent efficiency is very good, but it is less than 100%. That does not mean that indirect designs should never be used, but they should be carefully considered to make sure that energy is not wasted just because someone wants indirect lighting.

Energy can, and should, be saved in lighting by using one or more of these techniques because lighting is the single largest user of electricity in many buildings. So use lower levels where justifiable, highly efficacious sources, highly efficient luminaires, and automatic controls to turn off (or dim) unneeded luminaires—or some combination thereof. Any improvement will help a little.

Under most codes, new installations are required to meet current rules, so one should plan on using only energy code-compliant hardware and controls for new work. When doing renovation work, it is often not clear exactly what is required. It is common for jurisdictions to require total replacement and full compliance if more than 50% of the existing luminaires are to be replaced, but that would have to be verified jurisdiction by jurisdiction. One cannot go wrong—except with the owner's money—by bringing a project up to full compliance, but owners will sometimes resist such extensive work if the applicable code gives them an out.

8.9 Lighting design by layers

This is where real lighting design starts. In any space, in an ideal world, there would be at least four layers of light:

1. ambient (general);
2. task;
3. focal (accent);
4. decorative (some people omit this as a valid layer, but the author chooses not to).

In addition, in many situations, the task, focal, and decorative layers might have numerous sub-divisions within them due to the desire for variable effects. The ultimate example of this is theatrical lighting, where each individual lamp is separately controlled in order to achieve a broad range of effects and changing effects throughout the course of a production.

In architectural lighting, it is rarely practical to control each lamp—mostly due to the very high cost involved—but real systems vary considerably from incredibly simple to extremely complex, depending on the needs of the project and the budget.

All designs come down to the design and implementation of one or more of these layers. Effective designs can be simple or highly complex, but the key is to find a way to relate the lighting design to the interior design (and/or the architecture) and the use of the space. One would never light a high-end jewelry store the same way that one would light a warehouse discount store, a factory, or a concert hall. Good lighting design should "fit" the design concept of the space or building while also meeting functional needs. Even though good lighting might draw attention to itself, it should never do so at the expense of the overall design.

8.9.1 Ambient (general) lighting

Ambient (general) lighting is used to make spaces safe to move around and to provide for the typical basic functions that are required—office, warehouse, factory, etc. By definition, ambient lighting is intended to be relatively even throughout the space, although there are exceptions where this is not true. In the past, this would have meant "general task lighting" as well, meaning that the ambient lighting system would be all that is required in a typical office setting. But, today, as we drive down energy use farther and farther, this is beginning to change to a situation where the ambient lighting truly is minimal and additional lighting is provided for specific tasks. Ideally, ambient lighting should be controlled by daylight and/or occupancy sensors, but there are situations where owners would not want to see the lights going off and on when a space is unoccupied—it might not be a good plan in some retail stores, for example, where it can be important to see the merchandise from outside the store, which will not work if the lights are turned off.

8.9.2 Task lighting

Task lighting is provided specifically for tasks—writing, reading, typing, sewing, cooking, horse-shoeing, taxidermy, surgery, etc. This can be provided by many different types of fixtures with many different sources, but it must be pointed out that it accomplishes nothing from an energy savings point-of-view if it does not save energy. So we tend to avoid incandescent task lighting to the greatest degree feasible (with surgical lights being a prime exception—even these are being replaced with LEDs now). Today, there are many good compact fluorescent fixtures for task lighting on the market but also a number of even better LED fixtures. The latter would be preferred. Whatever the task lighting is, it must be separately controlled from the ambient lighting. Even if the ambient light is to be left on for some reason, there certainly is no reason to have task lights on if no one is there. Task lights can be controlled by occupancy sensors.

8.9.3 Accent (focal) lighting

Accent (also known as focal) lighting is lighting that is used to draw attention to artwork, displays, and signs in many different environments, and merchandise in retail stores. Accent lighting can be successful only if it is at least five times the brightness of the surrounding ambient level; 10 times is much preferred. This is another argument to keep ambient levels low. Accent lighting in a 30fc ambient environment would have to be 300fc—which can be done easily enough, but doing so would be wasting energy. It is better to lower the ambient level to 10fc and to use 100fc accent lighting. But, again, this is a design and perception question, and the installation will not be successful if users believe the space to be "too dark" no matter what the levels are. Accent light can be accomplished from above a ceiling, by using recessed adjustable fixtures (single, multiple, and even linear); from below a ceiling by using surface-mounted adjustable fixtures (including track light), and without ceilings by using suspended adjustable fixtures (including track light). Focal lighting requires the use of controlled beams, which can be provided from the lamps themselves (PAR, MR, and AR, for example) or by the fixture's internal optics (A and capsule incandescent lamps, or LEDs, for example).

8.9.4 Decorative lighting

Decorative lighting is lighting that does not actually light a space in a functional sense, but which provides decorative effects such as sparkle or high brightness (usually associated with special materials—polished metals, translucent glass and plastic, stone, iron, etc.). Decorative luminaires can be used to provide vertical illumination (as in the example of the table lamps in the waiting room noted previously) or to increase apparent brightness in a space. Decorative lighting can be used to reinforce or complete a design concept as well.

8.10 Daylighting

The most energy effective way—free!—to light a building interior is with the sun—by allowing daylight into the space in such a way as to make electric lighting completely unnecessary. In reality, this is actually quite difficult to do, especially if buildings are very thick (more than 50′ 0″ or 60′ 0″ or so), but it can be implemented at the perimeter even in very thick buildings. Typical high-rise office buildings built since the 1970s are often 100′ 0″ thick (and about 200′ 0″ long—to achieve the "standard" 20,000gsf per floor required by the real estate market). It is difficult to get daylight to penetrate very much more than 25′ 0″ into a building (unless ceilings are very high), so half of such a footprint cannot be daylit. But that does not mean that it would not be a good idea to use daylight in the rest.

How is daylight introduced into a building? Through glass windows, preferably clear glass windows. (Many architects use tinted glass in glassy buildings, but tinted glass usually compromises the effectiveness of daylighting.) This often requires lots of glass, but that varies according to latitude and the amount of sun in the local climate. Daylighting a building in sunny Phoenix, Arizona would require much less glass than in very cloudy Portland, Oregon.

There is a risk of over-lighting from daylighting. On sunny days, daylight commonly reaches 10,000 fc in the exterior. If the windows in a building let 50% of the light through, that is 5,000 fc—far more than anyone would want in most spaces, so it will probably be necessary to use shading devices at certain times of day in daylit buildings.

The most effective way to get daylight deep into buildings is to use light shelves, which are devices that are built below the ceiling and above the view window, with another window above the shelf. The top surface of the shelf bounces light deep into the building, increasing overall daylighting effectiveness. Of course, such devices are costly and drastically change the architectural look of a building, so they are infrequently used.

Skylights (a horizontal, or sloping, opening in a roof) can be used too, but there is a risk of getting too much direct sun into the building. If skylights are used, it is usually advisable to use non-clear glazing to limit the direct sun penetration. Products are available with light tubes to direct light from a skylight into a space some distance below; such devices are commonly used in one- and two-family dwellings to transfer the daylight through an attic space (or even an upper floor).

Roof monitors, with or without light wells, are another excellent technique for bringing daylight into the core of a thick building. A roof monitor is a shaft that extends above the roof, with windows (sometimes called clerestories, depending on the exact configuration) in one vertical side. The south side is commonly used, but north sides are used too, especially for art school studios, where north light is highly desirable. Light wells are simply vertical openings between floors inside buildings, which can be used to spread daylight from skylights or roof monitors.

8.11 Luminaires and controls

There are thousands and thousands of different luminaires on the market, from $5.00 track heads to $100,000.00 gold-plated pendants to no-limit full-custom designs. Much of the work of a lighting designer is to understand the general types of luminaires and to recognize which manufacturer makes which type of product. It is always helpful to see luminaire samples in person to develop an appreciation for the differences between $50.00 luminaires and $500.00 luminaires.

Luminaires fall into four broad categories:

1 direct (all light downward, with no light projected above 180 degrees);
2 indirect (all light upward, with no light projected below 180 degrees);
3 direct/indirect (most light downward with some light upward);
4 indirect/direct (most light upward with some light downward).

And three basic types:

- suspended (hanging from a ceiling or open structure);
- surface-mounted (attached to a ceiling or structure);
- recessed (into a ceiling, wall, floor, or other surface).

The myriad ways these can be combined result in a dizzying variety of products to choose from; from surface-mounted direct to recessed indirect; suspended direct/indirect to surface-mounted indirect/direct. And from tiny (the end of single fiber for fiber optics) to room sized, as in a luminous ceiling.

And there are multiple "grades" of products, from architectural grade at the high end, to specification grade in the middle, to economy (sometimes called commodity) grade at the bottom. Recessed LED downlights illustrate this quite well: some single manufacturers make low-end fixtures that cost roughly $50.00/each, mid-range fixtures that cost $150.00/each, and high-end fixtures that cost $300.00/each—from the same company! One can only imagine the complications when comparing one manufacturer to another.

The author designs and specifies extensively in both commercial and residential projects, but he sometimes struggles with residential products. In the residential world, price is all-important and all the products tend to be relatively low in quality (although they are often sold at high mark-ups). In the commercial world, low-quality products are available but so are mid-range and high-end products and high mark-ups are rarely seen. (Most commercial products are priced to include 5–10% for the sales representative and 10% or so for the distributor; the contractors add on more, of course.) As a result, the author usually specifies commercial products for residential projects, often at higher cost (justified by higher quality and better performance).

As an industry, lighting breaks down into "very large" players and small players, with few in the middle. There are only four major conglomerates in the United States in the lighting business: *(Note: the inclusion of the following manufacturer names is in no way any kind of endorsement of their products; this information is provided simply to provide the reader with an orientation to the makeup of the US marketplace.)*

- *Acuity Brands Lighting*: This conglomerate, with headquarters in Conyers, Georgia (near Atlanta) markets fixtures under the following brand names: American Electric; Antique Street Lamps; Carandini; Gotham Lighting; Healthcare Lighting; Holophane; Hydrel; Juno Lighting Group; Lithonia; Mark Architectural Lighting; Peerless; Winona Lighting (and six more brands for controls or components)
- *Eaton Lighting*: This is another large conglomerate with headquarters in Peachtree City, Georgia (also near Atlanta) that markets fixtures under the following brand names: Ameritrix; AtLite; Consumer Products; Control Systems; Corelite; Ephesus; Fail-Safe; Fifth Light; Greengate; Halo; Halo Commercial; iLumin; InVue; io LED; Iris; Lumark; Lumiere; McGraw-Edison; Metalux; MWS modular wiring systems; Neo-Ray; Portfolio; RSA; Shaper; Streetworks; Sure-Lites; Zero 88; (and 15 controls and components brands)

- *Hubbell Lighting*: This is another large conglomerate with headquarters in Columbia, South Carolina that markets fixtures under the following brand names: Alera; AAL (Architectural Area Lighting); Beacon Products; Columbia; Compass Life Safety; Dual-Lite; Hubbell: Control Solutions; Hubbell: Industrial; Hubbell: Outdoor; Kim Lighting; Kurt Versen; Litecontrol; Prescolite; Precision Paragon (P2); Progress Lighting; Security Lighting Systems; Sports Lighter Solutions; Sterner; Whiteway
- *Philips*: This is a very large Dutch conglomerate that is also one of the large lamp manufacturers and which now markets in the US under the following brand names: Chloride; Color Kinetics; Daybrite; Gardco; Hadco; Ledalite; Lightolier; Lumec; Optimum; Stonco; (and six controls and components brands)

Here are some other brands that are worth knowing about (this list represents only the author's opinion, which is based on his direct experience with all of the companies on the list). All companies are in the US unless noted otherwise:

3G Lighting
Advent/SPI
A-Light
American Glass Light
Architectural Lighting Works (ALW)
ARK
Artimede Lighting (Italy)
AXIS Lighting (Canada)
Baldinger: high-end decorative luminaires
Bega Lighting US
Beta-Calco Inc. (Canada)
B-K Lighting
Boyd
Bruck Lighting (sister company to Wila Lighting)
Condaz
Contech
Delray
DeltaLight (Belgium)
Designplan
Donovan Lighting
Edison Price
Electrix
Elliptipar/The Lighting Quotient
ELP
ERCO (Germany)
Fabbian (Italy)
Finelite
Flos (Italy)
Fluxwerx
Focal Point
Hess America (Germany)
iLight Technologies
Intense Lighting
iTRE (Italy)
Kenall
Kirlin Lighting
LBL
Linear Lighting
Litelab
Liton
LSI
Lighting Services Inc. (LSI)
Lucifier Lighting
Luminis
Manning
Martini (Italy)
National Cathode
Nessen
Nexxus
OCL (Original Cast Lighting)
Peachtree
Pinnacle
Louis Poulsen (Denmark, now a part of Targetti)
Prisma
Prudential
Pure Lighting

Selux (Canada)
SistemaLux
Solavanti (Italy)
Spectrum Lighting
Targetti
Tech Lighting
Translite-Sonoma
Visa Lighting
WE-EF
Wila (sister company to Bruck Lighting)
H.E. Williams
XAL (Xenon Architectural Lighting) (Austria)
Zumtobel US (Germany)

8.11.1 Troffers

What is a **troffer**? A troffer is a large-scale recessed luminaire, usually designed for use in a suspended lay-in (grid) ceiling. The earliest troffers were 2′ × 4′ units, but many other variations are available today. Now, such fixtures are available in 2″, 3″, 4″, 6″, 12″, 20″, 24″, and 48″ widths and 24″, 36″, and 48″ lengths. Some unusual round troffers are also available, from 20″ diameter up to 72″. Troffers come in a number of different versions:

- *Flat white lensed*: This is an LED-only product that may be edge-lit (with LEDs around the edges) or lit by a large number of low-output LEDs directly behind the lens. These luminaires are smooth and simple but they tend to have poor optical properties.

Front

LED board
Driver compartment
LED section

Figure 8.11.1.A
2′ × 4′ LED flat-lensed troffer

- ***Prismatic-lensed***: This is the lowest cost and simplest luminaire in the industry. It consists of a housing (a white painted box), 1–6 fluorescent lamps (T5 or T8) or an array of low-output LEDs, and a flat prismatic lens. There are several different types of lenses, but the standard, and lowest cost, is the A12 pattern. It is highly advisable to specify $\frac{1}{8}$″-thick minimum A12 lenses to avoid lens sag over time. The least costly luminaire (at $65.00 or so for commodity grade), of all of the options, is a 2′ × 4′, 3-lamp (F32T8/735) luminaire with a standard A12 lens (not $\frac{1}{8}$″ thick) and a single ballast. All other types and sizes (whether smaller or bigger) are more costly. These are available in commodity and "spec" grade.

Figure 8.11.1.B
2′ × 4′ prismatic-lensed troffer

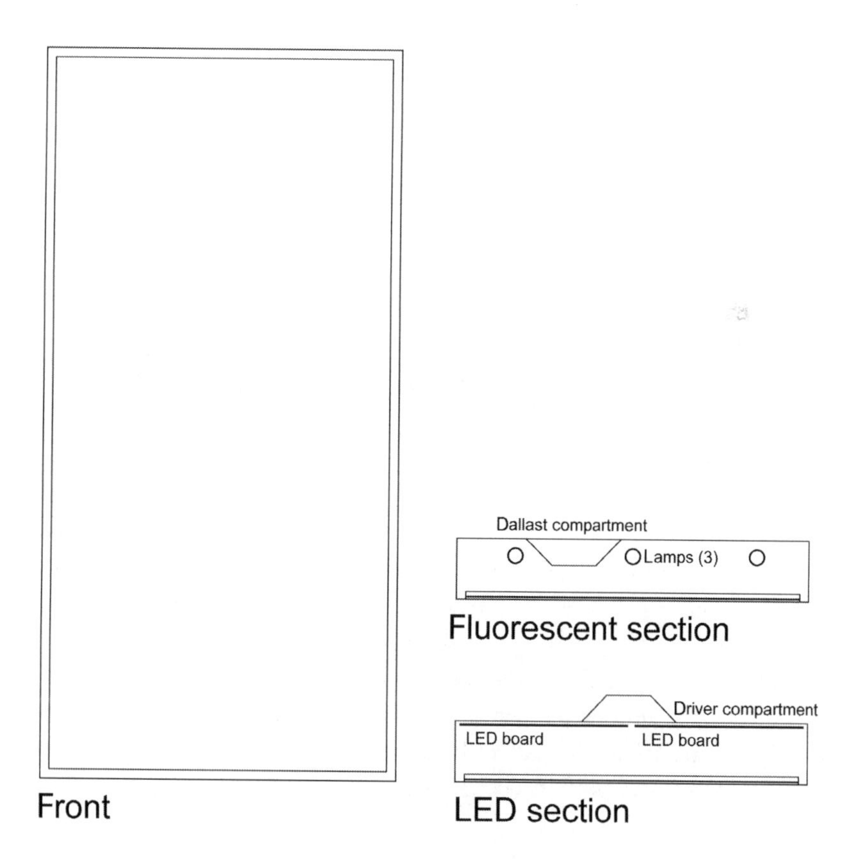

- *Deep cell* ***parabolic***: These are available in 12″, 20″, and 24″ widths, and 24″ and 48″ lengths. The face louvers are available in 2″, 3″, and 4″ depths. This technology was invented back in the 1960s in an effort to decrease glare by reducing surface brightness at the ceiling. These fixtures are designed to emit light in a controlled 90°-wide pattern, from 45° downward from one side of the fixture sweeping around to 45° downward from the other side. The only difference between a really good deep parabolic louver and a bad one is the tightness of this distribution: really good fixtures have very clear lines at the edge of the light beam and bad fixtures have fuzzy lines. But all of them follow the 45° lines. When VDTs (video display terminals) became dominant in offices in the 1980s, these luminaires were used to reduce glare on the shiny curved screens, which were highly subject to reflected glare problems from luminaires in the ceiling. Using deep

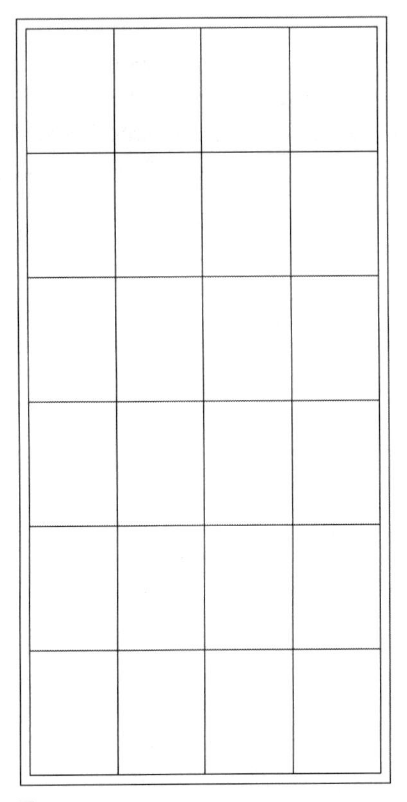

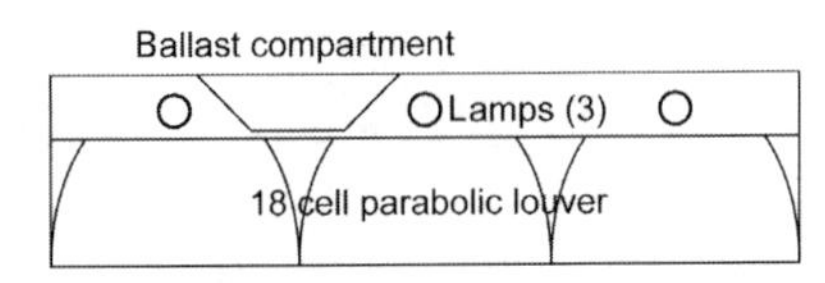

Figure 8.11.1.C
2′ × 4′ deep parabolic troffer

parabolics does reduce glare, but it also results in dark ceilings and dark upper walls, or what is called the "cave effect." This means that rooms illuminated with deep parabolic troffers tend to appear to be darker than they really are, mostly due to the low percentage of vertical footcandles. Luminaires like this are also usually moderate in efficiency (50–60%). It is quite common to put two ballasts into a 3-lamp luminaire to enable 1-lamp, 2-lamp, or 3-lamp operation, from two switches. This luminaire is extremely common in commercial office spaces but it should no longer be specified.

- *"Basket" troffer*, also known as *recessed indirect troffers*: These are available in 12″ and 24″ widths, and 24″ and 48″ lengths. The purpose of these luminaires is to reduce the cave effect without going to fully lensed luminaires (which are considered highly unattractive, even though they are actually quite effective). In the original design, the basket was below the main luminaire housing, or below the ceiling, which provides a better result because it reduces the brightness on the reflectors in the top of the luminaire housing. But such a design is costly to manufacture and ship and the "basket" soon moved up into the main luminaire housing. In recent years, a new generation has been developed with added lenses to reduce the brightness (the brightness of the reflectors in basic basket troffers is actually higher than the brightness of prismatic lenses; even though this actually increases glare, most people seem to prefer the look of basket troffers to lensed troffers). The most common luminaire basket troffer is a 2′ × 4′

Figure 8.11.1.D
2′ × 4′ basket troffer

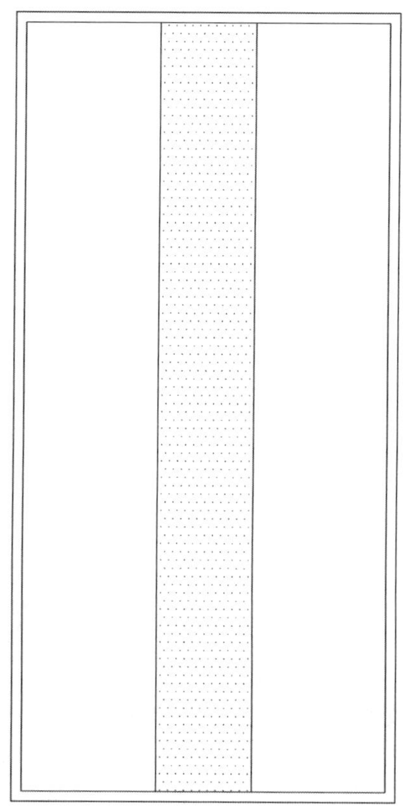

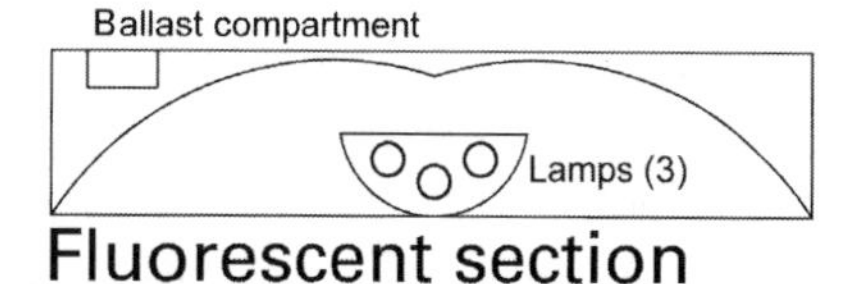

luminaire with three lamps (F32T8/735) with a perforated basket to provide minimal direct downlight and a single ballast. All other options—T5 lamps, different baskets, multiple ballasts, etc.—add costs. There are two reasons not to use these luminaires. First, they usually have moderate efficiency (50–70%), and second, they tend to collect dirt and trash in the open baskets. It is quite common to put two ballasts into a 3-lamp luminaire to enable 1-lamp, 2-lamp, and 3-lamp operation, from two switches. Available variations include 1-sided basket, 2-sided baskets, and 4-sided baskets. True basket troffers are not available with LEDs but LED luminaires that appear to be lensed-indirect troffers are quite common.

Lensed-indirect troffers really started to catch on by 2010 or so, when prices (for 2 × 2′s) dropped to $125.00 or so and the inefficiency of other types began to make energy code compliance a challenge. Many fixtures now exist in this category, including LED versions from a number of manufacturers.

- *Volumetric troffer*: Volumetric lighting was developed by Lithonia (a brand of Acuity Brands Lighting) to reduce the cave effect, increase vertical illumination, and increase luminaire efficiency all at once. Lithonia's luminaire was called the RT5 and it used only F28T5 and F54T5 lamps. It is a lensed luminaire that looks a little bit like a double basket troffer. The luminaires are available in 12″ and 24″ widths and 24″

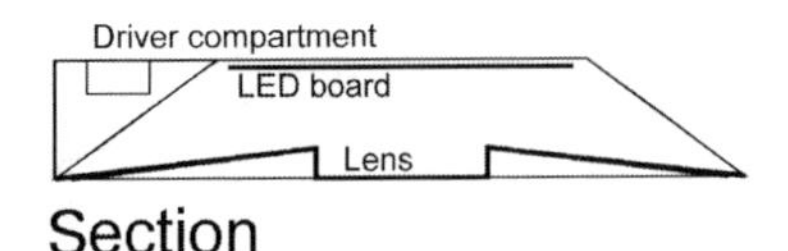

Figure 8.11.1.E
2′ × 4′ LED "indirect" troffer

and 48″ lengths. The most common luminaires are the 24″ × 24″ 2-lamp (F17T5) with step ballast and the 24″ × 48″ 2-lamp (F28T5 or F54T5) with step ballast. Each lamp has its own reflector and lens, so it is impractical to use two ballasts to achieve multiple levels; this is why Lithonia developed the "step ballast" which allows for high/low switching of both lamps simultaneously in a 2-lamp luminaire. Since the original 2-step ballasts, there are now ballasts with multiple steps too. These luminaires are extremely effective in all three of their goals. The cave effect is entirely gone in rooms illuminated with RT5 luminaires, and vertical illumination is dramatically higher. Most rooms illuminated with RT5 luminaires appear to be brighter than they really are. Finally, the standard RT5 luminaire is 89% efficient, which is far higher than most other products in the industry. This luminaire is available in an LED version now.

- *Specialty troffers*: In typical large sizes—2′ × 2′ and 2′ × 4′—Columbia Lighting offers the "Zero Plenum Troffer" (even in an LED version now) that is designed for environments with little to no plenum space above the ceiling. Typical troffers vary in depth from 3.5″ to up to 7″ for some deep parabolics, and standard practice requires lifting the luminaire over the ceiling grid for installation. Columbia's ZPT luminaire is only 1.50″ tall, which is less than the thickness of standard suspended ceiling grid framing members.

For a number of years, some manufacturers have made 6″-wide troffers with both lens and parabolic louver options. In recent years, the trend has been to see smaller

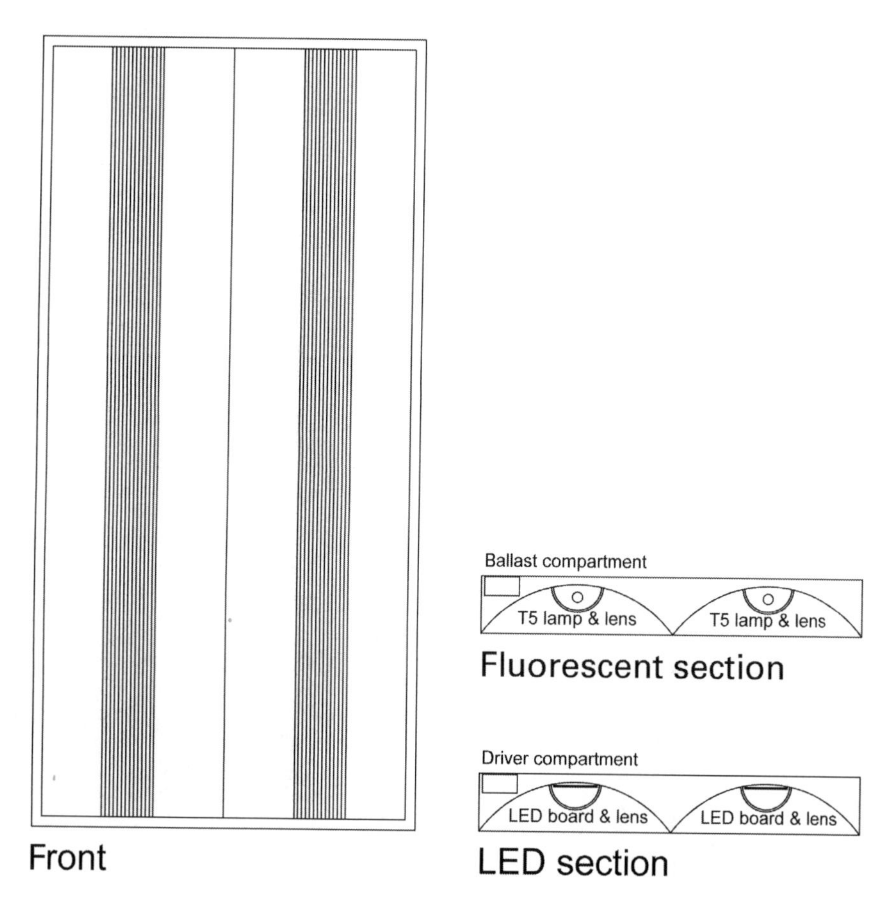

Figure 8.11.1.F
2′ × 4′ volumetric troffer

and smaller fixtures, and now we can get 4″, 3″, 2″, and even 1″ aperture troffers, also with lenses and louvers (usually non-parabolic louvers). Anything smaller than 2″ in width is LED-only, but LEDs have replaced fluorescents in nearly all of these products. There are even products available that *are* the ceiling grid, in both $\frac{15}{16}$″ and $\frac{9}{16}$″ widths; the luminaire lenses are about $\frac{3}{8}$″ below the plane of the grid but the upper part of the luminaire (including the heat sink) becomes part of the ceiling grid. (The author has these in his office.) Narrow luminaires like this tend to be used in continuous runs, and can be recessed into walls as well as floors. These luminaires are also usually available for pendant hanging without ceilings, as downlights, uplights, or both in a slightly taller housing. LEDs have made even smaller luminaires practical, and sizes are down to 20 mm (about 0.8″).

- *Round troffers*: Large-scale (2′ to 6′ diameter) lensed round troffers have been around for many years, but smaller semi-recessed luminaires have made their way to market more recently. In these semi-recessed luminaires, the housing is above the ceiling as in a conventional troffer, but the trim, or part of the trim, is below the ceiling. The latter allows for light to spill directly onto the ceiling around the perimeter of the luminaire, increasing apparent brightness and decreasing contrast between the luminaire and the ceiling surface.

Figure 8.11.1.G
2″-wide linear troffer

Driver compartment
LED board
White lens
Front
Section

8.11.2 Recessed downlights

What is a recessed downlight (sometimes called can light or pot light)? A downlight is usually a fully recessed housing for one or more fixed or adjustable lamps. As such, this is a very broad category that covers everything from tiny fiber optic star feature luminaires up through 16″-aperture luminaires with 400 watt metal halide lamps. Generally, these luminaires break down by lamp categories, as follows:

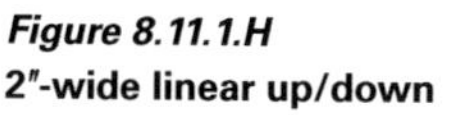

Figure 8.11.1.H
2″-wide linear up/down

- *HID downlights* (mercury vapor, high-pressure sodium, metal halide, and ceramic metal halide) are still available for virtually all HID lamps, from the smallest 20 watt T4 and MR16 CMH to 400 watt ED17 metal halides, but it is almost unimaginable that one would need to specify an HID downlight today.
- *Induction recessed downlights* are available from several major manufacturers for Philips QL lamps, in all wattages. They tend to be large downlights with apertures starting at 7″ or so. But, again, it is almost unimaginable that one would need to specify an induction downlight today.
- *Compact fluorescent downlights* are still available for virtually all compact fluorescent lamps (excluding the biax, or "long," compact fluorescent lamps), in 1-, 2-, or

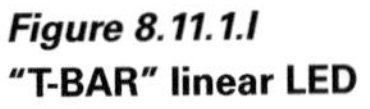

Figure 8.11.1.l
"T-BAR" linear LED

Remote driver

LED board with lens

Front

Section

even 3-lamp versions, but, again, it is almost unimaginable that one would need to specify a CFL downlight today.

- *Incandescent downlights* are still available for many incandescent lamps (single lamps and multiple lamps), including A, R, BR, ER, PAR, MR, and AR types. There is a clear separation between "line-voltage" (120V) and "low-voltage" (6V or 12V) luminaires. For very low budget projects, it is still common to see incandescent luminaires with LED lamps but that is usually a significant compromise in performance (mostly in efficiency and lifetime). The standard of the industry for many years was the 6″ aperture (ceiling opening), mostly because large incandescent lamps—R40, PAR38, etc.—require such a large luminaire; today, when using LEDs, there is little reason to use such large luminaires and the standard is closer to a 4″ aperture. But lots of work is being done using 2″ apertures too when ceilings are relatively low (say less than 10′ 0″ or so).

- *LED downlights* are available in the widest variety of all: from 1″ to 8″ apertures; round, square, and rectangular; lensed on the surface to deeply regressed, and from one to dozens of watts; a few hundred lumens to several thousand lumens, and from a few hundred centerbeam candlepower to 100,000 CBCP. New products are arriving monthly, if not more often, and prices are dropping rapidly.

Figure 8.11.2
Glass-trimmed LED downlight

8.11.3 Track lighting

Track lighting is a simplification of theatrical lighting that was developed for use in museums and retail stores so that luminaires could be moved easily for re-aiming as display needs change. (An older style in some department stores was the "pull down" downlight, where an adjustable sub-housing is actually pulled down and out of the ceiling, and aimed at the display.) It is a simplification because each lamp is not individually controllable; instead, all lamps on a common circuit are controlled together. Track lighting can be recessed (most costly), surface-mounted (most common), or suspended.

Track lighting comes in low voltage and line voltage (and line voltage comes in both 120V and 277V). It comes in 1-circuit (standard), 2-circuit, and 3-circuit, and in many different finishes (white, black, aluminum, etc.). Track lighting includes KableLite

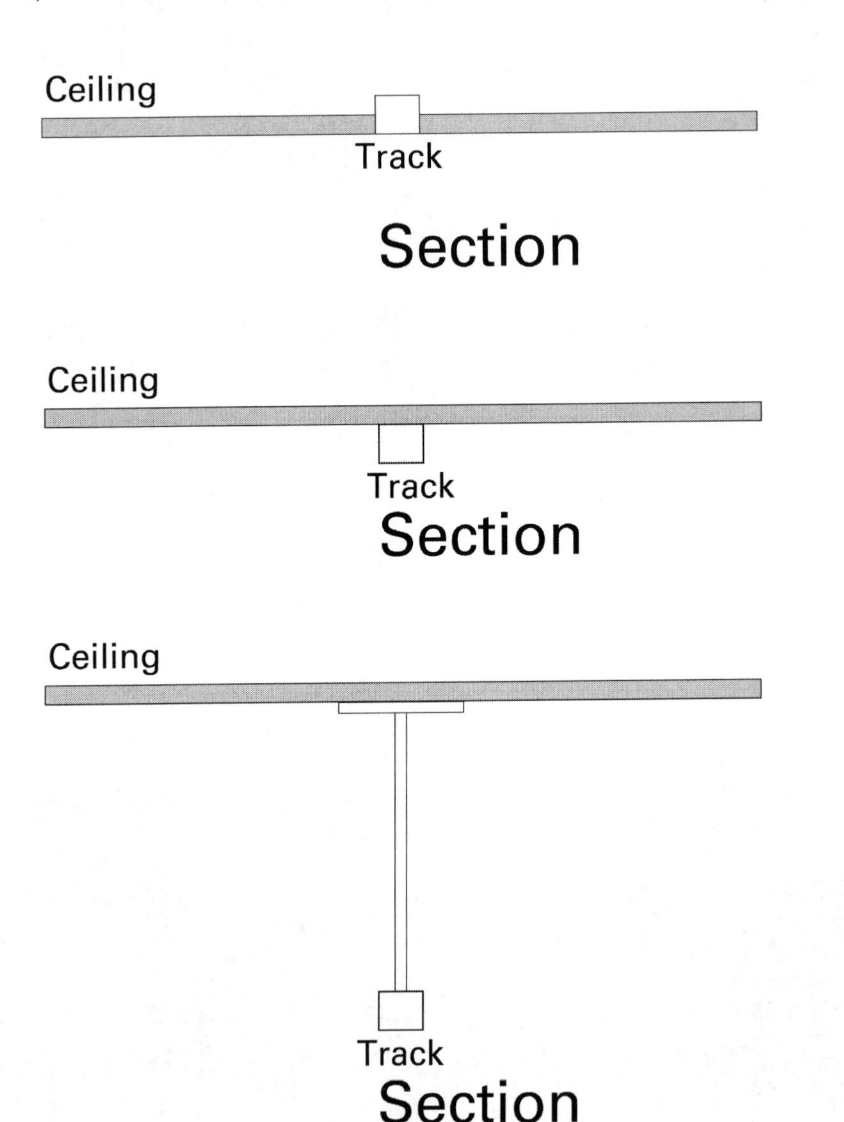

Figure 8.11.3.A
Recessed track

Figure 8.11.3.B
Surface-mounted track

Figure 8.11.3.C
Suspended track

(where the 12V power is carried by the parallel cable supports stretched between walls or other supports, with lamps mounted between the cables) and numerous different kinds of other low-voltage systems, many of which can be curved.

There are thousands of different head types available for incandescent, compact fluorescent (including biax), linear fluorescent, HID, and LED lamps and LED sources. Virtually anything is possible with track lighting. Of course, most people who have track lighting (even most stores) rarely actually move the heads around, which defeats the purpose of having track lighting.

In quality terms, there are two basic levels: commodity and commercial. Some manufacturers will describe their track as "museum grade" but that is just a small improvement on commercial. The one significant difference between commodity and commercial is that commodity systems combine the physical support of the luminaire to the track with the electrical contacts, and the commercial systems separate those two functions. This is important because having the electrical contacts involved in the luminaire support can cause arcing, which can cause luminaires to get welded in place (like incandescent lamps in non-porcelain sockets). Naturally, commercial grade track is considerably more costly than commodity grade track (by a wide margin—at least 200%, probably 300% or even more).

8.11.4 Pendants

Pendants are luminaires that hang from something—the roof structure or a ceiling—and they can take a very wide variety of shapes and sizes. They can use virtually any type of lamp. They can cost anywhere from $100.00 to "the sky's the limit." They can be ugly and functional or beautiful and not functional or beautiful and functional. They can be individual luminaires or continuous runs of linear luminaires. They can dominate a space or virtually disappear into the environment. They can be highly efficient or highly inefficient. They can produce strong shadows and strong contrasts or virtually no shadows or contrasts.

8.11.5 Custom luminaires

Even though there are literally thousands upon thousands of products to choose from, the perfect luminaire is not always available for a given design. When this happens, two paths are available: customization of a cataloged product and full-custom design.

Many manufacturers, especially manufacturers of decorative and semi-decorative products, are very open to customization of their designs: different finishes, colors, sizes, shapes, trims, etc. and there is no reason not to take advantage of these capabilities, if the owner can afford it and does not mind paying for it. This will rarely involve significant technical changes to the product though, because changes like that often involve re-testing (with UL or ETL) a luminaire which is a very costly process.

When is a full-custom design justifiable? When an appropriate product is not available and when the owner does not mind paying for a full-custom design. Here are a few examples:

1 *Historic 1876 church*: In late 1998, the author got the opportunity to redesign lighting in the large sanctuary of a church that was completed in 1876. Originally, there were two very large gas chandeliers in the space (which were documented in a drawing in a newspaper article); in the early 1890s, those chandeliers were replaced with two large round rings of gas jets with heat-removing hoods above them (and new vents in the ceiling to the attic). The hoods had short vertical sides (a foot or so) and an upper cone that closed the 8-foot diameter of the ring down to 2 feet or even less. In the 1920s these gas luminaires were removed, and the first electric lighting was installed. The latter consisted of six octagonal decorative plaster enclosures (about 3 feet across) for all indirect lighting running down the middle bays of the ceiling and five smaller (about 2.5 feet across) similar enclosures running along the side bays of the ceiling. The larger center bay luminaires had three 500 watt PS35 incandescent lamps in them, in special reflectors, and the smaller side bay luminaires had four 300 watt PS25 incandescent lamps in them. This system was still in use in 1999 but the maximum light level that was achievable (on a bright sunny day) was only 3 fc or so in the middle of the pews. The owner wanted to increase the light level and put in a more energy efficient system. The author determined that 200,000 lumens were needed for general lighting in the space, preferably from two large fixtures at the locations of the original chandeliers. It was not feasible to replicate the original gas chandeliers and achieve such high light output, but it was feasible to design full-custom fixtures that resemble the shades from the 1890s gas fixtures. The shape of the shade was turned upside-down and the bottom cone was made using faux alabaster acrylic. There are 12 50 watt biax lamps (dimmable) on top of each fixture for up-lighting and eight 50 watt biax lamps (dimmable) inside each fixture for downlighting, and the light level increased at the middle of the pews to roughly 25 fc. These fixtures cost about $12,000.00 each, but there were only two and the owner was happy with the cost. In addition, 12 recessed adjustable 300PAR56 downlights were added to light the pulpit, lectern, altar, piano, baptismal font, and the choir. The latter fixtures can be re-lamped and aimed from the attic and the large pendants have permanent power winches to lower them to the floor for maintenance. Because the owner did not want to lose all of the 1920s decorative housings, custom compact fluorescent (eight 42 watt) pans were made for the smaller side bay fixtures and those fixtures were re-installed. The overall project was a big success.

2 *Historic state park inn*: In 2000, the author got the opportunity to relight the lobby, main lounge, and dining room at a 1930s historic state park inn. One original pendant luminaire remained in the lobby, and that luminaire was removed, cleaned, reworked, repaired, and re-lamped using compact fluorescent lamps. (The original 1930s carved Lalique glass panels in this luminaire were carefully restored.) It would not have been feasible, or affordable, to try to copy the original luminaire for the lounge and dining room, but it was practical, and affordable, to design a new pendant, taking inspiration from the original luminaire. The new pendants use compact fluorescent lamps, perforated metal, and flat glass to evoke the original design. Despite some technical problems with the fixtures (the manufacturer installed ballasts in incorrect locations and left out some critical heat removal features—all corrected by the manufacturer), the net result was very

successful. (These luminaires cost $3,500.00 each and they are roughly 26″ diameter by roughly 12″ high.)

3 *1940s church*: in this case, another overly dark church sanctuary needed more light but the exposed wood structure made it almost impossible to add more locations for luminaires (without having excessive amounts of ugly exposed wiring), so the author designed a very high-output luminaire to take the place of the existing luminaires. This consisted of eight 50 watt dimmable biax lamps in vertical cylinders, using art glass and aluminum trim with recessed downlights in the bottoms of the luminaires. (The original idea was to use induction lamps for the downlights but that proved to be too difficult to do in a full-custom luminaire and the final design uses dimmable 42 watt CFLs.) The owner insisted upon changing the cylindrical shape to octagonal (greatly increasing the cost in the process) and then insisted upon adding gothic arch trim (despite the fact that the building is not a neo-gothic design), but the luminaires were manufactured and installed, at a material cost of $5,000.00/each. Despite some technical problems (these luminaires were also constructed incorrectly and had to be modified by the manufacturer), the overall installation was a success.

When doing full-custom design, it is highly advisable to spend the money to have a complete mock-up manufactured for testing and inspection by the designers and owner. All three of the examples above were too small to do this and there were no mockups. Mockups could have exposed the technical problems with examples (2) and (3), saving everyone lots of time and trouble (and money for the manufacturers).

8.11.6 Luminaire applications

Troffers are used almost exclusively for ambient lighting, although some of the new very small high-tech luminaires can do a great job at merchandise and task lighting too. Whenever small luminaires, or luminaires that use point-source lamps (incandescent, HID, or LED), are used for ambient lighting, it is necessary to follow the triple-overlap approach.

The triple-overlap approach requires placement of luminaires so that every point in the space (within reason) is covered by at least three luminaires; this minimizes shadow problems while assuring relatively even coverage. This can require a large number of luminaires (at very close spacing) if a narrow beam luminaire is used, so wide beams should be used instead. In some environments (lobbies, meeting rooms, restaurants, etc.) large-scale troffers can be very unattractive, so recessed downlights, or other small-scale luminaires, are often used instead. If recessed downlights are used for ambient lighting, they would also usually be placed in a regular pattern to result in even illumination, and using the triple-overlap approach. (When recessed downlights are used for other purposes—wall washing or accent lighting—the spacing might not be regular.) Pendants of many different types can be used for ambient lighting as well. Track lighting can also be used for ambient lighting, if the triple-overlap approach is followed.

In very basic designs—simple offices, basic meeting rooms, gymnasia, etc.—the ambient lighting is usually designed to be the task lighting too. This simply means that the level throughout the space is raised to the task level (say 30 fc for offices). This is done mostly because it can be inexpensive. As concern grows about energy use, this will probably start to change, and we are likely to see ambient levels dropping to something closer to 10 fc, as long as each workspace has sufficient separate task lighting.

In other environments—churches, schools, stores, restaurants, homes, etc.—the issues are more complicated because accent lighting and even decorative lighting become important parts of the overall design.

Even in a simple office, though, this is all about design. And atmosphere. And "feel" and "comfort." What is the intended sense of the space? Bright? Dark? Flat? Textured? Dull? Dramatic? The answers to these questions lead to lighting design concepts, and lighting design concepts in turn lead to hardware selection (or design in some cases). It might be tempting to start from the assumption that a law firm's reception area should be illuminated with recessed downlights, but how would one select the size, the trim, the lamping, and the placement? Only by knowing how the space is intended to be seen and experienced can these questions be answered. Every space should be approached from these basic questions. Many of them will have very quick answers. A typical storage room needs only functional ambient light—but one has to be careful about the placement of luminaires in relation to shelving and the items being stored. A large public restroom might need only functional ambient lighting (plus good mirror lighting) but which of the many available techniques should be used? Lines of miniature troffers? 2′ × 4′ prismatic troffers? Downlights? Wall-washers? Wall-slot, which is a variation on wall-washers? Answers to these questions are not so simple.

All designers have preferences—for materials, shapes, sizes, colors, and lighting—and those preferences play a role too. But do not fall into the trap of doing something because that is the way it was done the last time. Especially in light (no pun intended) of the rapid development of LED technology, much will change in the very near future, so habits should be questioned even more than usual.

At the beginning, think about shadows and brightness. Lighting the law firm's reception area ceiling from cove lights will minimize shadowing but it will increase the apparent brightness in the room due to the very bright ceiling. Using deeply regressed (regressing is moving the lamp above the ceiling plane) LED downlights will greatly decrease the apparent brightness but it will drastically increase shadowing. Perhaps both are needed? Or not? What about pendants or other decorative luminaires? What about decorative downlight trims? Table lamps? Sconces (wall-mounted luminaires)?

When using any luminaire close to a wall—especially round recessed downlights—it is important to consider how the luminaire puts light on the wall or some other vertical surface. Recessed round downlights, even wall-washers, tend to put scallops (a sharp curve with much less light above than below) of light on the wall. This effect can be manipulated into looking intentional, and even attractive, but random or irregular scallops of light on walls tend to look sloppy and unattractive. But all other luminaire types have light output patterns that affect nearby walls, or columns, or pendant light luminaires, or other objects too, and the issue should not be ignored.

Unfortunately, the budget will end up determining much of what is possible. The least-cost way to light anything is to use 2′ × 4′ lensed troffers, and that cost is roughly $1.00/sf, installed. Changing to lensed-indirect-look LED 2′ × 2′ troffers will increase that to roughly $2.00/sf. Obviously, this is not an insubstantial budget increase, but this is a *small* increase as lighting increases go. Changing to all downlights could increase the budget anywhere from 400% to, well, there is really no limit. Adding decorative luminaires could be enormously costly. Oftentimes, it is necessary to carve out a few special areas for "enhanced" lighting and to tolerate "ambient only" approaches in utilitarian spaces. Rarely is it possible to achieve truly excellent lighting throughout a project. But it is necessary to understand the budget boundaries so as to make the most of what is available. Ignoring the budget could simply end up destroying the entire design.

8.11.7 Controls

Lighting controls fall into three categories, each of which will be covered separately:

1 integral (to luminaires);
2 wall box (switches, dimmers, and sensors);
3 systems (switches, dimmers, and smart controls).

8.11.7.1 Integral controls

Integral controls are switches or dimmers that are built into luminaires, which is the norm for table lamps, torchieres, and some ceiling fan light kits. Depending on the quality of the device, these can last for decades or for months. Dimmers that are integrated into luminaires tend to be unreliable. This category could include "in cord" switches and dimmers as well.

8.11.7.2 Wallbox controls

This is the most common form of lighting controls, by far. Wall box controls are switches and dimmers located on the wall, usually in a recessed box. These can be found in quantities from one to several in individual or combined boxes and in many different styles and quality levels, including a wide variety of trim options.

8.11.7.2.1 Switches

Switches can be used for low-voltage (12V or 24V) and line-voltage (120V or 277V) lighting, and the basic types of switches include:

- standard toggle;
- illuminated toggle;
- paddle;

- illuminated paddle;
- keyed;
- pilot light.

Pilot light switches are rarely used for lighting; they are often used for devices such as exhaust fans. The purpose of the pilot light is to make it obvious that the device that is controlled by the switch is turned on.

The handles for switches are nearly always thermoplastic and come in a number of different standard colors: white, black, brown, gray, ivory, red, and sometimes more. The wall plates for these devices are available in thermoplastic in all the same colors, in galvanized steel, in stainless steel, and in brass, wood, ceramics, etc. Some of the thermoplastic devices are available with snap-on face plates that eliminate visible screws. Wireless switches are possible too these days, but only in conjunction with a control system.

8.11.7.2.2 Wall box dimmers

Wall box dimmers are available in two basic types: rotary and slide, but in hundreds of different specific configurations. They are also readily available in combination switch and dimmer devices. Rotary dimmers are the old style, and even though they are reliable, they are relatively unattractive. Slide dimmers come in many different types and styles. Dimmers are rated in maximum wattage and typical sizes include 300 watts, 600 watts, 1,000 watts, 1,500 watts, and 2,000 watts, but the larger sizes are rarely used today given very low LED loads. (Back when 150 watt incandescent lamps were commonly used, large dimmers were very common. Today, a powerful LED fixture might use 35 watts.) When high-capacity dimmers are installed side-by-side, it is usually necessary to "de-rate" the dimmers, meaning that their capacities are reduced. This de-rating is due to the potential of over-heating due to close proximity, and, again, this is rarely relevant today given small dimmer loads.

8.11.7.2.3 Sensors

Occupancy, vacancy, and daylight sensors are available in wall box versions and surface-wall and surface- and recessed-ceiling mount versions (even including some wireless devices). These devices provide automatic control to turn off or dim the lights when daylight is adequate or when spaces are unoccupied. Occupancy sensors turn on the lights when someone enters a space and they turn off the lights after the space is empty (according to some programmed delay time); vacancy sensors do nothing when someone enters the space (it is necessary to turn the lights on manually with a switch) but they do turn the lights off after the space is empty.

Daylight harvesting, as it is called, is very practical and reasonably affordable and can be implemented on many different projects with high daylight penetration. Using daylight harvesting does mean that the luminaires have to be grouped into zones that step away from the windows; in most situations unless there are elaborate light shelves, daylight penetration is unlikely to exceed 20′ 0″ or so, so the luminaires within that 20′ 0″-wide zone must be on separate control zones.

8.11.7.3 Systems

Control systems can be anything from a special controller in a wall box up to multiple rack systems that can control all of the lights in a major performance venue (such as a concert hall, football stadium, convention center, etc.). Systems can incorporate astronomical clocks (to account for sunrise and sunset with or without daylight-saving time adjustments), pre-programmed schedules (for occupied and unoccupied, for example), automatic controls (including occupancy, vacancy, and daylight sensors), and pre-set scenes.

These systems fall into the categories of:

- simple;
- architectural;
- theatrical.

8.11.7.3.1 Simple systems

Simple systems are small systems that fit into large recessed wall boxes. Systems like this can usually handle a couple of circuits and up to eight zones with several pre-set scenes, which can manage lights, automatic projection screens, and even room blackening shades in a typical conference room. All programming is done from the controller in the room.

8.11.7.3.2 Architectural systems

Architectural systems are used in conference rooms, meeting rooms, convention centers, schools, churches, restaurants, retail stores, etc. These systems have extensive capabilities and can be operated from numerous different types of controls, from simple 4-button low-voltage switches in wall boxes to elaborate wireless hand-held devices. Most architectural systems use pre-set scenes (a scene is a predetermined arrangement of all the lights controlled by the system, thereby allowing all of the lights to be set from a single button) and most systems can accommodate several different such scenes. For a conference room, the scenes might be "PowerPoint," "talk," "video," "video conference," and "meeting"; for a restaurant, the scenes might be "breakfast," "lunch," "dinner," and "cleaning"; for a church, the scenes might be "traditional worship," "contemporary worship," "music," "wedding," "funeral," and "cleaning." Most systems also include programmable fade rates, so that scenes can change instantly or over such a long time period that no one even notices the change. (Long fade rates are very useful in restaurants for changing from "lunch" to "dinner"; it is far better to change gradually over 30 minutes or so than to have the sudden "dimming" that all of us have noticed in some restaurants at the end of the lunch period or the beginning of the dinner period.)

Control systems interfaces (devices to operate the systems) can include wall panels, wall devices, multi-button wall devices, wireless hand-held remote controls, and even linkage to PDAs. These control systems can also integrate HVAC controls, audio systems, video systems, security systems, and even the internet (for news, weather, etc.).

Frequently, control systems are used when systems become too complicated for easy use from wall box devices. A conference room with pendant lights over the table, wall-washers for a white board, general area recessed downlights, and video conferencing lights, would require four individual controls, which might be manageable. The author once worked on a medium sized retail store that was so complicated there were 35 separate zones of lighting control on an 8,000 sf sales floor! No one could manage 35 wall box dimmers in a situation like that. A large, high-end residence could easily have several dozen groups of light fixtures to be individually controlled.

These systems have changed dramatically over several decades, starting in the 1980s. Prior to that, large-scale dimmers were physically large, requiring lots of space and cooling due to their high heat production. In the 1980s, the solid-state (electronic) dimmers became available, which made it feasible to install them in cabinets similar to panelboards (see Chapter 9). But the systems were still very costly. (The system for the complicated retail store cost more than $15,000.00.) The next step in the evolution of this technology was to make it feasible to put dimmers and relays (remote-controllable switches) near the loads and not in cabinets with extensive line-voltage wiring; this greatly reduced costs. That $15,000.00 system in a distributed arrangement might cost half as much. But these systems still required highly trained personnel to modify them; in other words, if the restaurant wanted to change the settings for "dinner," an expert had to be called in (at high cost) to re-program the system. The most recent development is that the systems are now easily programmable, and it is no longer necessary to pay for an expert to come in to make changes. So the cost and ease of use have now reached the point where the systems are applicable to far more projects—smaller restaurants and retail stores, even offices.

8.11.7.3.3 Theatrical systems

Theatrical systems are the ultimate in lighting control because each instrument (as luminaires are called in the lighting world) can be controlled individually. This also includes color changing (also being seen on a limited basis in the architectural world), fading, dimming, flashing, strobing, and even moving lights. LEDs have not yet completely taken over the theatrical world mostly due to the very high CBC that can be required (over 100,000 sometimes), but they are gaining rapidly.

8.12 Lighting calculations

There are three main areas of calculation in lighting design: zonal cavity (for ambient lighting), accent (for display lighting), and energy use. Each will be covered separately.

At the current time (2017), most zonal cavity calculations are done by using specialized software on computers (from independent software developers or from one of the large lighting conglomerates), but it is still useful to learn to do the calculations manually in order to have a better understanding of what happens when changes are made to ceiling height, reflectances on surfaces, etc. Accent lighting calculations can also be done in the software, but it is cumbersome and it is more common to see hand calculations. Energy calculations are computer-form based (especially when using ComCheck reports from the U.S. Department of Energy), but it is still required to count luminaires by hand, work out building areas, etc.

When zonal cavity calculations are done using software, the output is far more sophisticated than a single average footcandle number (as in a manual zonal cavity calculation), and it usually consists of a **point-by-point** plot and/or an **iso-footcandle** plot. The software is also capable of calculating on vertical surfaces, ceilings, etc., depending on the sophistication of the input model.

Some software packages are also capable of generating photorealistic renderings of spaces. All such software uses data files for luminaires that are called "IES files" because the IES developed the format. If a manufacturer can provide an IES file for a luminaire, that luminaire can be calculated in any model. If the manufacturer cannot, or will not, provide an IES file, then the luminaire cannot be calculated accurately by any software. (Manufacturers are usually happy to provide their IES files, although they sometimes prefer to receive a model and do the calculations themselves when sensitive new products are involved.)

8.12.1 The zonal cavity calculation for ambient lighting

There is a simple formula that is used to calculate average ambient lighting on a horizontal workplane:

$$TL = \frac{A \times fc}{CU \times LLF}$$

where:

TL = total lumens
A = room area
fc = footcandles
CU = Coefficient of Utilization (a function of the specific luminaire in a specific room)
LLF = Light Loss Factor (a combination of lumen depreciation factor, dirt depreciation factor, room surface dirt depreciation factor, and lamp burnout factor)

Coefficient of Utilization is based upon Room Cavity Ratio (RCR). The equation for RCR is:

$$RCR = \frac{5h\,(L + W)}{L \times W}$$

where:

h = distance (in feet) from the workplane to the bottom of the luminaires (in most situations, the workplane is 30″ above finish floor—standard desk height—but 0″ is used for sports and other special situations)
L = length of the space (in feet)
W = width of the space (in feet)

Once RCR is know, the CU can be determined from information provided by the luminaire's manufacturer. The information that is published by the luminaire manufacturers includes CUs for all potential situations, from RCR = 0 to RCR = 10, and at various combinations of room reflectances. The standard assumptions for the latter are:

- pcc (ceiling cavity reflectance) = 0.80, or 80% (with possible options for 0.70, 0.50, 0.30, 0.10, and 0.00)
- pw (wall reflectance) = 0.50 (with possible options for 0.70 and 0.30)
- pf (floor reflectance) = 0.20

The software that is used for lighting calculations can use any reflectance values.

It is vitally important to understand that room shape and size can have a very powerful effect on the effectiveness of any particular luminaire in any particular space. A luminaire that has a CU = 0.76 (equivalent to 76% efficiency) at RCR = 2 could have a CU = 0.41 at RCR = 8, which means that the luminaire is only 41% efficient at RCR = 8—a drop in efficiency of 35%, just by changing room proportions! Given high efficiency requirements in modern energy codes, this is a key issue.

The zonal cavity equation can also be written this way:

$$fc = \frac{CU \times LLF \times TL}{A}$$

The equation can be used in the first form to determine how many luminaires are needed to achieve a given light level (fc) and the second form can be used to determine the average light level (fc) based on a given quantity of specific luminaires.

Here is how to use the formula for the first case (to determine how many luminaires are needed):

1. Determine the desired footcandle level.
2. Determine the room's dimensions and calculate room area ("A").
3. Determine the general type of luminaire to be used (or, better yet, the specific luminaire).
4. Determine the workplane height.
5. Determine "h," the cavity height, based on the workplane height and the mounting height of the luminaires (to the bottom).

6 Calculate the RCR.
7 Determine the reflectances of the floor, walls, and ceiling.
8 Determine the CU based on the RCR and reflectances determined above and LLF (use 0.70 if LLF is unknown).
9 Calculate total lumens using the first form of the equation.
10 Determine the number of lumens per lamp in the selected luminaire by looking up the lamp in a standard lamp catalog (or on the lamp manufacturer's web-site).
11 Determine the number of lumens per fixture by multiplying the number of lamps by the lumens per lamp.
12 Determine the number of luminaires required by dividing the total lumens (step 9) by lumens/luminaire (step 11), rounding to an even number.
13 Complete a luminaire layout in the room, adjusting the number from step 12 as required to result in an even pattern.
14 Verify light output of the final layout. This can be done by re-running the equation (in the second form) or by simple mathematical ratios. In the latter case, if step 12 says 5.4 luminaires and step 13 results in six luminaires, then the average light level will increase by a factor of $\frac{6}{5.4}$, or 1.11, if six luminaires are used.

Here is a specific example of how to do this for the room in Figure 8.12.1.A.

1 Light level = 46.5 fc (given, based on Category R, *The Lighting Handbook*, 10th Edition)
2 Area = 30′ × 16′ = 480 sf (given)
3 Fixture type: 2′ × 4′ recessed prismatic LED troffer with 4,865.4 delivered lumens (given by the author for purposes of the example)
4 Workplane height = 30″ (given)
5 Cavity height = 10′ (ceiling height = fixture mounting height) – 2′ 6″ (workplane height) = 7′ 6″.
6 RCR = [5(7.5′)(30′ + 16′)]/(30′ × 16′) = 3.6

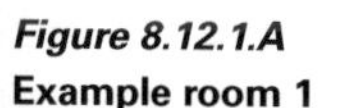
Figure 8.12.1.A
Example room 1

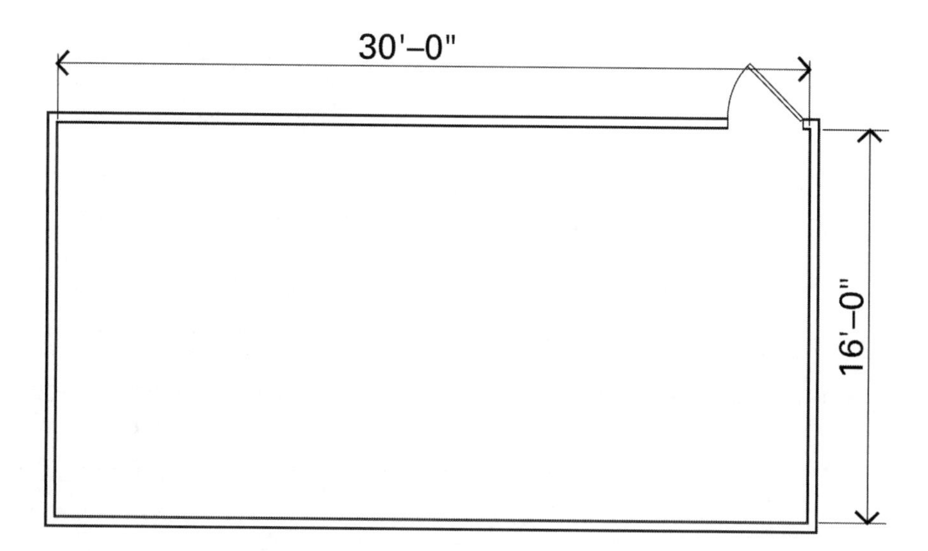

PHOTOMETRICS

2TL4 48L FW A12 EZ1 LP840, 4865.4 delivered lumens, test no. LTL26934P10, tested in accordance to IESNA LM-79.

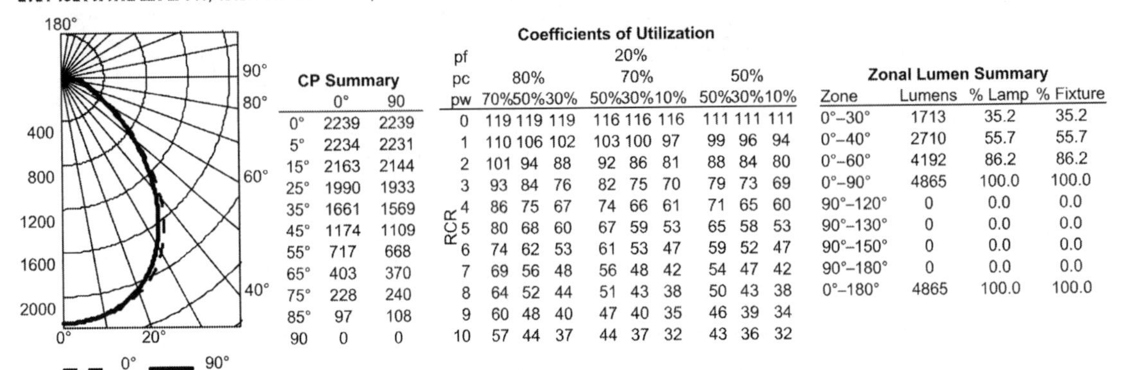

CP Summary

	0°	90
0°	2239	2239
5°	2234	2231
15°	2163	2144
25°	1990	1933
35°	1661	1569
45°	1174	1109
55°	717	668
65°	403	370
75°	228	240
85°	97	108
90	0	0

Coefficients of Utilization

pf	20%								
pc	80%			70%			50%		
pw	70%	50%	30%	50%	30%	10%	50%	30%	10%
RCR 0	119	119	119	116	116	116	111	111	111
1	110	106	102	103	100	97	99	96	94
2	101	94	88	92	86	81	88	84	80
3	93	84	76	82	75	70	79	73	69
4	86	75	67	74	66	61	71	65	60
5	80	68	60	67	59	53	65	58	53
6	74	62	53	61	53	47	59	52	47
7	69	56	48	56	48	42	54	47	42
8	64	52	44	51	43	38	50	43	38
9	60	48	40	47	40	35	46	39	34
10	57	44	37	44	37	32	43	36	32

Zonal Lumen Summary

Zone	Lumens	% Lamp	% Fixture
0°–30°	1713	35.2	35.2
0°–40°	2710	55.7	55.7
0°–60°	4192	86.2	86.2
0°–90°	4865	100.0	100.0
90°–120°	0	0.0	0.0
90°–130°	0	0.0	0.0
90°–150°	0	0.0	0.0
90°–180°	0	0.0	0.0
0°–180°	4865	100.0	100.0

***Figure 8.12.1.B* Prismatic-lensed LED troffer CU**

7. Floor reflectance = 0.20; wall reflectance = 0.50, and ceiling reflectance = 0.80 (given)
8. Figure 8.12.1.B shows the CU figures for a Lithonia 2TL 48L FW A12 EZ1 LP840 troffer. As you can see, at pf = 20% (0.20), and pc = 80% (0.8), and pw = 50% (0.50), the value at RCR = 3 is 84 (actually 0.84 or 84%) and the value at RCR = 4 is 75 (or 0.75 or 75%). The value for RCR = 3.6 (as in the case here) must be estimated by using mathematical interpolation, which is done as follows: 3.6 is 60% of the way between 3.0 and 4.0, so we need the CU at 60% of the way between 75 and 84. So we take 60% of 9 (the difference) and get 5.4. Round that down to 5.0 and add it to 75 to get 80, or 0.80. So the CU for this fixture at RCR = 3.6 is 0.80.
9. Total lumens = (46.5 fc × 480 sf)/(0.80 (CU) × 0.70 (LLF)) = 39,857 lumens.
10. Lumens per luminaire = 4,865.4 (given)
11. Number of luminaires = 39,857 lumens/4,865.4 lumens/luminaire = 8.2
12. 8.2 luminaires cannot be used, so it will have to be eight. Using eight, the layout is as shown in Figure 8.12.1.C.
13. The average light level is 8.0/8.2 × 46.5 fc = 45.4 fc.
    Or fc = [8(4,865.4)(0.8)(0.7)]/480 = 45.4

Now for some comparisons:

A Raise the ceiling 2′ 0″

In this case, everything remains the same except RCR and CU. RCR is now = [5(9.5′)(30′ + 16′)]/(30′ × 16′) = 4.6, so CU is now equal to 0.71 or 71%. So the new calculation is:

Total lumens = (46.5 fc × 480 sf)/(0.71 (CU) × 0.70 (LLF)) = 44,909 lumens

This is 11.3% more than we had in Example 1. This equates to 9.2 luminaires, so the average level for the eight-luminaire layout will be 8/9.2 × 46.5 fc = 40.4 fc. So the average level will drop by 5 fc just because the ceiling went up two feet! Nothing else changed.

What happens when different luminaires are used with the original ceiling height? Here are three more examples.

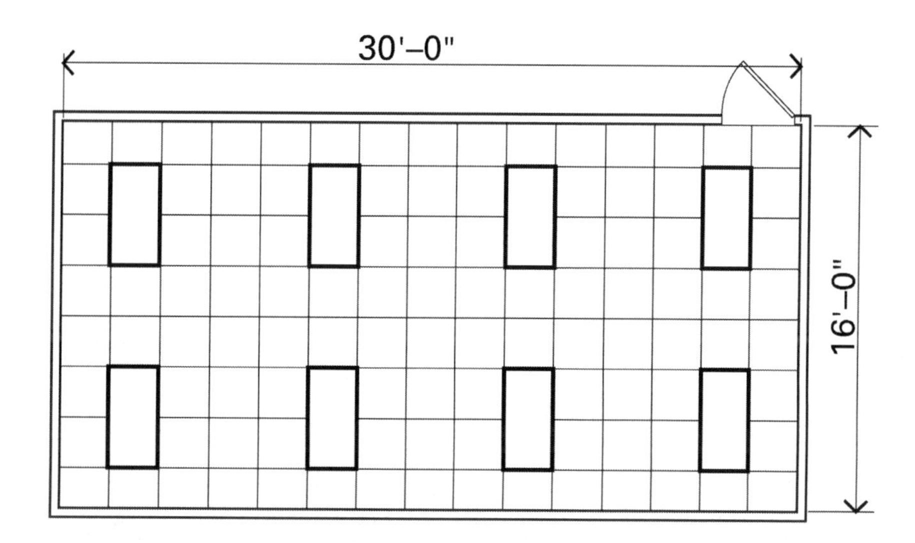

Figure 8.12.1.C
Example 1 layout

PHOTOMETRICS

2VTL4 40L ADP LP835, 4211 delivered lumens, test no. LTL24782P4, tested in accordance to IESNA LM-79

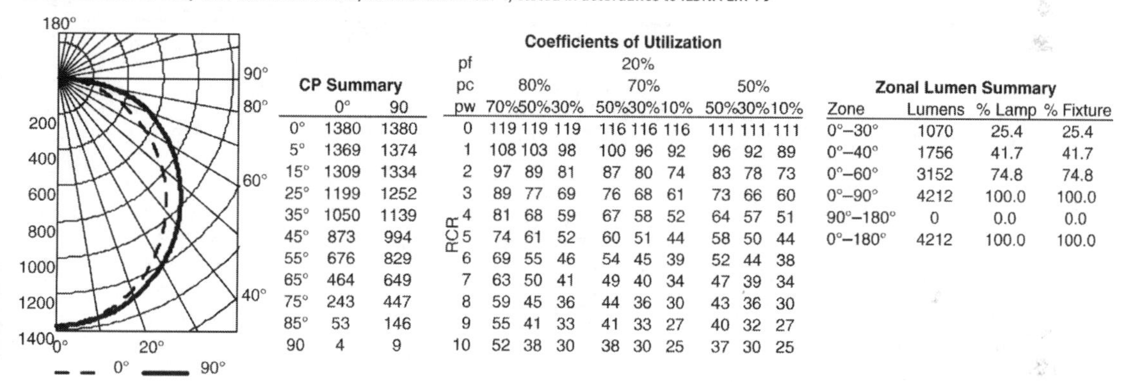

CP Summary

	0°	90
0°	1380	1380
5°	1369	1374
15°	1309	1334
25°	1199	1252
35°	1050	1139
45°	873	994
55°	676	829
65°	464	649
75°	243	447
85°	53	146
90	4	9

Coefficients of Utilization

pf	20%								
pc	80%			70%			50%		
pw (RCR)	70%	50%	30%	50%	30%	10%	50%	30%	10%
0	119	119	119	116	116	116	111	111	111
1	108	103	98	100	96	92	96	92	89
2	97	89	81	87	80	74	83	78	73
3	89	77	69	76	68	61	73	66	60
4	81	68	59	67	58	52	64	57	51
5	74	61	52	60	51	44	58	50	44
6	69	55	46	54	45	39	52	44	38
7	63	50	41	49	40	34	47	39	34
8	59	45	36	44	36	30	43	36	30
9	55	41	33	41	33	27	40	32	27
10	52	38	30	38	30	25	37	30	25

Zonal Lumen Summary

Zone	Lumens	% Lamp	% Fixture
0°–30°	1070	25.4	25.4
0°–40°	1756	41.7	41.7
0°–60°	3152	74.8	74.8
0°–90°	4212	100.0	100.0
90°–180°	0	0.0	0.0
0°–180°	4212	100.0	100.0

Figure 8.12.1.D
LED volumetric troffer CU

B 4,111 lumen 2′ × 4′ LED volumetric troffers (Figure 8.12.1.D)

From Figure 8.12.1.D, the CU will be 0.72. Therefore, TL = (46.5 × 480)/(0.72 × 0.70) = 44,286 lumens, or 10.5 luminaires. So our eight-luminaire layout will result in 8/10.5 × 46.5 fc = 35 fc average (or about 77% of the output of the lensed troffers).

C 1,500 lumen 4″ LED downlights

From Figure 8.12.1.E, the CU will be 0.92. Therefore, TL = (46.5 × 480)/(0.92 × 0.70) = 34,658 lumens or 23 luminaires (at 1,509 lumens/luminaire). So we need to change the layout to 20 luminaires, which will provide an average level of 20/23 × 46.5 = 40.4 fc.

D Decorative pendants ("schoolhouse" type glass pendants with 100 watt metal halide)

The CU is unknown for the HID pendants. In a case like this, use 0.50 to be reasonably close. Therefore, TL = (46.5 × 480)/(0.50 × 0.70) = 63,771 lumens or 7.4 luminaires (at 8,600 lumens/luminaire). So six luminaires will provide an average level of 6.0/7.4 × 46.5 = 38 fc.

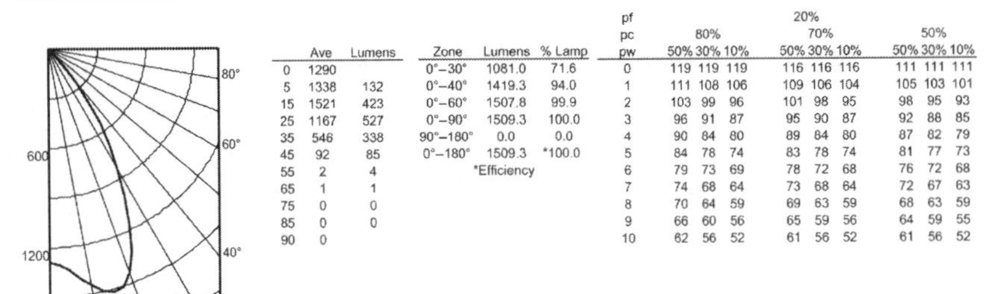

EVO 35/15 4AR LS	INPUT WATTS: 17.3, DELIVERED LUMENS: 1509, LM/W=87.2, 1.08 S/MH, TEST NO. LTL27786

	Ave	Lumens
0	1290	
5	1338	132
15	1521	423
25	1167	527
35	546	338
45	92	85
55	2	4
65	1	1
75	0	0
85	0	0
90	0	

Zone	Lumens	% Lamp
0°–30°	1081.0	71.6
0°–40°	1419.3	94.0
0°–60°	1507.8	99.9
0°–90°	1509.3	100.0
90°–180°	0.0	0.0
0°–180°	1509.3	*100.0

*Efficiency

pf	20%								
pc	80%			70%			50%		
pw	50%	30%	10%	50%	30%	10%	50%	30%	10%
0	119	119	119	116	116	116	111	111	111
1	111	108	106	109	106	104	105	103	101
2	103	99	96	101	98	95	98	95	93
3	96	91	87	95	90	87	92	88	85
4	90	84	80	89	84	80	87	82	79
5	84	78	74	83	78	74	81	77	73
6	79	73	69	78	72	68	76	72	68
7	74	68	64	73	68	64	72	67	63
8	70	64	59	69	63	59	68	63	59
9	66	60	56	65	59	56	64	59	55
10	62	56	52	61	56	52	61	56	52

Figure 8.12.1.E
Downlight CU

Luminaire	Quantity	fc	% of goal	Installed cost	Multiplier	Adjusted multiplier
A 2 x 4 prismatic troffers	8	45.4	98%	$1,200.00	1.00	1.02
B 2 x 4 volumetric troffers	8	35.8	77%	$2,600.00	2.17	2.81
C 1500 lumen 4" downlights	20	40.4	87%	$4,000.00	3.33	3.84
D Metal halide pendants	6	38.0	82%	$5,600.00	4.67	5.71

notes

1. % of goal means percentage of target footcandles achieved
2. Adjusted multiplier is the multiplier divided by the % of goal

Figure 8.12.1.F
Zonal cavity cost versus performance comparison

Now, we will add costs to the mix. The range in quantity of luminaires is 8 to 20 and the luminaires that are required in greater quantity are also more costly, so the range in installed cost is quite wide (more than 4:1; see Figure 8.12.1.F).

Again, the purpose here is not to say "do not use downlights or decorative pendants"; instead, the purpose is to explain the wide variation in cost that can occur in order to develop an understanding of cost-benefit and cost-quality. Unless a designer has an idea of what is affordable across the entire project, it is nearly impossible to determine where special features and improved lighting can be used.

The manual zonal cavity method can even be used to estimate accent lighting for certain applications. For example, some years ago the author designed a shoe store that was to be illuminated by using only 4" aperture recessed adjustable luminaires with 75 watt PAR16 narrow flood lamps (this was the author's decision in collaboration with the owner); there would be no ambient lighting at all. So how could he figure out how many accent luminaires would be needed? By running the zonal cavity calculation for some average level to determine total lumens, which could then be used to determine how many PAR16 lamps to use. The technique worked well and the store's lighting was highly effective.

Finally, there is a simplified approach to this that can be done much faster and more easily. Back to the original formula:

$$TL = \frac{A \times fc}{CU \times LLF}$$

If we assume that CU = 0.7 (usually close, but on the low side if not) and LLF = 0.7 (also on the low side these days), CU × LLF = 0.49 ≈ 0.50. So the formula becomes:

$$TL = \frac{A \times fc}{0.5}$$

Then, multiply both sides by 2, and it becomes:

$$2 \times TL = A \times fc$$

which then becomes:

$$fc = \frac{2 \times TL}{A}$$

Most of time, using this simplified approach will get close enough to establish rough luminaire quantity, which is very useful when designing ceilings, etc. There is no need to have CU or RCR information to do this, but it is necessary to have a sense of the needed lumen output per fixture and/or the fc goal.

8.12.2 Accent (focal) lighting calculations

Accent lighting calculations might appear to be more complicated than zonal cavity calculations because it is necessary to use basic trigonometry to find beam sizes. It is necessary to understand beam sizes so that the accent lighting relates appropriately to the object being illuminated. A good example would be the face of a speaker at a lectern. First, the face should *not* be illuminated from a single source (no matter what the source or the direction) unless strong shadows are desired, which is usually not the case. Second, two sources usually work well, spaced on either side, above, and in front of the face. Third, the ratio of brightness on the face to the background illumination is key; remember the 5:1 minimum brightness ratio? It would be better to target 10:1. If it is assumed that the background level is 10 fc, that means that there should be 100 fc on the face. Each source would need to do half of this, or 50 fc. Fourth, how large does the beam of light need to be? Not very large because the goal is to illuminate only the face and head. If it is assumed that the beam diameter should be 2′ 0″ (a round figure, no pun intended), how big does the beam need to be if the source is 15′ 0″ away from the face? Also, what is the beam intensity that is needed to provide 50 fc?

In order to answer the last question, the following equation is used:

$$fc = \frac{CPC}{d^2}$$

where: CPC = centerbeam candela (or centerbeam candlepower—indicated CBCP); d = the distance from the lamp to the object (in feet).

So if one wanted to see 50 fc on the face from 15′ away, that would require CPC = 50 fc × 15′ × 15′ = 11,250. Light falls off with the square of the distance, so the CPC at 30′ would be CPC = 50 fc × 30′ × 30′ = 45,000. (For what it is worth, 45,000 CBC is a *big* number that is not easily achievable.)

But this is true only at centerbeam. The primary beam in situations like this is defined so that the level at the edge is 50% of the level at centerbeam. So in the two cases above, if it is 50 fc at centerbeam, it will be 25 fc at the edge of the beam.

But what is the angle of the beam that is needed to provide a 24″-diameter beam at 15′ 0″ from the face? This is where trigonometry comes in.

Trigonometry (sines, cosines, tangents and their inverses) only works for right triangles where one of the angles is 90 degrees. So we start by drawing a center line through the light beam to create two right triangles. In such a case the distance "d" would be the long side of the triangle; the distance "r" (half of the diameter) would be the short side of the triangle, with the hypotenuse as the third side. The length of the hypotenuse is not usually critical (see Figures 8.12.2.A and 8.12.2.B).

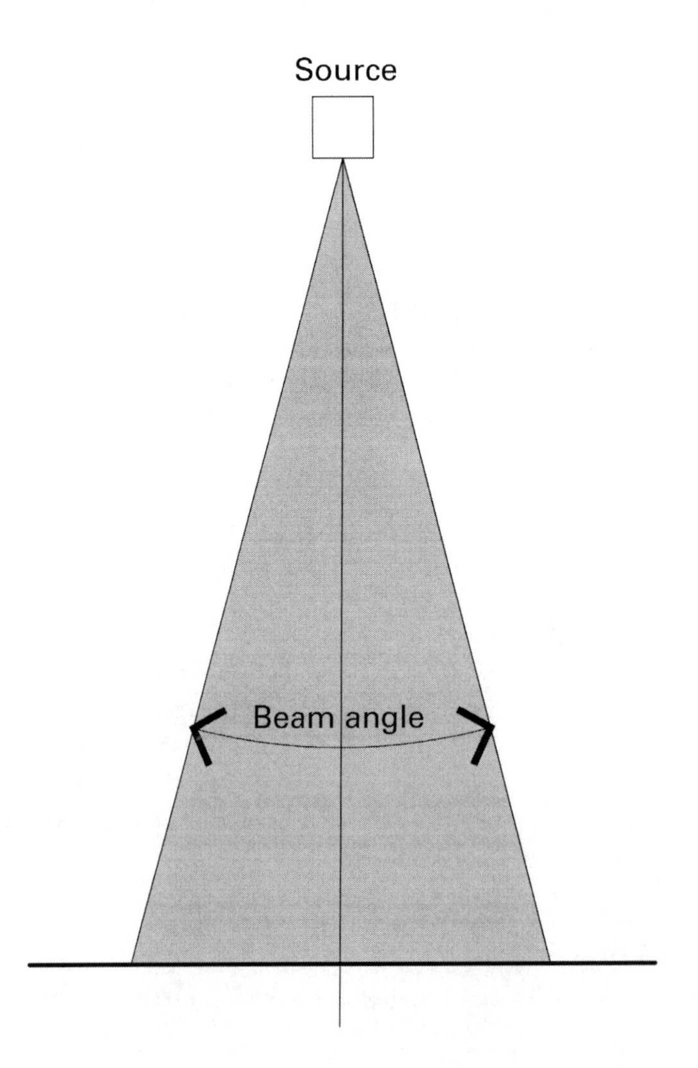

Figure 8.12.2.A
Beam angle

Figure 8.12.2.B
Calculation triangle

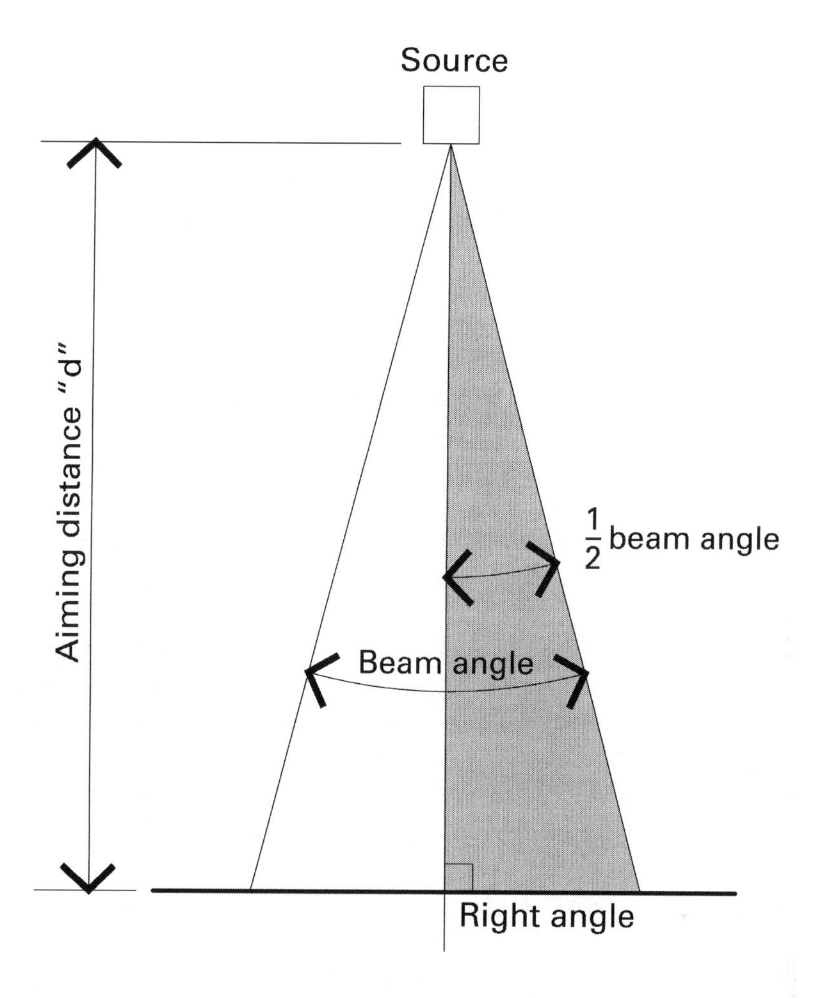

Using the right triangle, the acute angle at the top (half of the beam from the lamp or luminaire) is found as follows:

$$\Theta = \text{arcsine } (r \div d)$$

or

$$r = d \times \sin\theta$$

or

$$d = r \div \sin\theta$$

Back to 50 fc on the face from 15′ away. We are trying to find the angle, and we know r (12″ = 1′ 0″—half of the diameter) and d (15′ 0″), both given. So use the first equation:

$$\Theta = \text{arcsine } (r \div d)$$

$$\Theta = \text{arcsine}\ (1 \div 15)$$

$$\Theta = \text{arcsine}\ (0.067)$$

$$\Theta = 3.82$$

Remember that this angle is *half* of the beam though, so the beam angle that we need is $2 \times 3.82 \approxeq 8$ degrees.

Is this workable? That is hard to say, but it is unlikely. In the old days when incandescent and halogen lamps were used for things like this, it was simpler to look in the lamp catalogs to see what the available sources could do. (For example, one might find a 75 watt MR16 lamp with a 15 degree beam and 10,000 CBC—not too far from what is needed in this example.) With LED luminaires, there are actually more options than there ever were before, especially CBC. So now, it makes sense to look at LED luminaire cutsheets to see what beam angles and CBCs are available to find a decent match. This will never be perfect though—the beam angle will be a little too wide or narrow or the CBC will be a little too high or low, or some combination of both.

Also, in a perfect accent lighting world, this is very carefully worked out. In an art gallery, each art work has lighting that is designed specifically for that piece, which can mean that many different beam angles and CBCs are needed in a single space. One has to be careful here because more complexity makes operation more difficult for the owner and it can make maintenance extremely challenging. One of the advantages to new LED products is that many of them are now available with low-cost field changeable optics, which means that it is feasible (easy, really) to change a 25° beam to a 50° beam, for example. Of course, if that change is made CBC will drop considerably, so one still has to be careful—it might be necessary to use two luminaires instead of one where a wide beam is needed.

That is actually all there is to basic accent lighting calculations, but this only applies where the source is aimed at the object on the centerline and perpendicular to the object—head on and on center. In reality, few accent lights are aimed that way. Instead, the most common arrangement is for the source to be on the centerline but to be aimed at an angle; this usually means a source that is mounted above the item being illuminated and aimed downward—which changes round beams into ovals. When two sources are needed, as in the face example, they are usually located at an angle to each side, which adds even more complexity.

The rule-of-thumb for such aiming is to keep the angle between 30 and 45 degrees from vertical; the main reason for the 30° limit is to avoid extremely elongated ellipses (ovals) of light on the object (unless that is specifically desirable) and the main reason for the 45° limit is to limit shadowing from viewers (if one stands between a low-aimed light and the object being illuminated, one casts a large shadow).

Aiming these lamps at angles does complicate the math considerably. We will take one angle first, meaning that the lamp is angled vertically but not horizontally. Determining fc at the center is the same at it was before, so determining the fc at the side edges of the beam is the same too (the beam width does not change as the angle changes). But determining the fc at the top and bottom of the beam requires calculating at two additional distances; this last is by far the most complicated calculation in lighting design. Note the various items, distances, and fc's in Figures 8.12.2.C, 8.12.2.D, and 8.12.2.E.

Here is the list:

Θ = aiming angle (from vertical)
Φ = lamp beam angle

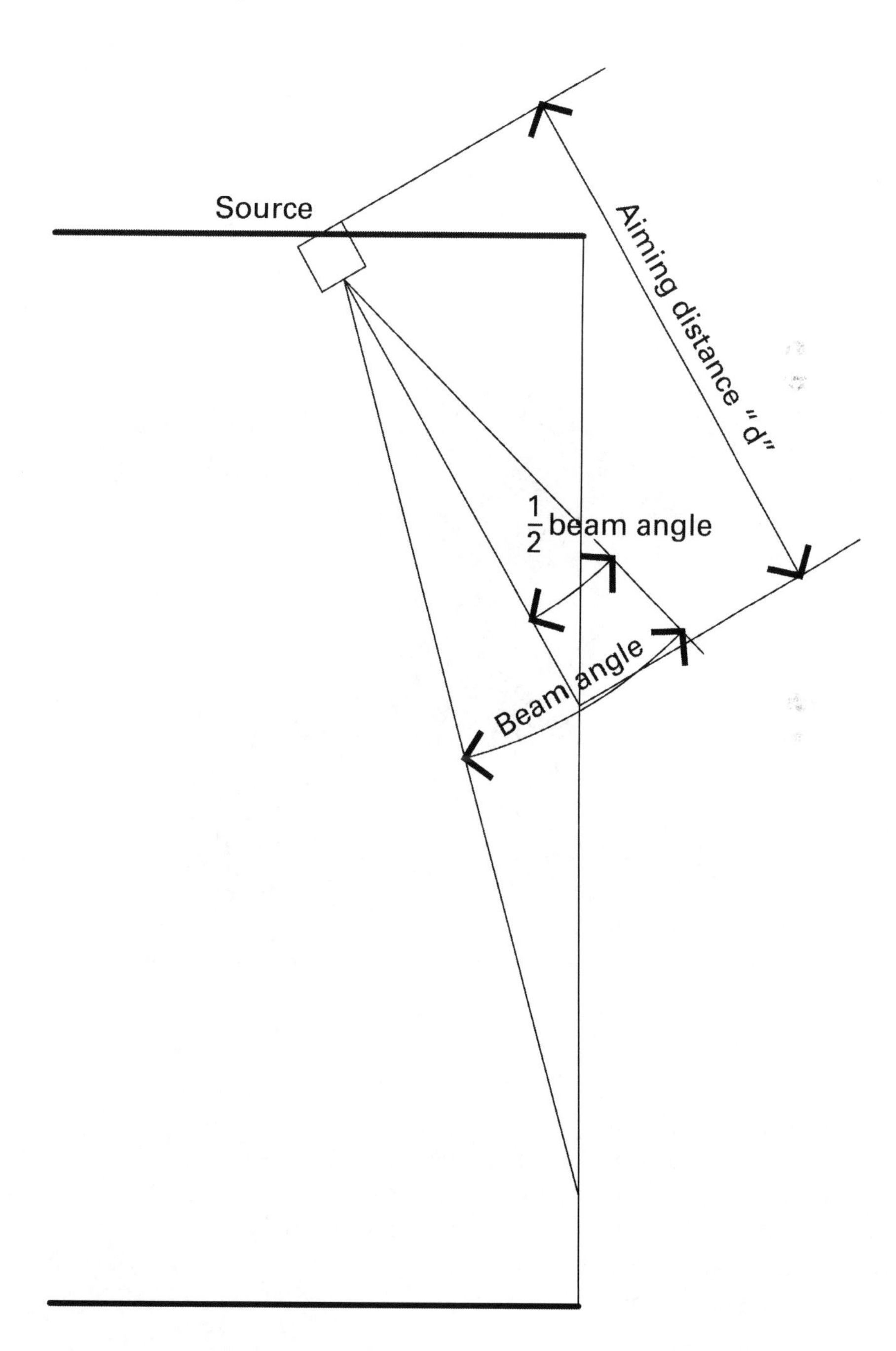

Figure 8.12.2.C
Vertical aiming angle

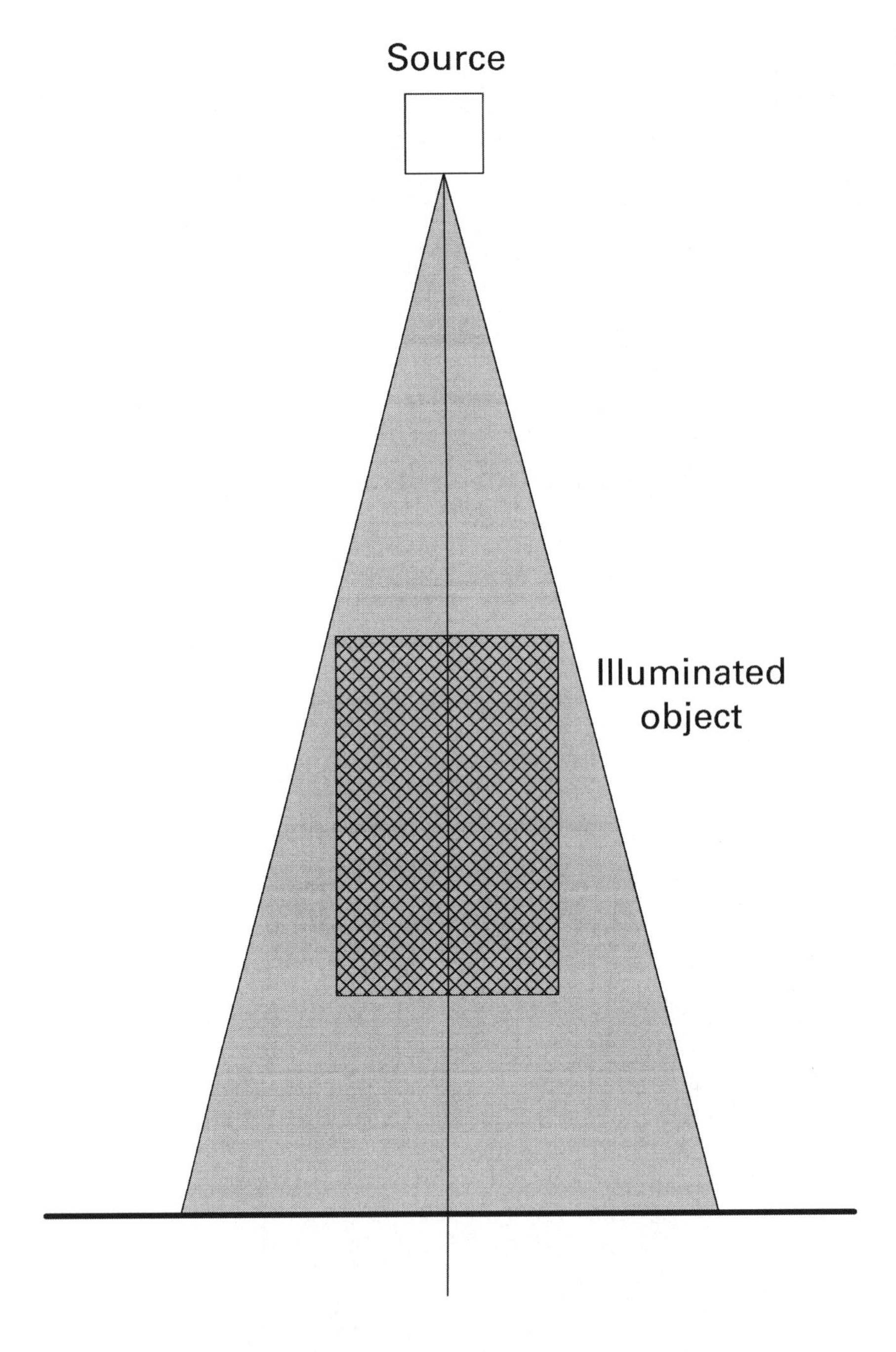

Figure 8.12.2.D
Coverage

s = the setback distance from the wall

d1 = distance from the lamp to the top beam edge intersection with the object

d2 = distance from the lamp to the centerbeam intersection with the object

d3 = distance from the lamp to the bottom beam edge intersection with the object

d4 = vertical distance from d1 to d2

d5 = vertical distance from d2 to d3

r = half diameter (radius) of the primary beam
dia = diameter of the primary beam (= 2 × r)
fc-c = footcandles at centerbeam
fc-e = footcandles at primary beam side edge
fc-et = footcandles at primary beam top edge
fc-eb = footcandles at primary beam bottom edge

Here are the calculations for each of these

Θ = given
Φ = given
s = given
$d1 = (s \div \sin\Theta) - d4$
$d2 = s \div \sin\Theta$
$d3 = (s \div \sin\Theta) + d5$
$d4 = r \text{ (at d1)} \div \tan\Theta$
$d5 = r \text{ (at d1)} \div \sin(\Theta - \Phi)$
$r = d2 \times \sin\Phi$
$dia = 2 \times r$
$fc\text{-}c = CPC \div (d2 \times d2)$
$fc\text{-}e = 0.5 \times fc\text{-}c$
$fc\text{-}et = 0.5 \times [CPC \div (d1 \times d1)]$
$fc\text{-}eb = 0.5 \times [CPC \div (d3 \times d3)]$

Here is an example using the Philips ALU111MM50W G53 12V 8D 50 watt AR111 lamp having CBC = 23,000 and an 8° beam, aimed at Θ = 30° from vertical (refer to Figure 8.12.2.E).

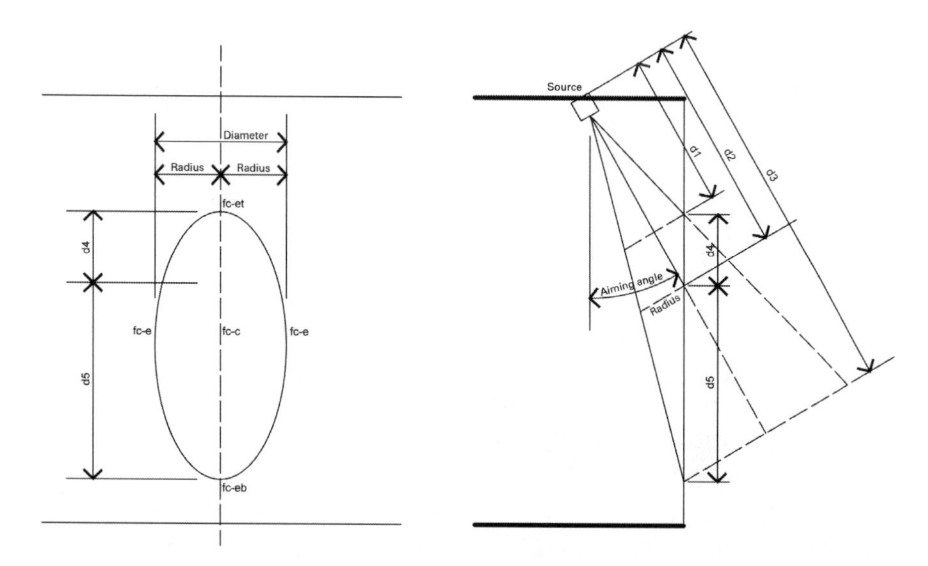

Figure 8.12.2.E
Full calculation

$\Theta = 30$
$s = 3'$
$d1 = (s \div \sin\Theta) - d4$
$d2 = 3' \div \sin 30 = 6'$
$d3 = (s \div \sin\Theta) + d5$
$d4 = 0.42' \div 0.577 = 0.73'$
$d5 = 0.42' \div \sin (26) = 0.95'$
$r = 6' \times \sin 4 = 0.42'$
$dia = 2 \times 0.42' = 0.84'$
$fc\text{-}c = 23{,}000 \div (6' \times 6') = 639\,fc$
$fc\text{-}e = 0.5 \times 639\,fc = 319\,fc$
$fc\text{-}et = 0.5 \times [23{,}000 \div (5.27' \times 5.27')] = 414\,fc$
$fc\text{-}eb = 0.5 \times [23{,}000 \div (7.04' \times 7.04')] = 232\,fc$

In order to limit the brightness range across the beam, all of these numbers must be understood. The brightest spot will be fc-et or fc-c, depending on the specifics of the source being used and the distances and angles involved. The dimmest spot will always be fc-eb.

So the range in this elliptical beam will be from 232 fc at the bottom to 639 fc at the focal point to 414 fc at the top to 319 fc at the sides. That is a ratio of 2.8:1, which is not too bad. It would be nice to stay at, or even below, 2:1, but that can be extremely difficult. Much more than 3:1 will result in obviously uneven lighting. This would be effective accent lighting at an ambient level of 23 fc to 63 fc or so.

This is a very small beam (at 8° that should be expected), at only 0.84′ wide by 1.77′ high, but it could be appropriate for a small sculpture or a large piece of jewelry in a highly dramatic setting.

If we were to change to a Philips ALU111MM 50W G53 12V 24D, 50 watt AR111 with a 24 degree beam and CBC = 4,000, this would be the result:

$\Theta = 30$
$s = 3'$
$d1 = (s \div \sin\Theta) - d4$
$d2 = 3' \div \sin 30 = 6'$
$d3 = (s \div \sin\Theta) + d5$
$d4 = 0.42' \div 0.577 = 0.73'$
$d5 = 0.42' \div \sin (26) = 0.95'$
$r = 6' \times \sin 12 = 1.25'$
$dia = 2 \times 1.25' = 2.5'$
$fc\text{-}c = 4{,}000 \div (6' \times 6') = 111\,fc$
$fc\text{-}e = 0.5 \times 111\,fc = 56\,fc$
$fc\text{-}et = 0.5 \times [4{,}000 \div (5.27' \times 5.27')] = 72\,fc$
$fc\text{-}eb = 0.5 \times [4{,}000 \div (7.04' \times 7.04')] = 40\,fc$

Now the range is 40 fc at the bottom to 111 fc at the focal point to 72 fc at the top to 56 fc at the sides, for a ratio of 2.7:1, which would be OK. This would be highly effective accent light at an ambient level of only 4.1 fc and reasonable up to about 8 fc. This lamp, under these circumstances, could light a mannequin or a human-sized statue or something else of similar size.

The next level of complication arises when we consider side-to-side aiming angles, which can vary from 0–45° or so. The math for that is extremely complicated; suffice to say, for all of these calculations, estimates of the distances involved will work well in most situations. The simplest way to account for these side angles is to apply an "increase" factor due to the side aiming. If the angle is 45°, the increase would be the square root of 2 = 1.41; at 30°, it would be 1.15, and for 15° it would be only 1.04. To give a specific example, if the distance is known to be 12′ and the side angle is 30°, then use 12′ × 1.15 = 13.8′ ≈ 14′ and move on from there.

This covers basic lighting calculations. These manual calculation methods really are quite useful and applicable to day-to-day practice for quick checking, if nothing else.

8.13 Lighting applications

In this section, basic approaches will be explained for various types of projects, including correctional, educational, governmental, health care, historic, hospitality, industrial, office, recreational, religious, residential, restaurant, and retail. Each one will be covered separately:

8.13.1 Correctional lighting

Correctional facilities include federal and state prisons (of various security levels) and local jails (county, city, parish, town, village, etc.). Lighting in these facilities is highly utilitarian and the most important factors are durability and tamper-resistance. If exposed fasteners must be used on luminaires, they must be special tamper-proof fasteners that cannot be removed without special tools, but even this must be avoided in very high security environments. (It must be kept in mind that incarcerated persons have lots of time on their hands and it might be reasonable to take a week to remove a screw; in non-correctional environments, that would never be a concern.) There are specialty manufacturers for this type of product with virtually unbreakable lenses (usually heavy gauge polycarbonate) and extra-heavy duty steel housings. Aluminum might not be tough enough for this kind of application. Luminaires should be recessed if at all possible to make them more tamper-proof. Controls are likely to be centralized and automatic.

Even though lay-in ceilings are very common in most project types, they are unlikely in correctional projects due to security issues. Hard ceilings (usually gypsum board) or exposed structure (especially concrete) are more likely and luminaire selection needs to take this into account.

8.13.2 Educational lighting

Educational facilities cover a broad range, including pre-school (maybe even day-care), elementary, middle, and high schools, and various types of higher education facilities (trade schools, community colleges, private colleges, and public and private universities).

These facilities can include nearly all types of functional spaces, including corridors, stairways, elevators, dining, kitchen (both institutional kitchen and training kitchen), office, shop, lab, classroom, toilet room, locker room, pool, gym, mechanical room, art room, break room/lounge, music performance room, music practice room, theater, library, lecture room, smart classroom, shooting range, etc., so there is a need for a very wide variety of lighting design approaches and associated technology—luminaires and controls.

Generally speaking, most spaces will be on the simple side with just one layer of lighting to serve for both ambient and task lighting; this might be even be true in classrooms, although it is very common to have grouped control so that portions of the room, or portions of the lighting, can be controlled separately. In many jurisdictions, lighting levels are regulated in classrooms (50 fc maintained average in Indiana) and food preparation areas are nearly always regulated (70 fc maintained average in Indiana).

Particularly in higher education, architectural high design is as likely as not, so sophisticated lighting design should be expected too—large-scale gathering spaces, atriums, etc.

Ceiling types in educational projects could include exposed structure (concrete, wood, steel, etc.), hard ceilings (plaster or gypsum board), lay-in ceilings (extensive in most schools), and specialty ceilings—metal, wood, and other materials.

8.13.3 Governmental lighting

There are many different governmental entities in the US: the federal government, 50 state governments, and thousands of local governments. At the federal level, design and construction of most facilities is handled by the GSA which has extensive facility requirements, which are non-negotiable. The most meaningful from a lighting point-of-view is that offices, even now, are required to be illuminated to a maintained average of 50 fc (way too high in the author's opinion). But the Department of Defense does not use the GSA so different requirements might come into play.

At the state level, each state is different. In Indiana, the main lighting regulations relate to commercial kitchens (70 fc in all food preparation areas), K-12 classrooms (50 fc average), day-care centers (50 fc in most areas), and long-term care sleeping rooms (15 fc average); all of these regulations come from the Indiana Department of Health, and not through the building, electrical, or energy codes. Anything is possible and it is necessary to be very careful to verify state and local requirements. Under most energy codes, which often have stringent limits on allowed wattage, there is an exemption for areas that have "other statutory" requirements for lighting, such as the items noted above. In those cases, whatever wattage is required to meet those rules should be acceptable, energy code or no energy code (but do not forget to check the local rules).

Just like educational facilities, government facilities can include almost any conceivable type of function—the whole educational list plus aircraft hangars, armories, barracks, correctional facilities, etc. It would be an exaggeration to say that budgets are always higher for government projects, but it is fair to say that budgets tend to be higher. This is true for two reasons: first, the government is not interested in building or owning short-term low-quality facilities (usually), and, second, the government usually wants to minimize maintenance, which requires higher quality, and more costly, systems across the board.

In terms of lighting design, this means that selected products should be high-quality (even if relatively high in price), durable, and well-suited to a long lifetime. The design approaches can vary from extremely simple to extremely complex, even including explosion-proof equipment in some cases (maintenance facilities and embassies come to mind). The entire universe of options should be available, to be tailored to the needs and budget of each individual project.

Ceiling types in governmental projects could include exposed structure (concrete, wood, steel, etc.), hard ceilings (plaster or gypsum board), lay-in ceilings (extensive in most projects), and specialty ceilings—metal, wood, and other materials.

8.13.4 Health care lighting

Much current research in lighting is related to health care, especially circadian rhythm issues, Alzheimer's disease treatment, etc. This is also related to the recent discovery of slow-acting photo-sensitive ganglion cell receptors on the retina, which seem to be related to circadian rhythm, sensitivity to blue light, etc. Just as in governmental work and higher education, a wide variety of lighting applications will be found in health care settings—plus imaging rooms, nurses' stations, med rooms, operating rooms, med gas rooms, etc.

Recently, the American Medical Association issued a report that recommends no "blue" outdoor lighting (largely due to the concern about blue, or bluish, lighting and sleep patterns). But this has been met with great controversy and passionate argument. Broadly speaking, there is a tendency in the industry to increasing color temperature—making lighting bluer, so to speak—and that flies in the face of this recommendation to reduce blue light. Part of the desire to increase color temperature is the fact that LEDs are more efficient at higher color temperatures because it takes more phosphor in the lenses to produce lower color temperatures. Human vision is also more sensitive to blue light (makes sense, eh?) so we tend to believe that we see more light as light gets bluer. And, lastly, metal halide is by far the most widely used source in existing exterior lighting and it is bluish.

The concepts are important but what really matters is what works in the project at hand. If an owner wants a warm and cozy waiting room, it would be a big mistake to use 4,000 K LED lighting; on the other hand, one would never light an operating room with 2,700 K LED sources. (Actually, many procedures in operating rooms are done with the primary room lighting turned off these days; those lights are used mostly for prepping and cleaning the room.) At the most basic level, the author believes that humans naturally relate "warm" light to low light levels and "cool" light to high light levels; at dawn, as the sun breaks over the horizon spilling yellow and even orange light (very warm), it is still pretty dark, relatively speaking. But when the sky is 5,000 K or 6,000 K—very cool—it is much brighter. So it makes sense to keep color warmer as levels are lower and cooler as levels go up.

Systems are being used now to mimic this effect in spaces automatically, especially in Alzheimer's treatment units—low level and warm in the morning; neutral and bright by late morning; cool and very bright by mid-day, reversing through the afternoon back to low level and warm by evening. The main thing that separates this from nature is that "very bright" indoors might mean a few hundred footcandles while very bright outdoors could mean 10,000 fc.

Especially in residential health care facilities (and outpatient facilities for stress-inducing treatment, such as chemotherapy), there is a strong desire for spaces to be quiet and comfortable, home-like almost. So colors tend to be warm and materials tend to be natural—wood, stone, tile, etc. The lighting should follow the lead of the space design: warm to warm and cool to cool. If a changing white scenario is to be attempted in the lighting design, the color scheme in the space should be highly neutral—neither obviously warm nor cool—so that the character of the space does not change dramatically as the lighting changes.

Ceiling types in health care projects could include exposed structure (concrete, wood, steel, etc.), hard ceilings (plaster or gypsum board), lay-in ceilings (extensive in most projects), and specialty ceilings—metal, wood, and other materials.

8.13.5 Hospitality lighting

Hospitality projects include hotels and motels, maybe casinos and restaurants. Restaurants will be treated separately in this text. As to casinos, an important purpose of the design is to eliminate all windows to assure that occupants cannot tell time according to outdoor conditions.

There are two separate aspects to most hotel projects: public areas and guest rooms. Guest rooms tend to be illuminated in ways that are similar to single-family homes to make guests feel like they are literally at home—table lamps, sconces, some low-output downlights, etc. Public spaces can be large and highly designed so the lighting is often used to accentuate the experience of arrival, lounging, eating, drinking, relaxing, etc. Many hotels incorporate conference centers, where room lighting might more closely resemble office lighting, and restaurants, coffee shops, and even small retail stores. Ballrooms are often size-adjustable with movable walls, so lighting control systems have to be highly flexible to accommodate varying room configurations.

Ceiling types in hospitality projects could include exposed structure (concrete, wood, steel, etc.), hard ceilings (plaster or gypsum board), lay-in ceilings, and specialty ceilings—metal, wood, and other materials. Lay-in ceilings are somewhat less likely in this project type, except for back-of-house areas.

8.13.6 Industrial lighting

Needless to say, lighting for industrial projects tends to be highly utilitarian. In the 19th century, industrial facilities were built with extensive windows and skylights (saw-tooth roofs were very popular with the north-facing facets being all glass) but more recent facilities tend to be entirely windowless. For many years, this meant that fluorescent or HID systems were used to light the entire space to task levels. That has changed somewhat, and now it is more common to see separate low-level ambient lighting with localized high-level task systems in individual work areas. This is more energy efficient and it can make spaces look more attractive than endless seas of luminaires in vast spaces.

Ultimately, what matters in industrial projects is performance. If the lighting does not make it possible to build or operate whatever needs to be built or operated, it will have failed and it could well be dangerous too.

Ceiling types in industrial projects could include exposed structure (concrete, wood, steel, etc.) in most areas with possible hard ceilings (plaster or gypsum board), lay-in ceilings, and specialty ceilings—metal, wood, and other materials—in offices, lobbies, etc.

8.13.7 Office lighting

When the author started practicing in 1981, the standard of the industry in office lighting was the 4-lamp (F40T12) prismatic-lensed 2′ × 4′ fluorescent troffer in quantities required to achieve 100 fc average everywhere. That sounds awful today and it actually was awful at the time. But it was the standard anyway.

By the mid-1980s, desktop computer screens were becoming commonplace and glare control became important; early CRT screens were curved, dark, and shiny—excellent reflectors for bright spots (e.g. luminaires) overhead. The deep cell parabolic fluorescent fixture was invented to reduce glare and had become the "standard first upgrade" for office lighting by the late 1980s. Also by the late 1980s, the 4-lamp troffer had given way to the 3-lamp troffer (which is more efficient) and the T8 lamp was available. The biax lamp was available in the early 1990s.

Even now, standard office lighting is still prismatic-lensed 2′ × 4′ troffers, with deep parabolic 2′ × 4′ troffers as the first upgrade, and basket troffers as the second upgrade. This has finally begun to change and we are seeing many 2′ × 2′ LED troffers now (most commonly the faux indirect look), and that is what the author has throughout his firm's office (except his personal office, which has LED fixtures that *are* the $\frac{15}{16}$″-wide ceiling grid). Progressive office design includes removal of continuous ceilings, using clouds and exposed structure in many areas, which has led to more use of indirect/direct lighting. LEDs have brought smaller luminaires with better optics to the market, so it is feasible to do better indirect/direct lighting than ever before. But it is still considerably more costly than troffers in the ceiling.

The most common trend in office light at the moment is the use of narrow linear fixtures across ceilings, suspended below structure or ceilings, or even in walls.

The most common width is probably 4″ but 2″ is gaining rapidly; smaller products (1″ or even less) are sometimes used where higher budgets allow. The most common aperture for recessed downlights is probably 4″ but 2″ luminaires are catching on in this area too.

Up to the mid-1980s office ceilings were usually installed at 8′ 0″; then the standard was 8′ 6″ for many years, and some "premium" buildings have 9′ 0″ as the standard. But today, the trend is to push ceilings as high as possible or to eliminate them entirely.

As noted previously, the author has been designing office lighting to 30 fc average for decades and he expects to continue to do so for the foreseeable future. But some people still want higher values, while others will accept lower values. Whatever the target level might be, it is vital to remember that what matters most is the perceived brightness and not the actual brightness.

8.13.8 Recreational lighting

This category includes indoor sports facilities, such as gyms, racquetball courts, pools, bowling alleys, etc. Glare control is important in these facilities to minimize reflections of bright luminaires on shiny floors or the surface of a pool (especially critical for the latter). Pools in particular are often illuminated using indirect techniques, but it is important to remember to include some small amount of glare for the psychological impact. Since the 1970s, most of these facilities have been illuminated with HID sources (mostly metal halide) and with some high-powered fluorescent systems, but any of them can be done with LEDs today.

These facilities commonly have exposed structure without ceilings, but ceilings are commonly seen in courts, pools, etc. Spaces tend to be very high, so accessibility for maintenance should be considered when selecting and locating luminaires.

8.13.9 Religious lighting

This category includes churches, mosques, and temples and their related facilities, and the required lighting can be highly theatrical, highly decorative, highly dramatic, or all of the above. The architecture can be dramatic as well—usually in the form of very high spaces—and a solid understanding of accent lighting is key. A typical sanctuary chancel in a church might include a choir, a piano, an organ console, lectern, pulpit, and sometimes a baptismal font; each of those would require at least two aimable luminaires (adjustable recessed downlights, canopy-mounted track heads, track mounted track heads, etc.) and the final aiming is crucial. One should plan for a couple of hours on site with the electrical contractor to set up the aiming properly. In a space that is used for weddings, it is a nice feature to have separate aisle lighting to accentuate the wedding party walking down the aisle, but such hardware is not needed for other events. If a space is used frequently for funerals, additional aimable luminaires for a casket might be needed.

Control systems are also key for worship spaces so that scenes can be changed easily from multiple locations (near doors for ushers, at the pulpit, sacristy door, etc.).

But these facilities can also include fellowship halls, gyms, dining rooms, classrooms, offices, etc., so many different luminaire types might be needed. Ceilings tend to be gypsum board (public spaces) or lay-in (offices, classrooms, corridors, etc.) and higher than they would have been 20 years ago.

8.13.10 Residential lighting

This category includes one- and two-family dwellings (houses) and multi-family projects (apartments and condominiums) from a few rows of houses to hundreds of apartments in large high-rise buildings. In large projects, the lighting tends to be much like hotels—more dramatic public spaces with higher quality hardware more sophisticated treatment and very home-like private spaces. The smaller the project is, the more likely it is to be illuminated like someone's house.

But how does one illuminate a house? This has changed greatly over the years. When electricity first came to houses, there might have been wall sconces and pendants in most rooms. After World War II, as housing developed very rapidly in most part of the US, it became common to put on-ceiling glass fixtures in bedrooms, hallways, and a few sconces here and there. A bit later again, the surface-mounted fixtures largely disappeared from bedrooms, to be replaced by switched receptacles to use with table or floor lamps. In more recent years, recessed downlights have become popular in houses, even in bedrooms.

Like everything else about lighting, this is all about atmosphere—the look and feel of the space. If someone is going to read in the living room, it is necessary to have light for reading—or in the den, bedroom, dining room, or anywhere else. It is necessary to have higher light levels in kitchens for food preparation, in laundry rooms, and maybe in the garage if there is a workshop.

Ceilings in residential environments tend to be mostly gypsum board; heights were very high in the 19th century (10′ to 14′), rather low for most of the 20th century (8′to 9′), and trending upward again early in the 21st century. Ceilings 9′ and 10′ high are not usual these days, even in basements. Ceilings like this tend to be attached directly to the framing above (usually wood joists, rafters, or roof trusses), so it is necessary to coordinate recessed fixtures with the framing.

8.13.11 Restaurant lighting

Restaurants are a form of retail store and they are commonly associated with hospitality sites. They range from crude food shacks to elaborate fine dining establishments, and they have lighting from the simple to the sublime. Nearly all restaurants need control systems to manage changing scenes throughout the day (and for morning prep and night cleaning). Light levels are usually regulated in food preparation spaces but possibly nowhere else, and some restaurants are quite dark. (It is not uncommon to see older folks in restaurants using small flashlights or smart phone flashlights to read menus, the author included.)

Color accuracy is very important because low CRI lighting can make food look very unappetizing. (A grocery store near the author's home has metal halide lighting throughout much of the store, including the check-out lanes, and red meat tends to gray and gross in that area.) Dimmability is usually very important.

The nature of incandescent and halogen lamps is that they color shift to yellow and then to orange as they dim lower and lower, and Edison lamps are always orange-yellow. The author was very happy when LED luminaires came along that do *not* change color when they dim, and then he got in trouble on a restaurant project because the owner wanted the lights to turn orangish at low levels; the solution was to add orange color gels to the LED luminaires, making them always orangish. Due to the common desire for this color shifting at low light levels, LED manufacturers have figured out how to duplicate the effect. It is usually called "warm dim" or "warm glow" and the LEDs go from 3,000 K to 1,800 K as they dim. The author commonly uses sources like this in restaurants now, but it is something that should be discussed for each project. All manufacturers cannot do this, but there are several good options for those who can.

Current restaurant design trends are for high spaces with minimal acoustically absorbent material—loud in other words. Low light levels can help to hold down noise a little, but only a little. If ceilings are used, they are often colored; the heights can vary from 9′ upwards. Kitchen ceilings are most cost effective at 8′ 5″ (just below the tops of the hoods) but they are often higher if they are visible from the dining spaces or the bar.

8.13.12 Retail lighting

This category includes stores of all types: big boxes to tiny boutiques; jewelry to lawn mowers; clothing to autos; and everything in between. A high-end mall in Indianapolis has all kinds of outlets: Apple store (ultimate minimalism); a Tesla store (including full sized cars); Saks Fifth Avenue; high-end dining and many other stores. Lighting for retail is usually all about the product, and making the product look so good that it will sell itself. How is this done?

There are various different ways, and different retailers have different approaches. Low-end big box stores are illuminated mostly with only the ambient layer (with some accent lighting in produce and maybe task lighting in jewelry) and the main concern is that customers can see everything easily and that the system is not costly to operate. A high-end jewelry store is more or less the opposite: low ambient lighting with bright accent lighting on the products in the display cases to evoke luxury and comfort.

Particularly in an enclosed mall, there is a technique for lighting a retail store that can improve sales, and that is to create the brightest spot in the store on the back wall. The reason for this is that human eyes naturally look to the brightest spot in the visual field, no matter what it might be, and this is involuntary. So by putting a bright spot in the back, that brings eyes to the back of the store; the only way for the person who is looking at the back of a store to re-orient vision to the mall or another store is to move visually from the back of the store to the front, seeing much of the merchandise in the store on the way. That is a good way to force customers to look at merchandise, but it is so subtle that most people do not even realize that it is happening.

Ceilings in stores can be almost anything imaginable and at heights from very low to very high—it all depends on the look and the merchandise.

8.14 Egress and emergency egress lighting and exit signs

Code requirements for egress and emergency egress lighting are widely misunderstood (by code enforcement officials and designers alike), but they are not all that confusing.

First, egress lighting is required in all occupied spaces whenever they are occupied, but this is covered by standard room lighting so it is not usually challenging.

Second, emergency egress lighting and exit signs are required for the egress path whenever two or more means of egress are required. When exit signs are required, it is also required that an occupant be able to see two exit signs from any given point along the egress path (this can cause lots of exit signs in corridors that go around a lot of corners). It is sometimes said that emergency egress lighting and exit signs should be provided in all egress paths, whether required or not, but that does not make sense. If that were true, wouldn't there need to be exit signs and emergency egress lighting everywhere? In every room? This is a little bit confusing because the IBC does say that emergency egress lighting is required throughout, but then it has a series of exceptions that make it clear that it is only required when two or more means of egress are required.

The most common condition where two exits are required is when occupant load exceeds 49 (there are important exceptions so it is necessary to be careful); this can be the case in a conference room that is larger than 735 sf (15 sf/occupant × 49 occupants); or in an open office that is larger than 4,900 sf (100 sf/occupant × 49 occupants), or in any combined space, floor, or facility where the combined occupant load exceeds 49. (It is worth noting that this is how the egress path is defined as well; below the occupant load threshold, there is no egress path at all, which helps to explain why exit signs and emergency egress lighting are not required.)

When emergency egress lighting is required, the level must be 1 fc average (on the floor) with a minimum of 0.1 fc *along the egress path*. This must last for 90 minutes although the levels are allowed to drop by 50% at the end of the 90 minutes. The back-up power for this can be an on-site generator, batteries in individual luminaires, or centralized battery systems (called inverters).

There are two basic techniques that are commonly used:

1. Individual emergency-only luminaires, which are often wall-mounted along corridors and which are often called "bug-eyes." They are unattractive but low cost. It is difficult to meet the photometric requirements (1 fc average) using hardware like this.
2. The other technique is to use normal luminaires with internal batteries or connected to inverters. This is less obtrusive (although each battery requires a visible red test switch) and more effective but also oftentimes more costly than using bug-eyes.

The critical issues here are performance and maintenance. Performance is much better with approach (2) as is maintenance. Batteries are required to be tested monthly, which rarely happens with bug-eyes or even with in-luminaire batteries; inverters have internal computers that do this testing automatically, which eliminates most of the maintenance. But the owner has to be convinced that an inverter is affordable and that is not always easy to do.

8.15 Lighting design documentation

Documentation for lighting design has two parts: (1) design presentations and (2) construction documents. Design presentations are usually pretty basic unless photometric calculations and/or models are involved. Early lighting design presentations tend to include mostly images or drawings of luminaires (usually culled from "cut sheets" or "data sheets" or "product sheets" from manufacturer web-sites) coupled with descriptions of effects—wall washing, accent lighting—and a discussion of performance. Performance usually includes footcandle targets, possibly brightness ratios, and control schemes. There might be hand sketches or models to illustrate the design approach as well. Later in the design process, this will get more detailed and plans (and maybe elevations) will be drawn to shown luminaire locations, installation methods, and possibly control grouping.

For construction documents, the latter phase design drawings are completed to include *all* hardware (no matter where it is mounted), full control grouping, and full product specifications for all luminaires (usually on a Light Fixture Schedule). Controls might require additional diagrams and/or schedules if they are complicated.

There is much confusion in offices about how a lighting plan differs from a Reflected Ceiling Plan. First, the RCP is a drawing for the construction workers who build the ceiling(s) and it is *not* for the electrical contractor (EC) who installs the lighting. Second, the RCP should show everything that is attached to, or recessed into, the ceiling, including HVAC, low-voltage, etc. much of which does not matter for lighting. The lighting drawings are intended only for the EC, so they must show *all* elements of the lighting—ceiling recessed, surface ceiling, and ceiling suspended luminaires, wall-mounted luminaires, floor-mounted luminaires, and full control information. (Wiring can be shown on separate power drawings, but these are often combined together so that there are "power" and "lighting" pages in the construction document set.)

The complexity of projects varies from extremely simple (maybe just one luminaire type) to extremely complex (a recent new hospital project in Indianapolis is said to have had 650 luminaire types), and the drawings vary accordingly. But the basic techniques are always the same; identify, specify, and locate each luminaire and show how controls will work.

Finally, even if the ID selected luminaires and located them on drawings, when the construction documents have been completed the final lighting information should be found (nearly always) on the electrical drawings and that information should no longer appear on interior design drawings. This is done simply to make the drawings as clear as possible for the construction team.

Summary

This whirlwind overview is intended only to familiarize the reader with the basics of lighting design. There is much more to be learned and it is best to learn that by doing. There really is no substitute for experience, so dive right in and give it a try. There are plenty of resources out there to help: professional colleagues, local IES chapters, local IALD chapters, local universities, lighting reps, manufacturers, etc. And there is always the 10th edition of *The Lighting Handbook* from the IES, if you can afford to buy a copy.

Lighting design really does tie many of the other building systems together to make spaces and places beautiful and functional. To return to the beginning of this chapter, good lighting design can greatly improve a weak design and bad lighting design can ruin a brilliant design. And nothing is beautiful, or functional, if it is in the dark.

The subject used to be complicated by having to understand and know how to use several different sources: incandescent, halogen, fluorescent, compact fluorescent, metal halide, high-pressure sodium, cold cathode, and even induction. Today, there is essentially only one source: LEDs. But that one source can be as complex as all of the legacy sources combined—almost unlimited options for lumen output; more color variety; more optical options; anything goes controls; and on and on.

So, in a way, this is really the most exciting time in lighting design that we have seen to date. It always was exciting to see new technologies come along, but all of the legacy sources were blunt instruments when compared to LEDs. We did not use 105 watt 3-T8 2′ × 4′ troffers because we really needed 8,900 lumens in a 2′ × 4′ housing; we did it because it was easy. And the result was that we over-lit virtually everything. LEDs give us the option to dial everything down to the levels that we would really like to see, and not just pile it on to be safe. Of course, the closer we try to hit footcandle targets, and the lower those targets get, the greater the risk. If we missed an average level of 100 fc by 10 fc, no one would ever have noticed, but if we miss a 15 fc target by 5 fc, we could get into trouble for "underlighting" something. But all-in-all, this is a good thing because we will use less light, and less energy, in a more appropriate manner—is that not at least one definition of sustainable design?

Outcomes

8.1 Understanding basic indoor architectural lighting and daylighting.
8.2 Understanding human vision, especially color and acuity.
8.3 Understanding the history of artificial lighting, especially the various sources.
8.4 Understanding quantity of lighting—lumens, footcandles, and centerbeam candela.
8.5 Understanding general optics—the shaping and control of light.
8.6 Understanding quality of lighting, especially CRI and accuracy.
8.7 Understanding sources, especially the difference between the legacy sources and LEDs.
8.8 Understanding energy conservation and sustainable design in lighting.
8.9 Understanding lighting design by layers—ambient, task, accent, and decorative.

8.10 Understanding daylighting.
8.11 Understanding luminaires, including controls.
8.12 Understanding basic lighting calculations, zonal cavity, and accent.
8.13 Understanding lighting applications across many different project types.
8.14 Understanding egress and emergency egress lighting and exit signs.
8.15 Understanding lighting design documentation.

Notes

1 *The Lighting Handbook*, 9th Edition
2 *The Lighting Handbook*, 10th Edition, page 4.33
3 Ibid., page 32.8.
4 Ibid.
5 Ibid., page 32.10.

Chapter 9

What are power systems?

Objectives

9.2 To understand the basics of voltage, amperage, and electrical power
9.3 To understand the basics of electrical equipment
9.4 To understand the basics of over-current protection
9.5 To understand code requirements for working space
9.6 To understand code requirements for large equipment
9.7 To understand the basics for various low-voltage systems
9.8 To understand basic power for lighting
9.9 To understand power documentation

9.1 Introduction

Electrical power affects nearly everything that we humans do, at home, at work, while shopping, visiting the doctor, going to a museum or concert, sporting event, etc. So wiring is almost literally everywhere (even inside a computer) and it has a real impact on *all* design projects. The objective of Chapter 9 is to provide a basic understanding of terminology and technology so that interior designers can avoid negative impacts on their designs—the same point-of-view as for HVAC and every other topic in this book. As has been the case since the Preface, the overall objective is to assist both students and practicing interior designers in improving their design work, and even more importantly, their completed projects.

9.2 Electrical power: the volt, the ampere, and the watt

Electrical power is measured in **watts**. A watt is a small unit that is the product of **voltage** and **amperage**.

The most commonly understood form of electrical power is the rating system for old-style incandescent light bulbs, which are commonly available in 25, 40, 60, and 100 watt versions. In most parts of the US, the voltage that is supplied to a lamp socket is 120 volts, so 60 watts = 120 volts × 0.5 amperes. To be more precise, the unit of volt-ampere should be used instead of watt because volt-amp = watt only when something called the power factor (PF) = 1.0. For incandescent lighting and electric heaters, PF = 1.0, so volt-amp = watt. For other types of loads—motors, compressors, electronics, etc.—the PF is not equal to 1.0, so a volt-amp is not equal to a watt. In such cases, PF is most commonly less than 1.0 which means that apparent wattage is higher than volt-amps. This is not of great concern to most interior designers, but it can be critical in the design of some power systems.

This calculation applies to all single-phase, single-pole and 2-pole loads, but it does not apply to 3-phase, 3-pole loads. For 3-phase loads, there is a standard power correction factor that has to be applied, and that factor is 1.73 (the square root of 3 for those who might care). Here are some examples to show power capacity at various voltages:

Single-phase, single-pole: 50A × 120V = 6,000W = 6.0kW
Single-phase, 2-pole: 50A × 240V = 12,000W = 12.0kW
Single-phase, 2-pole: 50A × 208V = 10,400W = 10.4kW
3-phase, 3-pole: 50A × 208V × 1.73 = 17,992W = 18.0kW

9.2.1 Voltage ranges

In the US, virtually all electrical design work is controlled by NFPA 70, the National Electrical Code (2017 is the most recent version), which is the most dominant of all code books, nationwide.

The NEC defines three ranges of voltage:

1. High voltage, meaning >1,000V[1]
2. Low voltage, meaning <24V[2]
3. "Medium" voltage, meaning >600V but <1,000V[3]

There is no formal definition for voltage between 24 and 600 volts, but this is the range commonly referred to as "line voltage." For most practical purposes, everything in the power system will be line voltage according to these definitions.

9.2.2 Standard service voltages

In most areas of the US, there are four commonly used service voltages:

A 120/240V, single-phase ("2-pole")
B 120/240V, 3-phase ("3-pole," usually called "open delta" or "high leg delta")
C 120/208V, 3-phase ("3-pole," usually called "wye")
D 277/480V, 3-phase ("3-pole")

System A is what most people are familiar with at home. There are three wires coming in from the utility (overhead from a pole or underground). The voltage from each "hot" conductor, of two, to ground is 120V; the two phases are 180 degrees opposite one another, so when both are tapped simultaneously (as for an electric laundry dryer, an electric range, or an electric air-conditioner), the voltages add up: 120 + 120 = 240.

This system can be used for small commercial buildings too, but it cannot be used in buildings that require 3-phase power (see below) for a commercial elevator or other large-sized motors. Please note that this is not 230V and it is not 220V; the author has never encountered a utility company in the US that provides 220 or 230 volts. All of them provide 240 volts for 2-pole loads and 120 volts for single-pole loads.

At this point, it is necessary to talk about receptacles—wall plugs. Standard receptacles for portable appliances, including computers and many other items, are 120V but they can be either 15A or 20A and grounded (3-prong) or ungrounded (2-prong—rare these days). Receptacles are used for larger loads too at 120V, 208V, 240V, 480V single- and 3-phase, and even 277V. What matters here is that each size, or capacity, of receptacle is unique; in other words, each one has a unique plug configuration that

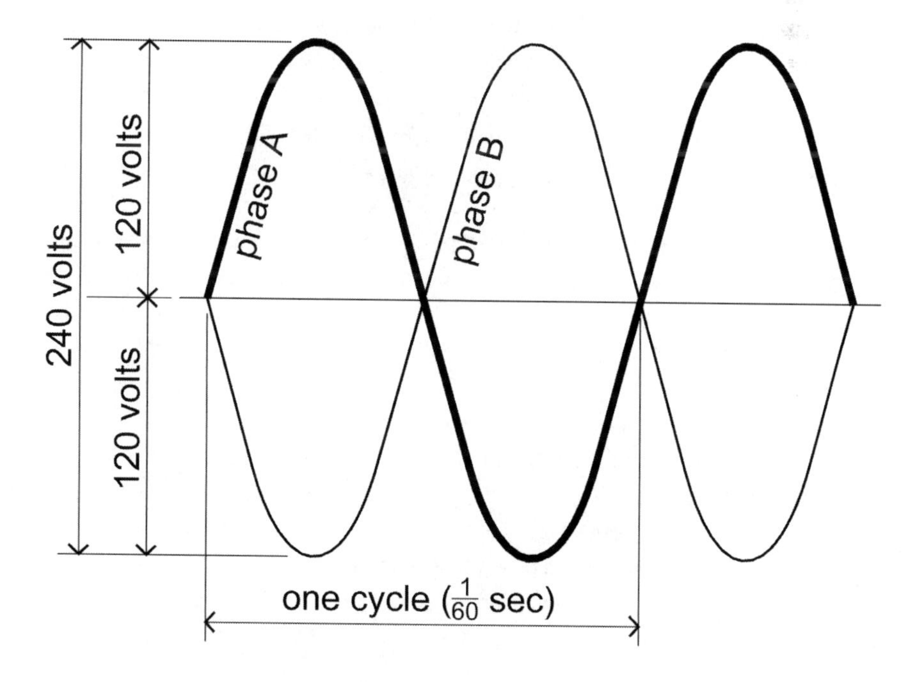

Figure 9.2.2.A
120/240V, single-phase system

requires unique over-current protection and even wire sizing. This does mean that the wiring for a 30A, 240V receptacle is different from that for a 60A, 240V receptacle and one cannot be used for the other. (Actually, the 60A circuit can be used for 30A if the over-current protection is changed, but that is not often done.) This matters because it is common parlance in the industry, even among electrical contractors, to say to provide a "220 outlet" but that statement is entirely inadequate in determining exactly how to do that. In order to provide that outlet, the amperage and phasing must be known and each circuit must be designed individually.

Next, the term "dedicated" circuit needs to be understood. Under the NEC, a standard 20A, 120V duplex (two outlets) receptacle is counted as 180 watts of load, so it is perfectly legal to put ten such receptacles on a single circuit. But it rarely makes sense to do that. In residential settings, some receptacles are required to have dedicated circuits, usually in kitchen and bathrooms. Dedicated circuit means that there is just one receptacle on that circuit. The reason for this is so that the circuit breaker does not trip when the blow dryer, or toaster, or microwave oven, blender, etc. is turned on. The capacity of a 20A circuit is actually 1,920 watts, so that circuit breaker should trip if two blow dryers are plugged into the same duplex receptacle. (It is possible to wire duplex receptacles so that each half is on a different circuit, but that is not often done—it is more costly, if that is not obvious.)

This confusion is encountered commonly in park and street improvement projects (outdoors) where the owner, usually a town or city, wants to provide wiring to vendors for street or park festivals, farmers' markets, etc. As soon as someone in a meeting says, "Let's provide some 220 outlets," the author is forced into an explanation of the wiring complications of that vague statement. And it is a rare owner who actually knows what is really needed: six 30A @ 208, single-phase; four 40A @ 208, single-phase, 20 standard 20A @ 120, etc.

There are two more receptacle terms that need to be understood: ground-fault and arc-fault. In ground-fault, if a receptacle were to get wet, it might fault to ground because the water could provide a conductive path between the hot and neutral or hot and ground. This is dangerous so the NEC has long required ground-fault protection for receptacles in areas that could be wet: kitchens, basements, crawl-spaces, and outdoors. This can be done in two ways: (1) by using ground-fault circuit breakers in panelboards or load centers; or (2) by using ground-fault receptacles. The latter is far more common and nearly everyone has seen one of them; these are the receptacles that have "trip" indicators and "re-set" buttons. Arc-fault is very different. Whenever an electrical connection is made (if the power is on), there is a small arc—essentially a tiny lightning bolt—that jumps across the gap between the two electrical parts just before contact is made or just as contact is broken. (Sometimes, you can see this when you unplug something from a receptacle.) This applies to 20A, 120V receptacles like everything else. If such a receptacle is located in a place where an object might inadvertently knock the cord loose but not completely out of the receptacle, it is possible for the arc to cause a fire. (The most obvious example would be a pillow falling off a bed.) So the NEC and the IRC require arc-fault receptacles in many areas in residential settings (basically everywhere a ground-fault is not required). This can be done only at the circuit breaker though, so the receptacles themselves are the same.

System B is an old system that is said to be rarely found these days (in truth, the author encounters it on a regular basis). This was the first 3-phase system used for commercial buildings. In this case, 3-phase power is achieved by putting the 3-poles in a triangular configuration, so that the peak voltages for two poles add up: 120 + 120 = 240. In reality, the third pole is usually at a higher voltage (usually 180V or so) and it cannot be used for single- or 2-pole loads, although it will work for 3-pole loads. This results in an inherently unbalanced system that is used less and less often. In fact, if possible, the author changes this system to system C during renovation projects if that is feasible.

System C is the most common system used for commercial buildings. Here, 3-phase power is different from the 120/240V delta system because the phases are deliberately kept separated—by 60° between the phases. That means that the peaks of the sine waves never cross (add up). The crossing point occurs where the voltage is equal to 104V, so adding two poles gives 104 + 104 = 208V.

Adding all three poles also gives 208V. This system can provide 120V for single-pole loads, 208V for 2-pole loads, and 208V for 3-pole loads.

System D is the same (essentially) as system C except that the base voltage is increased from 120V to 277V. In this case, the waves cross where the peak voltage is 240V, so 240V + 240V = 480V.

This system is used for equipment (usually at 2- or 3-pole) and for lighting (usually 1- or 2-pole, but not for incandescent lamps that must operate at 120V) in large facilities. The reason for using these higher voltages is that wire sizing is based on amperage, so that the same size wire can carry 100 amps at 480V as at 120V—or four times the total amount of power.

But it should not be assumed that all commercial buildings should have, or will have, a 277/480V power system. Even though the wiring for the 277 and 480V circuits will be less costly than at 120 or 208 volts, there is a downside. In all facilities, it is necessary to have 120 volts available for convenience receptacles (wall outlets). If the main power system is 277/480V, there must be transformers inside the building to provide 120/208V power for the receptacles. Such transformers are costly, large, noisy, and generate lots of heat, so it only makes sense to do a 277/480V system when the savings in the 277/480V wiring are larger than the additional costs to add transformers (and other equipment) and to pay for the space and additional cooling they require.

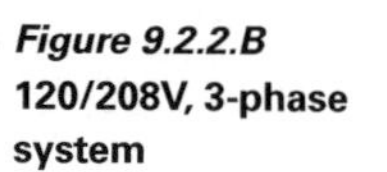
***Figure 9.2.2.B* 120/208V, 3-phase system**

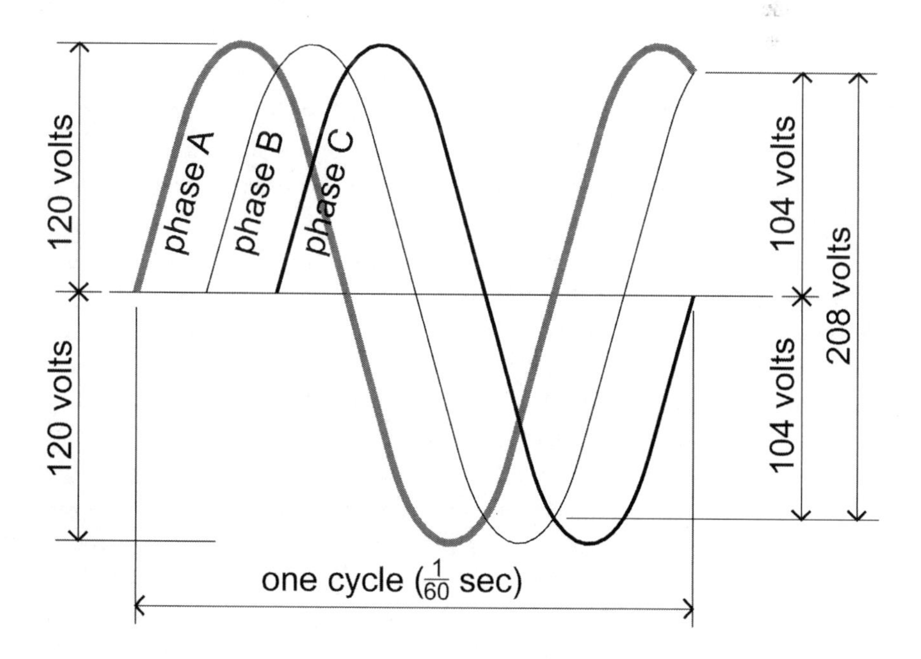

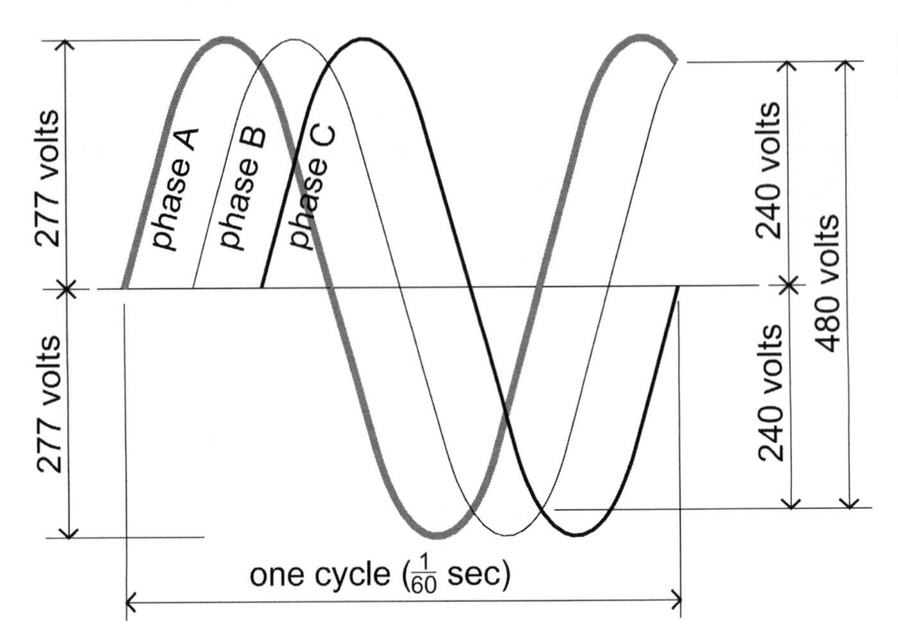

Figure 9.2.2.C
277/480V, 3-phase system

When is 3-phase power (at either 208, 240, or 480 volts) required? Only if there will be motors that require 3-phase power; although there is no strict cut-off, this is usually the case if there are motors that are larger than 10 horsepower or so. To provide context, a 20 hp motor is very small for an elevator (and 40 or 50 hp is not uncommon), and a 10 hp motor in an air-conditioner is equivalent to about 10 tons of cooling—not large at all. If there is a commercial elevator (residential mini-elevators are designed to run on 240V, single-phase power), a chiller, medical imaging equipment, or even RTUs that are larger than 10 tons, it is usually going to be necessary to provide 3-phase power, one way or the other.

9.2.3 Transformers

There are two different classes of transformers in building power systems: utility and non-utility.

All utility companies use transformers to reduce voltage from the distribution voltage (4,160V, 12,470V, or 13,200V usually) to one of the four systems already described. Many years ago, this was usually done with tub-style transformers hanging on poles (or standing or pole-mounted racks for very large ones) for all of the possible service voltages. But since the 1970s, ground-mounted transformers have become far more popular. For 120/240V 1-phase systems, there are usually very small (3′ × 3′ × 3′ or so) and installed on pads near streets or driveways. For 3-phase systems, they are usually much larger (6′ × 6′ × 7′ or so) but still installed on pads near streets or driveways. All of these transformers are liquid-filled, meaning that their internal parts are cooled by a combination of conduction (contact with the cooling liquid) and radiation (from the transformer casing—large ground-mounted transformers have obvious heat sinks for this purpose). Liquid-filled transformers are not more efficient than dry-type transformers, but they are much less costly. They are also more dangerous because if they explode, they can fling burning liquid all around them.

For 277/480V systems, as noted above, it is necessary to have transformers to lower the voltage to 120/208V. (This can be done for some large facilities by bringing both 277/480V and 120/208V directly to the building from the utility, but some utilities are not very fond of the idea and will not cooperate.) If the transformers are internal to the building, they are nearly always dry-type (meaning air-cooled, not liquid-cooled) transformers, mostly due to safety concerns. They are expensive ($15,000.00 for "medium" sized), big (3′ × 2′ × 4′ for "medium" sized), and very heavy (1,500 pounds or so for "medium" sized) and they can be noisy. They also generate substantial amounts of heat. The NEC requires such transformers to be "readily accessible," so they cannot be installed above ceilings or hanging in open spaces where workmen cannot get to them easily. Ideally, they should be floor-mounted, but they do take up substantial space. In a building with a 277/480V system, there could be one large transformer or several smaller ones, depending upon how the power system is configured.

Even though most transformers are used to reduce voltage, they work both ways and we sometimes see transformers that are used to increase voltage. These are called "buck-boost" transformers, and they can be either liquid-filled (not used indoors) or dry-type and they are available to go from virtually any input voltage to virtually any output voltage. A common application is to see a small buck-boost transformer in a church with a 120/208V, 3-phase power system to provide 240V single-phase power to the pipe organ. (Most mechanical equipment is now designed to operate from 200 to 250 volts, which has greatly reduced the need for special transformers.)

9.2.4 Surges, phase drops, brown outs, black outs, and grounding

A *surge* occurs when the voltage moves above the normal range. The voltage that is delivered from the utility is never constant—it changes continuously according to the loads on the system—and it can vary 5% from the standard rated voltage. That means that, under normal circumstances, a 480V service can vary from 456V to 504V and a 208V service can vary from 198V to 218V. (This is part of the reason that equipment is often rated for 200–250V.) If 505 volts, or more, comes through the 480V wiring or 219 volts, or more, comes through the 208V wiring, that is called a surge.

Many years ago, this was a minor issue because most equipment (and lighting) was not very sensitive to voltage changes. But modern electronic equipment can be very sensitive to such changes, so we see surge protection in many projects today. (You probably have surge protection at home or at your office, laboratory, or classroom, in the form of the plug strip that you plug your computer into.) Given the extent of electronics in modern buildings—servers, phone systems, computers, HVAC controls, lighting controls, fire alarm systems, security systems, audio/visual systems, etc.—it has become common practice in recent years to build surge protection into the service equipment for the building. This does not mean that one should expect to see existing integral surge suppression on the main service to an existing building; in fact, few buildings have such protection.

Phase drop means that the utility's source loses one or two phases—they simply go away and no power is available. For a 3-phase load, such as a motor, dropping a phase usually causes major damage to the motor—within a matter of seconds. Utility companies usually deny that they have phase drops, but it is a common occurrence nonetheless. The best protection for this is a "phase loss relay" for each large 3-phase motor (or load) which will turn off power to the motor if a phase is dropped (or if two phases are dropped). Such devices are commonly used. Again, it should not be assumed that such protection is already there in an existing building.

Brown out means that the utility's system struggles to keep up with the load, which could mean phase drops, voltage sags (the opposite of a surge), or short-term losses in limited areas (the utilities have fuses and switches throughout their systems that can fail individually or that can be operated deliberately to reduce load on the system).

Black out means that the power simply goes out completely in an area, anything from a small area (from a single sub-station transformer failure) to a large region.

There is one other issue about power quality that should be mentioned: grounding. Grounding is dedicated wiring in the system that is designed to shunt excess power and faults literally to the ground. Excess power occurs whenever a switch is operated—when the switch is moved to the "off" position, the power that was flowing from the source to the load is suddenly cut off and a little power is trapped in the system. That small amount of power is shunted to the neutral, which takes it back to the panelboard where it is transferred to the ground. Fault currents occur when something goes wrong—two bare wires contact (called a short circuit), something overheats, there is a surge, there is some kind of excessive noise in the system, etc. These currents are also shunted directly to the ground, mostly to improve safety in the system.

Fifty years ago, grounding was not all that important in a building because the systems in those buildings were not very sensitive to grounding issues and because there were fewer issues. All of our modern electronic systems, in addition to being convenient, wreak havoc in the power system, creating all sorts of problems that we simply did not have 50 years ago. All of these electronic systems are highly sensitive to differences in ground. In other words, your PC might not be bothered by a ground resistance of 15 ohms (3 ohms is very low; 30 ohms is high) but there could be a problem if it is connected to a printer that is on a circuit having a 3 ohm ground—it is the difference between 3 and 15 that is the problem. For a long time, this was not well understood and many people in the computer world demanded "isolated" grounds for their equipment. In fact, isolating the ground is likely to increase problems and not to solve anything. What is best is to get everything to the same, consistent, level of ground resistance, at the lowest level feasible. But it would be better to have the whole system at 8 ohms than to have part of it at 2 ohms and part of it at 6 ohms.

Dedicated ground **buses** used internally in **load centers**, panelboards, **distribution boards**, and **busways** are also commonly seen in data centers and even in server rooms; the latter devices are readily apparent—exposed and usually wall-mounted—so it should be obvious if there is a dedicated ground bus in a given space.

9.3 Equipment

The primary purpose of electrical equipment is to provide **over-current protection** for wiring and fault protection for safety. Power systems are configured to bring in one circuit to a building (up to six circuits under some special circumstances) and then to break that circuit down so that it is useable for individual loads in the building. At each point where the circuit is sub-divided into smaller circuits, there must be additional over-current protection.

Here is an example of a typical single-family dwelling:

- *Incoming service*: 200 amps at 120/240V, 1 phase (2 poles)
- *Service wire protection*: 200 amp, 2-pole main circuit breaker that protects the wiring from the utility to the panelboard or loadcenter
- *Branch circuit protection*: circuit breakers for each individual load, probably a 50 amp, 2-pole for an electric range; a 40 amp, 2-pole for an electric dryer, a 40 amp, 2-pole for the ACCU, a few 20 amp, 1-poles for lighting, and several 20 amp, single-poles for receptacles

This entire system is contained in a single cabinet, called a loadcenter or panelboard.

Here is an example of a relatively simple small restaurant:

- *Incoming service*: 800A at 120/208V, 3-phase (3 poles)
- *Service wire protection*: 800A, 3-pole main circuit breaker
- *Distribution protection*: from a distribution board, circuit breakers for four 225A, 120/208V, 3-phase panelboards
- *Branch circuit protection*: various 2-pole and 3-pole circuit breakers for kitchen and HVAC equipment and lots of 20A, 1-pole circuit breakers for lighting and receptacles

These two systems actually look similar, except that there would be five cabinets in the restaurant (a larger distribution board and four panelboards).

9.3.1 Switchboards

Switchboards are the largest type of equipment commonly used indoors in buildings, and they would normally be "service-rated," which means that they meet the NEC's requirements for connection by the utility company. Such equipment is available in only three frame sizes: 2,000A, 3,000A, and 4,000A, it is physically large, and it must stand on the floor. Switchboards are available with either copper or aluminum buses. (In the author's

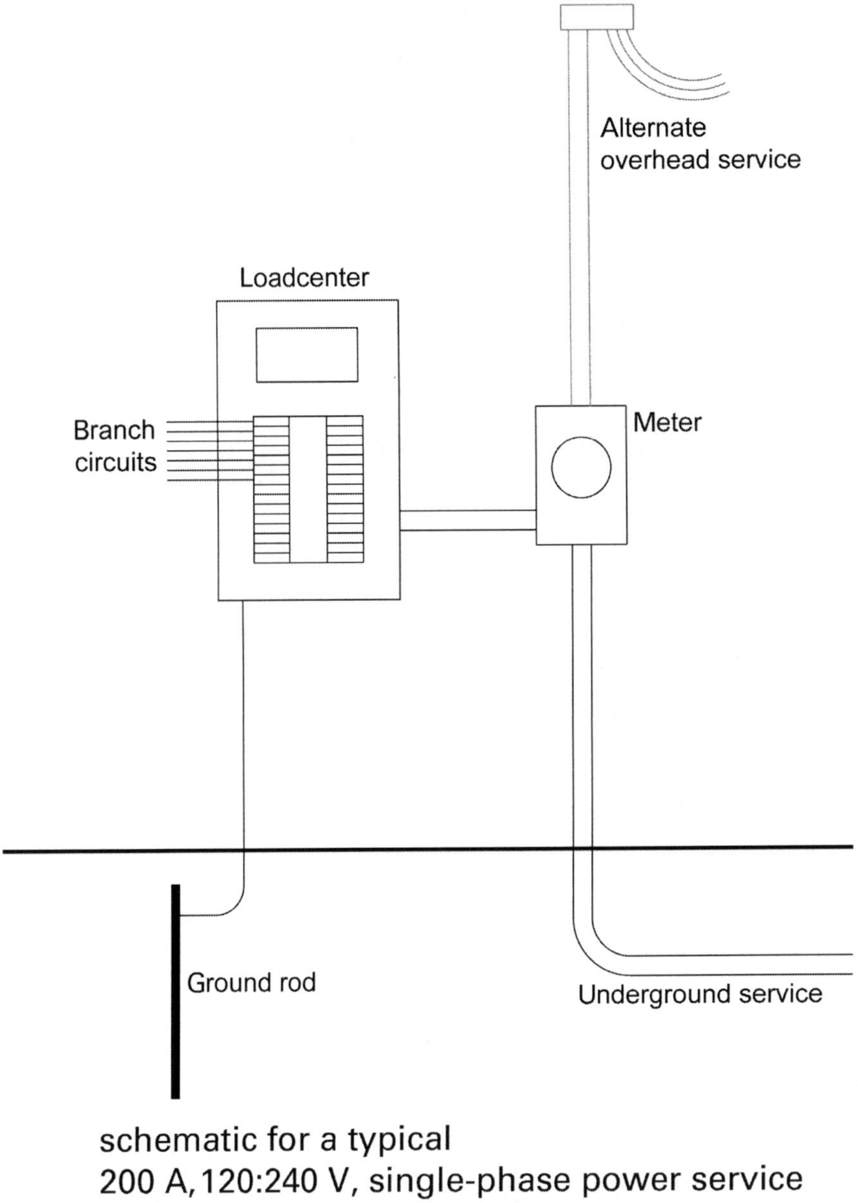

Figure 9.3.A
Single-family dwelling power system

opinion, lower cost and lower quality aluminum buses should be avoided if possible. Even though they are less costly, they are more prone to long-term problems and the relatively small savings do not truly offset the increased risk.) A typical 2,000A section would be about 36″ wide by 30″ deep by 84″ tall; a typical 3,000A section would be about 42″ wide by 36″ deep by 84″ tall, and a typical 4,000A section would be about 48″ wide by 42″ deep by 84″ tall. For a 2,000A switchboard, a main circuit breaker might be small enough to allow for some distribution in the same section as the main device, but that is not the case for 2,000A equipment using fused switches, or for 3,000A and 4,000A equipment. For the larger sizes, the main circuit breaker or fused switch would take up the

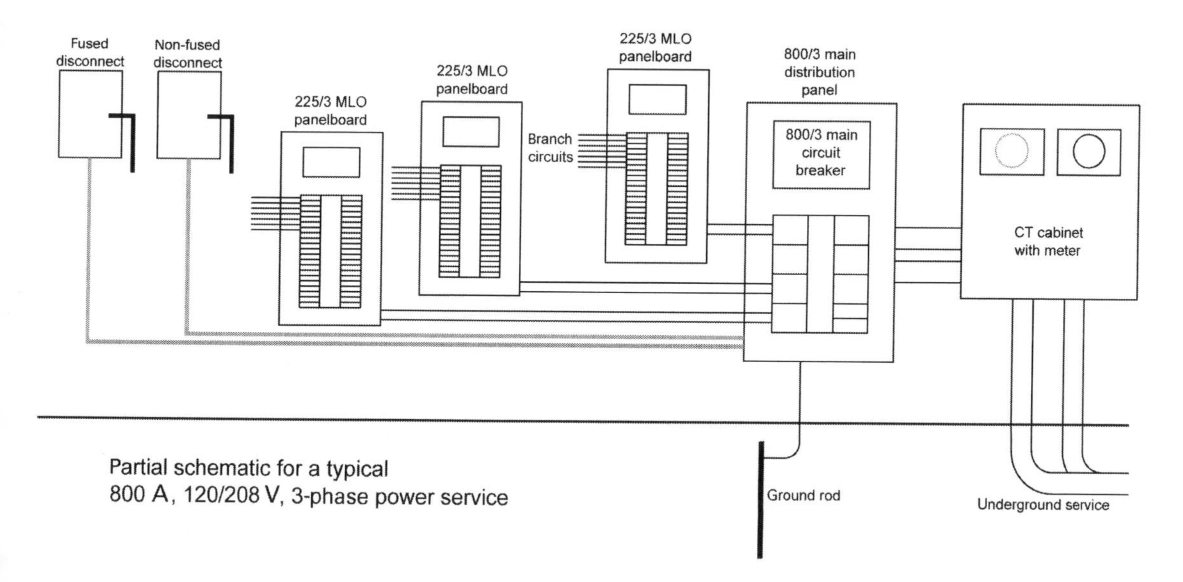

Figure 9.3.B
Small restaurant power system

entire cabinet, so additional cabinets are required for distribution (branch circuit) devices. One of the primary reasons to use all circuit breakers is that the equipment gets to be much smaller. (The author has worked on numerous large supermarket buildings, using both fused switches and circuit breakers. A typical main power distribution switchboard using fused switches would include eight sections or so and it would be almost 25′ 0″ wide; a similar all circuit breaker setup would be about 6′ 0″ wide.) The major US manufacturers for all of this equipment (switchboards, distribution boards, panelboards, load centers, safety switches, and circuit breakers) are Eaton (Cutler-Hammer), General Electric, Schneider Electric (Square D), and Siemens. All of this equipment is also available in indoor (usually painted steel), outdoor (painted steel, fiberglass, or PVC), and dustproof (stainless steel, fiberglass, or PVC) enclosures.

www.eaton.com/Eaton/ProductsServices/Electrical/ProductsandServices/ElectricalDistribution/index.htm

www.geindustrial.com/cwc/Dispatcher?REQUEST=PRODUCTS&famid=38&lang=en_US&omni_key=PrdSwtchBrdMO

www.schneider-electric.us/en/product-category/52100-lv-panelboards/?filter=business-4-low-voltage-products-and-systems

w3.usa.siemens.com/powerdistribution/us/en/product-portfolio/panelboards/Pages/panelboards.aspx

A more general classification is distribution boards, which includes both switchboards and smaller boards that might be floor- or wall-mounted. Equipment like this is likely to use all circuit breakers, and it ranges in size from 400A to 1,600A. The main difference between a distribution board and a panelboard (see 9.2.2) is that the distribution board is designed to accommodate large branch circuit breakers. Distribution boards might be service-rated if they are the termination point for the utility, but they do not need to be service-rated if they are installed downstream of a switchboard.

9.3.2 Panelboards

Panelboard generally refers to a branch circuit panelboard whose purpose is to provide over-current and fault protection for individual lighting, appliance, equipment, and receptacle circuits. Panelboards have to be service-rated if they are the termination point for the utility, but that rating is not required otherwise. Panelboards come in a wide variety of sizes and styles, with either copper or aluminum buses, but the most common type uses a back box (usually called a tub) that is 20″ wide and 5.75″ deep. Note that this tub will *not* fit into a standard wood or metal stud-framed wall, and 2″ × 6″ wood or 6″ metal studs must be used if the panelboard is to be recessed into the wall. The height varies considerably—roughly from 30″ to 84″—according to the size and configuration of the panelboard. Three-pole circuit breakers up to 100 amps are usually feasible, although some panelboards are configured for larger devices. Typical ratings are 100A, 225A, and 400A. In most panelboards, the circuit breakers are bolted to the buses, so that there is little or no risk of having the circuit breakers fall out of the panelboard.

9.3.3 Loadcenters

Loadcenters are lightweight and low-cost versions of panelboards, but they are available from 60A to 400A and they can be service-rated. They are commonly used in residential work. The main differences are that loadcenter tubs are 14″ wide and 4″ deep so that they will fit into standard 2″ × 4″ wood or $3\frac{5}{8}$″ metal stud walls and they usually have "push-on" circuit breakers. Push-on circuit breakers are attached to the buses only by friction and they are not bolted. Push-on circuit breakers should never be used in situations where there are significant vibrations because it is actually possible for the circuit breakers to fall out. Loadcenters are also available with either copper or aluminum buses.

9.3.4 Dry-type transformers

As noted under 9.2.3, dry-type transformers are required for nearly all 277/480V systems in order to provide for 120V at convenience receptacles. These transformers are available in a range of quality levels (mostly distinguished by the insulation class) but they all tend to look similar. They can be painted in the field, if necessary, but the nameplates should not be painted. They tend to hum at 60Hz (see 6.2), especially if they are heavily loaded and/or old, and the humming can be quite noticeable.

9.3.5 Safety switches (disconnect switches)

As noted previously, all wire must be protected for over-current, and the protection (fuse or circuit breaker) can be at either end of the wire. Sometimes, this protection is provided in the form of a fused switch that is used only for a single circuit (anywhere from a small piece of equipment all the way up to an entire service); such switches are usually called "disconnect switches" or even "disconnects." A good example is the load side wire for a transformer; if the panelboard that is fed by the transformer is not right next to the transformer, a fused disconnect is often used in place of a main circuit breaker in the panelboard.

Also, the NEC requires that the means to disconnect the service are visible and within reach of equipment, unless there is a lockable circuit breaker (or fused switch) in the loadcenter, panelboard, distribution board, or switchboard. In the case of HVAC equipment, such as ACCUs, this exception does not apply and a disconnect is required very close to the equipment. The reason for this is so that a workman can be sure that there is no power to the equipment that he, or she, is working on. Disconnects like these are usually non-fused switches because they are used only as switches for service purposes. (This is why there is a switch next to the ACCU at your house.)

9.3.6 Surge suppressor

As noted previously, surge suppression can be done locally, as in a plug strip, at a loadcenter, a panelboard, a distribution board, or a switchboard. Large devices that are used at panelboards, distribution boards, and switchboards are better because they can provide a higher level of protection while also monitoring themselves for failure. In other words, if the protection module for Phase A for a 2,000A switchboard has failed (due to the size or number of surges), it will actually go into alarm and notify anyone who is paying attention that it needs to be replaced. A typical plug strip simply fails and gives no indication that it does not work anymore.

9.3.7 Uninterruptible power system (UPS)

Many people have small-scale UPS units for their personal computers or small office servers. A UPS is a device that runs power through a set of batteries to a load, or loads. It is uninterruptible because the power is always flowing through the batteries; if the normal power source goes out, the unit simply runs on the batteries until the batteries are drained. Small units are usually boxes that sit on the floor, and they are rated in watts: 1,000W, 1,500W, 1,800W, etc. Due to the limitations of 20A, 120V, circuits (no more load than 1,920 watts per the NEC), it is not feasible to use large units with conventional wiring from receptacles. But large units can be used by placing the unit between a loadcenter, panelboard, distribution board, or switchboard and the loads. Large units can be very large. In addition to the load in watts that can be managed, the run-time for the batteries is also significant. For small residential units, the batteries might last for an hour or two; for large commercial units, the run-time is limited only by the number of batteries, but commercial systems usually have relatively short battery run-time (sometimes just a few minutes) because they are usually backed up by an on-site generator. Typical on-site generators will start and transfer power within 10 seconds of a power outage, so there is little reason to have lots of battery capacity in a UPS. These systems are very costly (tens of thousands of dollars for medium sized units) so they are usually limited to critical applications. They can take up substantial space; a typical 30 **kVA** unit might be 4′ 0″ wide by 3′ 0″ deep by 4′ 0″ high—and that is not a large unit.

9.3.8 Generators and automatic transfer switches (ATS)

For certain occupancies (911 call centers, some medical facilities (in-patient hospitals and outpatient surgery centers), security monitoring companies), it is necessary to have permanent on-site emergency power generation. Generators like this are available in a wide variety of sizes (from 2 kVA to 2,000 kVA), using gasoline (small units only), diesel fuel, natural gas, and propane. For uses where the generator is required, diesel fuel is usually preferred because the fuel is uninterruptible (if there is fuel in the tank—the tank has to be sized for a significant run-time (probably a few days) to make sure that there is time to re-fill the tank in an extended power outage). These units are available in two basic configurations: split package and unitary in an indoor or outdoor enclosure.

Split package units have the engine and generator inside the building and the radiator (or other cooling system) outside the building. Unitary units can be located inside a building with ventilation air ducted in for the radiator, or they can be located outside in an outdoor enclosure. Unitary units in outdoor enclosures are most commonly seen today.

These units are quite noisy when they run, and different outdoor enclosures can be specified according to noise sensitivity. If the unit is indoors, special acoustic treatment must be included for the walls (and floor above, if any) that enclose the unit.

For highly sensitive applications, it might be preferable to locate the generator inside to protect it from extreme weather. If the unit is critical but located outside, it might need special protection as well. (The author worked on a 911 call center project where a special super-strong roof was designed for the outdoor generator to protect it from flying debris in the event of tornado.)

In recent years, as we seem to see more and more power disruption issues (surges, phase drops, brown outs, and outright outages), many owners have decided to invest in on-site power generation, so we see more and more permanent generators that are not required by regulations. Also, some owners elect to have their buildings wired to accommodate portable generators that can be moved in on short notice to provide emergency backup power.

If the generator is permanent, it is necessary to add an ATS to the power system. This is a special switch that sits between the utility and the service switchboard, distribution board, panelboard, or loadcenter and the generator. It should be thought of as a three-way switch, with the load in the middle, the generator on one side and the utility on the other side. When the utility is providing power, the unit switches to the utility side and does nothing, except monitoring continuously to make sure that the utility's power is still on. If the utility's power goes off, the unit automatically turns off the utility side and switches to the generator side (after the generator has started running and the generator has stable output). An ATS can switch back to normal power automatically also.

For a portable generator, an ATS could be used, but it would only work as designed if the generator were already there (which it would not be, if it is portable). So, for portable generators, other systems are usually used. The two main options include a double-throw switch (which only works for 800 amp systems and smaller) or key-interlocked dual main switches in a switchboard or distribution board. The double-throw switch is exactly that: a double switch where up (or right) is "normal" and down (or left) is "emergency" with an "off" position in the middle. If the power fails, this switch would be set to "off"; then the generator would be set up and started, and then it would be manually switched to "emergency." The procedure would be reversed after the normal power is restored. Key-interlocked dual main switches are operated in a similar fashion: turn off the normal switch and remove the key (which means that everything is off); set up the generator, put the key in the emergency switch and turn it on. The procedure is reversed to go back to normal power.

The main interior design issue for an ATS is to make sure that there is adequate space for it. This can come up even for tenant projects in existing buildings, so it is important to keep it in mind.

9.3.8.1 Special considerations: large motors, UPS, run-time, and fire pumps

As noted under generators, an emergency power system has to be designed to accommodate all systems and special conditions. If there are large motors in the system (for HVAC equipment, elevators, pumps, etc.), those motors can have a large impact on the sizing of the generator. Similarly, some UPS designs affect a generator in different ways than other UPS designs, which can also affect the sizing and design of the generator. And, of course, the run-time for the batteries is a key element. Electric fire pumps (for sprinkler or standpipe systems—see 11.3) are a major issue in any power system with, or especially without, a generator, but the major issue for interior designers, in relation to fire pumps, has to do with space. There are several different types of ATSs too, and the selection of the type and its features can have a major impact on the budget.

9.4 Over-current protection

As noted previously, there are two basic forms of over-current protection: fuses and circuit breakers. While both are allowed by the NEC for virtually any application, circuit breakers are used far more extensively, primarily for the convenience of being able to re-set without having to locate and install replacement parts. That said, circuit breakers are supposed to be "exercised" annually, meaning that they should be turned off and then re-set at least once a year. Does everyone do that? Does anyone do that? Hardly anyone, if anyone at all. That does mean that circuit breakers are not always completely reliable. Also, equipment designs change over time, and it usually becomes difficult to get replacement parts for electrical equipment that is substantially more than 30 years old. Good electrical equipment can last for 50 years, or even longer, but it is usually advisable to replace it after 35 to 40 years to avoid problems with replacement parts.

Fuses are available in several different types: slow-blow (also called time delay), fast acting, and high speed, in sizes from fractions of an amp up to 4,000 amps. Some large fuses are bolted in their holders, but most fuses are held by friction. (Some readers might remember old-style screw-in fuses from panelboards in old houses; they are still around but they are becoming less common with every passing year.) The most important thing about fuses is that there must be spare fuses for every size and type on the premise at all times, so there must be a cabinet (a big cabinet for large systems using a number of sizes) with spare fuses in it in an accessible location. If a spare fuse is missing, the fuse cannot be replaced until someone locates a replacement.

Circuit breakers are also available in multiple types: standard, electronic, shunt-trip, ground fault current interrupters (GFCI), and arc-fault, to name a few. Standard circuit breakers trip when overloaded (over-current protection) or when there is a fault. Electronic circuit breakers work in the same way, but they can be linked to computer systems to monitor their status. Shunt-trip circuit breakers have electronic relays built into them so that they can be turned off by an electronic signal (this is required for elevator machines and for cooking appliances under a Type I hood in a commercial kitchen). GFCI circuit breakers are the same as those found in wall-mounted receptacles (having GFCI circuit breakers means that GFCI devices do not have to be used—but it also means that the circuit breaker has to be accessible for re-setting should something trip), and arc-fault is the special classification that applies mostly to sleeping rooms and dwelling units.

Circuit breakers are also available in a wide range of sizes: from 10A to 4,000A, with almost every size in between, and in bolt-on and push-on styles (the latter for small sizes only). Standard circuit breakers for panelboards are about 1″ high (or slightly less) but half-size circuit breakers are available for some panelboards and loadcenters so that more devices can fit into the same amount of space.

Even though most circuit breakers are located in loadcenters, panelboards, distribution boards, or switchboards, it is possible to use individual circuit breakers in enclosures (both indoors and outdoors) for special applications, such as where a fused switch might be used.

9.5 Working space

For worker safety, the NEC requires "working space" around all live electrical equipment. Generally speaking, this space is required to be 30″ wide (or the width of the equipment, whichever is greater), 78″ high (or the height of the equipment, whichever is greater), and either 36″, 42″, or 48″ in front of equipment, depending on what is on the other side.[4] The 36″ rule (Condition 1 in Figure 9.5.A) applies from 0V to 600V if there is non-grounded construction on the other side; non-grounded construction means lightweight construction, also known as not concrete or not masonry. The 36″ rule (Condition 1) also applies to grounded (heavy construction—concrete or masonry) construction if the voltage is 150V or less. The 42″ rule (Condition 2) applies to grounded construction for 151V to 600V, and the 48″ rule (Condition 3) applies to live parts to live parts, meaning when

Figure 9.5.A **Required electrical working space**

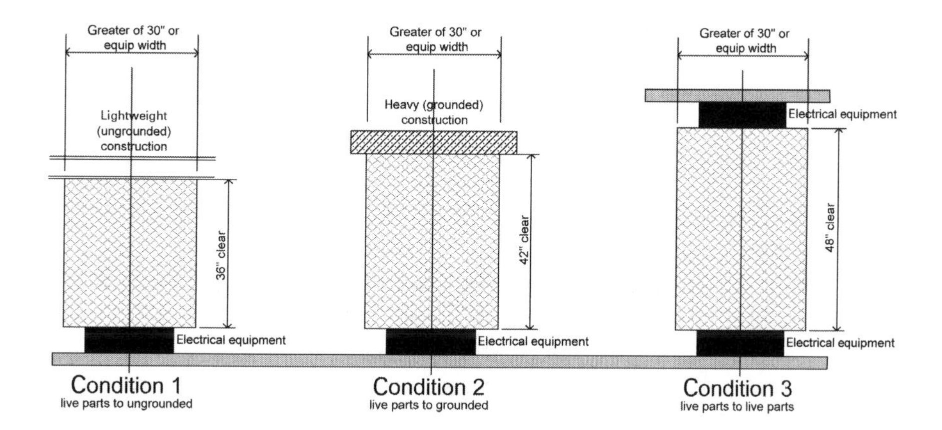

electrical equipment is facing other electrical equipment. (In any piece of electrical equipment, the live parts are not on the face of the cabinet, but it is prudent to use the face of the cabinet anyway when calculating these distances.)

The biggest challenge here is that these working spaces apply to "all electrical equipment,"[5] which is defined as "a general term, including material, fittings, devices, appliances, luminaires, apparatus, machinery, and the like used as part of, or in conjunction with, an electrical installation."[6] A literal interpretation of this would make it nearly impossible to provide a power system in any building. How many receptacles have furniture less than 36″ in front of them? How many light switches?

In reality, this is interpreted to mean that it is OK to have movable furniture encroaching into the working space for devices (switches, receptacles, etc.) but not for permanent equipment (including loadcenters, panelboards, disconnects, etc.) and to locate some equipment above ceilings (including junction boxes, disconnects, switches, etc.—as long as it is feasible to gain access to the equipment through the ceiling). The latter can be a major issue with hard ceilings, such as gypsum board, wood, metal, or plaster, because it is necessary to have access panels to get to any and all electrical equipment that might be above the ceiling. For new installations, the engineers try (or should try) to keep such equipment out of the area to the greatest extent feasible, but it might be difficult to move equipment in an existing building.

Working space requirements are important because they affect how much space electrical equipment takes up in a given situation, and it is necessary for the interior designer to work closely with the electrical engineer to work out the requirements. Interior designers commonly want to know if it is OK to conceal existing equipment, especially recessed panelboards or loadcenters in awkward locations. The answer is yes, usually, but it all depends on the details. In older buildings, branch circuit panelboards that were installed in concealed locations (usually electrical rooms) sometimes end up in occupied spaces when walls are moved around. Such panelboards usually have plain flush fronts, but most interior designers do not consider them to be decorative elements, so they want to hide them. A door can be placed in front of such a panelboard (within just a few inches) as long as the door opening provides a space at least 30″ wide centered on the panelboard that is also at least 78″ high (or higher if the panelboard is higher); when the door is closed, the panelboard cannot be seen. When the door is open, the working space is provided by the open door. The door must extend all the way to the floor though; it cannot be a cabinet door in the wall that is only slightly larger than the panelboard itself.

When planning space for electrical equipment, especially for multiple load-centers or panelboards, squarish large rooms do not work very well because the equipment is relatively flat (except for switchboards) and a large amount of unused space will be left over in the middle of the room. More often than not, electrical rooms should be relatively long and thin, providing lots of wall space.

What should a room look like to contain four standard panelboards? Each panelboard will be 20″ wide and there should be at least 4″ between them, so 24″ of wall is needed for each panelboard. For panelboards in corners, the center of the panelboard cannot be closer than 15″ to the side wall (due to the minimum 30″-wide working space), so the wall would have to be at least 15″ + 24″ + 24″ + 24″ + 15″ = 102″ wide, if all of the panels are on the same side. This room could be as little as 42″ wide, if the panelboards are surface-mounted and if the opposite wall is lightweight construction (see Figure 9.5.B).

If the panels are on both sides of the room, it could be 54″ wide and 60″ across (due to the 48″ between the panels that face each other). (See Figure 9.5.C.)

The narrow room would be 12 square feet in area and the squarish room would be 9.5 square feet. In this case, there is little difference.

How about a larger room to contain a 1,200A distribution board that is 36″ wide and 12″ deep (and 84″ high), a 1,200A ATS (also 36″ wide and 12″ deep but only 60″ high), and six standard panelboards? If this room were long and narrow, the length would need to be 36″ + 6″ + 36″ + 15″ + 24″ + 24″ + 24″ + 24″ + 24″ + 15″ = 228″ (or 19′ 0″) and

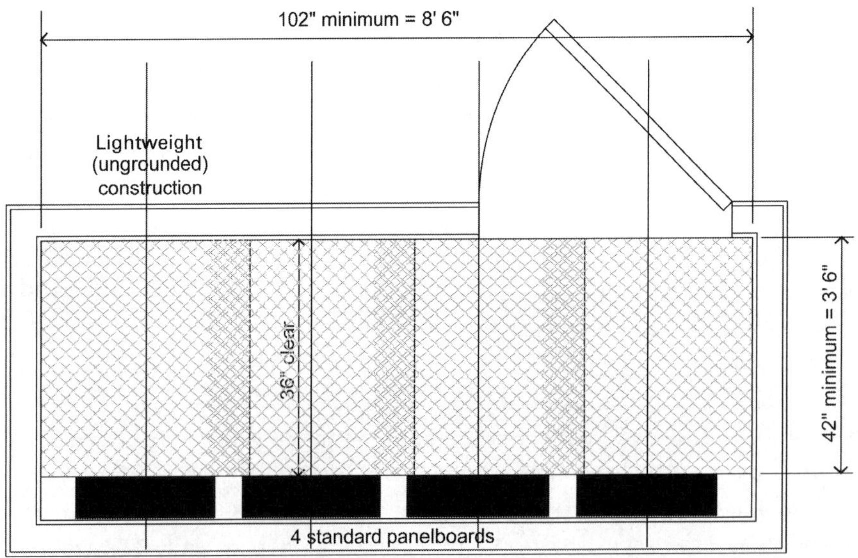

Figure 9.5.B
Example room 1

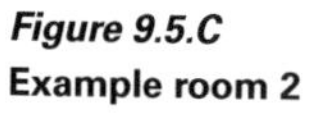

Figure 9.5.C
Example room 2

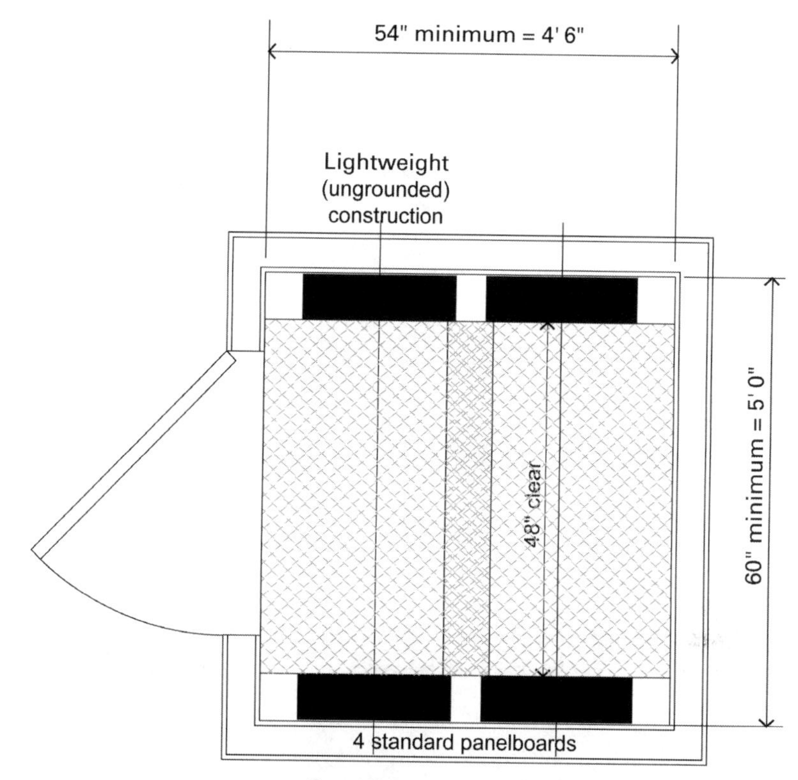

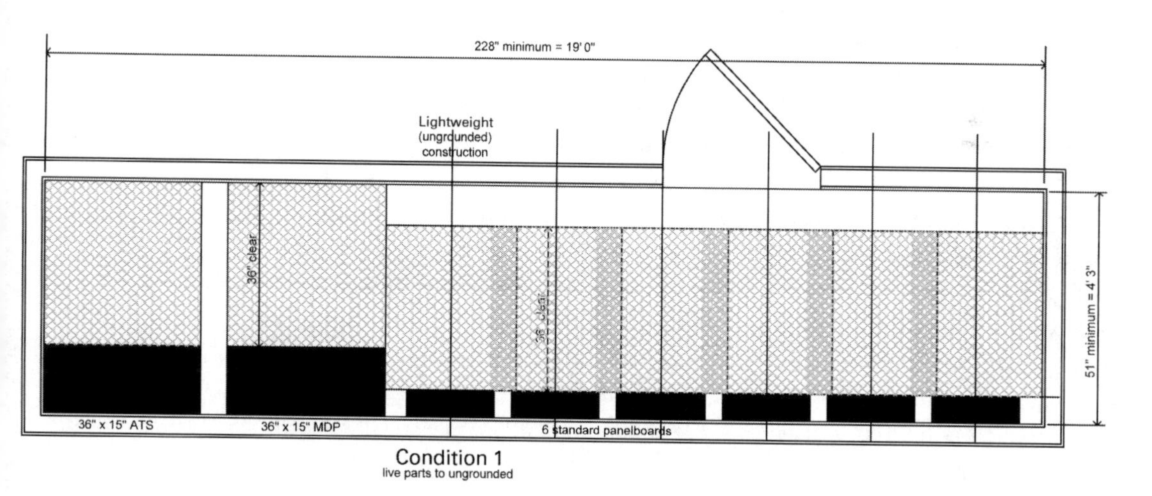

Figure 9.5.D
Example room 3

the width would have to be at least 48″ (if the opposite wall is lightweight construction). Such a room would have an area of 76 sf (Figure 9.5.D).

This room could also be 16″ (one more inch to avoid a conflict with panels on the side wall) plus 24″ + 24″ + 15″ (no panels at the other end), or 79″ by 86″ (22″ to clear the distribution board plus 24″ + 24″ + 16″, or 47 sf (Figure 9.5.E).

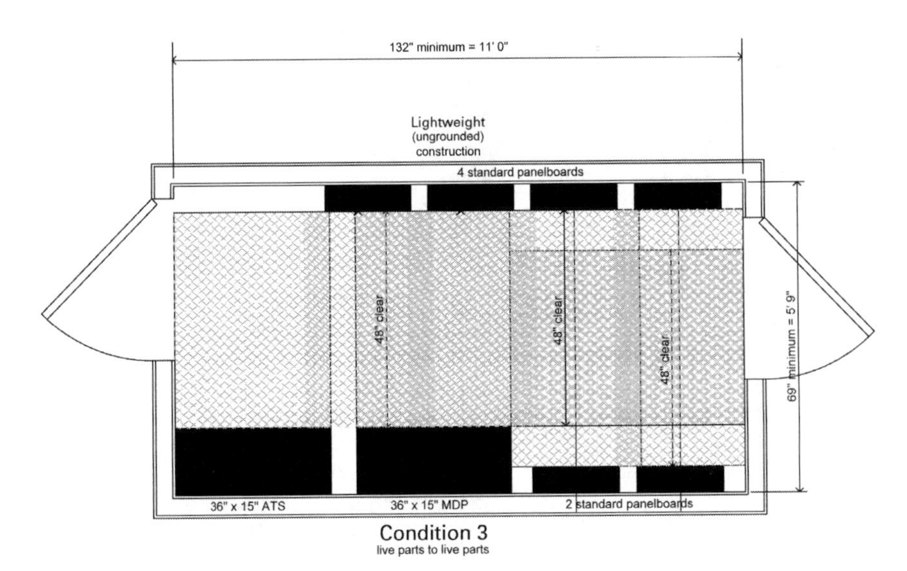

Figure 9.5.E
Example room 4

Obviously, the square is actually more space efficient, but the comparison is still valid. As more components are added, especially panelboards or loadcenters, rooms tend to become more rectangular.

Do not hesitate to get involved in equipment placement; this is your design after all, and there is no reason for an engineer or a contractor to ruin it due to lack of attention.

9.6 Large equipment

When electrical equipment is 1,200 amps (or larger) and 6′ 0″ wide (or wider), special code requirements come into play. The NEC requires two means of egress from such a room, with panic hardware on out-swinging doors at opposite ends that are at least 24″ wide by 6′ 6″ high, unless double working space is provided (96″ from live parts to live parts, for example), or unless a continuous and unobstructed way of egress travel is provided, or unless there is more than 25 feet between the door and the equipment.[7] There is not likely to be more than 25 feet between the door and the electrical equipment in most situations, so that exception is unlikely to apply. Double working space is easy, but it takes up more space which could be a problem. And no one knows what a continuous and unobstructed way of egress travel is because that is not the kind of language that is used in the building code. These requirements should not be in the NEC, because they are building requirements and not electrical requirements, but they are there and they must be followed.

9.7 Other low-voltage systems

Other low-voltage systems include fire alarm (see Chapter 2), voice (telephone), data (network), audio (sound), video, and security systems. Any or all of these systems can be configured to run on high-speed data wiring (most commonly level 6 unshielded twisted pair), but some video systems still use coaxial cable (like the thick cable used for cable TV at home).

Telephones can be done in two ways: old-fashioned copper wire (what everyone used to have at home and in other space) and data cable; the copper wire telephone line requires two pairs of wires, one pair for incoming voice and one pair for out-going voice, so it used to be common to see 15-pair cables for office telephones that had access to multiple lines. Telephones using data cable are usually called "voice-over-IP" which means voice via internet, which is all digital; voices are converted to digital bytes, transmitted across the network, and converted back to voice at the other end. This is becoming more and more common in the marketplace (the author even has it at his small office). One advantage to using digital telephones is that it also makes voicemail digital, so that it is easy to store and transfer voicemail messages to others in the system.

Data networks are usually built using what is called a "star" configuration. The cables are limited in length to 100 meters (about 300 feet), so nodes of network switches called DeMarcs (this stands for demarcation) are built so that every desk or wall outlet can be reached within 100 meters. If a space is too large for one DeMarc, multiple DeMarcs are connected together, usually using fiber optic cable. At some point, everything is routed back to a single server, a rack of servers, or even a room full of servers for a large installation. The servers are connected to the outside world using high-speed copper cable or fiber optic cable. For space planning purposes, this does mean that closets for DeMarcs must be spaced to meet the 100-meter limit for the cabling. A small DeMarc closet would be about 8 feet square.

Audio, video, and security systems can use network cabling as well, although video does still use separate coaxial cable as noted.

The main issue to note about all of this low-voltage cabling is that it must be specified to be "plenum-rated" if it is to be exposed in a return air plenum (or any other air plenum). This is simple to specify but it should be done only if truly needed because plenum-rated cable is substantially more costly than non-plenum-rated cable.

9.8 Lighting

As in so many topics here, lighting is a topic in itself, so it is covered in detail in Chapter 8. As far as power is concerned, most lighting is powered at 120V, single-phase; some at 277V, single-phase; some at 208V, single-phase; and even some at 480V, single-phase. The latest technology to come along is to power LED lighting at low voltage, directly from network cabling, which is called power-over-ethernet (POE). In POE installations, there is no line-voltage wiring for lighting at all.

9.9 Power documentation

As for lighting, power construction documents identify, specify, and locate every item in the power system (disconnect, switchboard, panelboard, circuit breaker, luminaire, duplex receptacle, etc.) plus all of the wiring feeding them, from the utility connection in.

These documents are usually divided into power and lighting sheets, plus Light Fixture Schedules, Symbols Legend, Panelboard Schedules, One-Line Diagrams (to explain the primary wiring arrangements), Grounding Details, and other details as needed for the particular project. A large system can include thousands of circuits, so the drawings can become very complex.

Outcomes

9.2 Understanding the basics of voltage, amperage, and electrical power and application to wiring in buildings, especially common voltage systems and their differences.
9.3 Understanding the basics of electrical equipment, such as loadcenters, panelboards, distribution boards, transformers, etc.
9.4 Understanding the basics of over-current protection, including fuses and circuit breakers.
9.5 Understanding code requirements for working space—36″, 42″, and 48″ and when each applies.
9.6 Understanding code requirements for large equipment—double working space or two means of egress, special door hardware, etc.
9.7 Understanding basic low-voltage systems, especially the need for DeMarc closets and the possible need for plenum-rated cabling.
9.8 Understanding basic power for lighting, including 120, 208, 277, 480V, and POE.

Notes

1 2014 NFPA 70, page 70–375
2 Ibid., page 70–514
3 Ibid., page 70–35
4 Ibid., page 70–40
5 Ibid., page 70–40
6 Ibid., page 70–30
7 Ibid., page 70–41

Chapter 10

What are plumbing systems?

Objectives

10.1 To understand the whys and hows of water usage in buildings, especially water conservation
10.2 To understand plumbing fixture types
10.3 To understand piping materials
10.4 To understand basic piping types and systems
10.5 To understand code issues related to the plenum
10.6 To understand code issues related to accessibility
10.7 To understand plumbing design documentation

10.1 Water usage in buildings

Water, even piped water, has been used in buildings for thousands of years—primarily for bathing but also for cooking, cleaning, and recreation. Modern indoor plumbing dates back to the latter half of the 19th century, when it became feasible to have pressurized domestic water piping and unpressurized sanitary waste piping indoors.

A few technical challenges had to be overcome to accomplish this:

1 How can pressurized water be prevented from spewing all over the place? *Answer*: valves and faucets.

2 How can the smell of sanitary waste water be kept out of the building?
   Answer: the water-seal trap.
3 How can solid waste material be moved into the sanitary waste water and out of the building?
   Answer: the flushing water closet, or toilet.

How much has changed since then? Remarkably little, really. We still use valves and faucets to control water flow; we still use water-seal traps to exclude sewer gases, and we still use flushing water closets. What *has* changed is the amount of water that is required to accomplish these goals.

Let us start with water closets. The amount of water required to flush a standard tank-type water closet in the 1970s was five or 6 gallons. Today, water closets flush with no less than 1.0 gallon ("low," for liquid only) and no more than 1.6 gallons ("high" for solids and liquid).

Urinals? In the 1960s in the author's elementary school, there were stall type urinals (probably about 12 side-by-side) in one of the boys' restrooms that never stopped flushing—there was a continuous flow of water across almost an entire wall. The flow rate for that cannot be calculated easily—clearly it was very high—but new urinals today flush with as little as 0.25 gallon and as much as 1.0 gallon, or no water at all for non-flushing waterless urinals.

Faucets? Sink and shower faucets used to use three, four, or even more gallons per minute. Today, faucets, even in showers, use 1.5 to 2.5 gallons per minute (gpm) or even less.

Why did this happen? Largely because the federal government got into the energy business in 1992 with the passage of the first Energy Policy Act (widely known as EPAct), which was updated in 1998, 2005, and 2007 and because we, collectively, have come to realize that water is a limited resource. The 1992 EPAct mandated a maximum flush rate of 3.5 gallons per flush (gpf) for water closets for commercial use from January 1, 1994 through January 1, 1997). For residential water closets, the limit was 1.6gpf starting in January 1, 1994. Urinals were limited to 1.0gpf, effective January 1, 1994, and faucets for lavatories and sinks were limited to 2.5gpf as well. The reason for this was to reduce water consumption in buildings to preserve water resources.

Initially, this caused many problems because most of the water closet manufacturers (except Toto, a Japanese company that was already in the US market in the mid-1990s) were not ready for such low flushing rates. Some people flushed the fixtures two (3.2 gallons), three (4.8 gallons), or even more times to get them to work, and lots of plungers were needed. Many older designs of water closets used small trap-ways (sometimes 2.25" or so) that require more water to force bowl contents through. It was necessary to redesign fixtures to have larger and less restrictive trap-ways in order to flush with dramatically less water, and it took the industry a number of years to catch up. Even today, some fixtures on the market do not flush very well with these low rates. This problem was so severe by 2002 that 22 plumbing industry organizations banded together to create MaP (Maximum Performance Testing), developing a special protocol for testing flushing of water closets (including the creation of fake "solid matter").

The original threshold was that a fixture had to be able to flush 250 grams of solid material, but that was soon raised to 350 grams. In the 2003 testing, the average fixture tested could flush 375 grams but that had improved to 675 grams by the 2008 testing. A number of fixtures on the market now can flush 1,000 grams, which is well beyond what should be required.

The most up-to-date testing summary available can be found at www.allianceforwaterefficiency.org/MaP-main.aspx. A searchable database can be found at MaP's website at www.maptesting.com/info/designers.html.

It is highly advisable to check this list before purchasing any water closet, unless there is good reason to know that it performs well.

This is no joke. Water closets that do not flush effectively are a serious problem and should be avoided at all costs.

Today, the 1.6 gpf limit remains in place, but the market has moved onto even lower flush rates. Dual-flush 1.6 gpf/1.28 gpf units are readily available, as are ultra low 1.0 gpf/1.28 gpf units. MaP tests these fixtures too. As noted previously, some flushing urinals are down to 0.25 gallon, and waterless urinals use no water at all.

How about valves and faucets? There have been relatively few technical improvements, but these include the washerless faucet and the ceramic cartridge faucet. A faucet is simply a special valve with a handle for easy operation. Valves have moving parts and moving parts wear out, especially when they are moved as frequently as they are in a faucet. Old-style faucets (which had mostly been replaced with washerless faucets by the 1970s) used rubber washers that had to be replaced periodically. Even though it was relatively easy to replace a washer in a faucet, it was still more work than not replacing the washer. So washerless faucets were born. The main differences between high-end faucets and inexpensive faucets is the quality and weight of the materials used. Highly durable faucets, which could last for 50 years, are made of brass (usually plated with chrome) but these are costly (typical high-quality commercial faucets rarely cost less than $500.00, not installed, and many of them are much more costly). Most of the faucets available at retail (through hardware stores and big box stores) are mostly plastic, with some lightweight metal parts here and there. So even though these faucets are much cheaper (as low as $50.00/each), they do not necessarily last very long—ten years is a long time for a product like this.

Single-handle faucets (as are seen in many kitchens) are simply double faucets with a single handle that can turn on the hot water, the cold water, or both simultaneously. Faucets that operate by touch have solenoid (electrically operated) valves that respond to changes in a small electric current running through the faucet spout and/or deck.

Water-seal traps in sanitary waste (usually called "sewer") lines operate by keeping a depth of water in the lower portion of a trap at all times, which prevents sewer gases from escaping the piping into the occupied space. When traps dry out because the water in them has evaporated (usually because water has not been added to them in a long time), the odor of sewer gas (which is quite distinctive) becomes very noticeable throughout the building. All traps are subject to drying out if not used, but some are more likely to dry out than others. Here is a list, in descending order, of likelihood of trap drying:

- Floor drains
- Infrequently used shower drains

- Infrequently used sink and lavatory drains
- Urinals
- Water closets

Some plumbing codes require trap primers for floor drains due to this problem. A trap primer is a very small domestic water line that is connected directly to the floor drain piping, just above the trap, which keeps a small amount of water running into the trap to keep it wet. Obviously, this wastes water, but it is code-required in some jurisdictions anyway. Another way to minimize the problem is to fill the trap with light vegetable oil, which evaporates much more slowly than water. If the oil gets washed out of the trap by flowing water, it probably was not needed in the first place because the trap is probably wet.

In Europe and Asia, waterless trap seals have been developed and they are being marketed very successfully. One product—the Wavin HEPVO—has been approved for use in manufactured housing (mobile homes and recreational vehicles) in the US, but approval for use in conventional buildings appears to be unlikely. This is a valve inside a PVC pipe that takes the form of a flat sleeve when there is no water present; it can seal out sewer gas but allow water to go through at any time. Despite the obvious advantages of such technology (no chance of traps drying out and the ability to install the device either vertically or horizontally), there is strong resistance in the US plumbing code community to allowing deviation from the water-seal trap approach. This might change in the future, but it could take a number of years. Using such traps would also conserve even more water.

Overall, water use reduction has been substantial for new fixtures that have been manufactured and installed since 1994, and especially since 1997. But there are no rules to require changing older, high water-using fixtures. Also, some people do not like the low flush fixtures and there are ways to modify them to use more water. It could take decades for all of the old fixtures to be replaced.

10.2 Fixtures

10.2.1 Water closets

Water closets come in two basic types: tank-type and flush-valve type and in vitreous china (fired china with an extremely tough vitreous glaze) and stainless steel (used mostly in correctional environments).

Tank-type fixtures are what most people have at home and what are commonly seen in light commercial applications, and they are nearly always vitreous china. (Stainless steel fixtures are likely to have **flush valves** for institutional use.) They function by storing water in a tank, which is dropped into the bowl when the flush handle lifts a plug in a hole in the bottom of the tank. The flushing mechanisms tend to be less than robust, and maintenance problems are typical, especially under heavy use. Tank-type fixtures are available in floor-mount and wall-mount types, and both of those are available with round bowls and elongated bowls. (Elongated bowls are required in commercial environments and they are becoming more common in residential environments.) Tank-type fixtures are available in a bewildering variety of styles and colors from many different manufacturers, some of whom are listed below:

www.americanstandard-us.com
www.craneplumbing.com
www.eljer.com
www.gerberonline.com
www.kohler.com
www.lacava.com
www.totousa.com
www.zurn.com

Flush-valve type fixtures are the type that is used in high traffic facilities—schools, theaters, arenas, stadia, airports, malls, government buildings, etc. Stainless steel fixtures (often with integral lavatories) are usually used in correctional environments. They are far more durable than flush-tank fixtures, and they are also available in floor-mount and wall-mount types. Few, if any, round bowl fixtures are available for flush

Figure 10.2.1.A
Flush-tank water closet

***Figure 10.2.1.B* Manual flush-valve water closet**

valves. Flush valves themselves are available in exposed manual, concealed manual, exposed automatic, and concealed automatic versions, all of which are highly durable. Automatic flush valves are available with line-voltage power connections (very costly) and with batteries. The batteries should last for several years and battery operation is the preferred type to use. Flush-valve fixtures are available in a number of types, but not in the amazing range found for tank-type fixtures.

When to use flush tanks or flush valves? Flush valves are far more durable and will stand up to frequent flushing much better than flush tanks, but they are also louder and should be avoided in noise-sensitive areas (such as restaurants—no one

Figure 10.2.1.C
Automatic flush-valve water closet

wants to hear lots of loud flushing during a meal). Flush valves are also more costly, especially if wall-mounted or if using an automatic flush valve. Wall-mounted flush-valve water closets used to be the commercial and institutional standard, largely to make floor mopping easier, but they are becoming less and less common. (Standard carriers for wall-mounted water closets are rated for a load of 500 pounds, but a person does not have to weigh 500 pounds to damage one. There are enough very large folks these days that wall-hung fixtures are avoided most of the time.)

Flush-tank fixtures are available with either a round or elongated bowl; elongated bowls are required for non-residential buildings and flush-valve fixtures are not available in round bowl. Standard and accessible heights are available in both types, and child-sized flush-valve fixtures are also available. The latter are sometimes required for licensed day-care facilities.

Lastly, flush-tank fixtures are available in a wider variety of styles and colors and are sometimes more suitable for high-design environments. The following is a list of major commercial manufacturers:

www.americanstandard-us.com
www.craneplumbing.com

www.eljer.com
www.franke.com
www.gerberonline.com
www.kohler.com
www.totousa.com
www.zurn.com

Here are some flush valve manufacturers:

www.sloanvalve.com
www.totousa.com
www.zurn.com

10.2.2 Urinals

Even though urinals are used mostly in men's rooms, various urinal designs are available for women. That said, the author has yet to encounter a female who has expressed interest in using such a fixture. Also, urinals are used in nearly all men's restrooms now, even in single user restrooms at convenience stores, coffee shops, etc. Urinals use less water than water closets and they are faster to use for men. Urinals are available in vitreous china and in stainless steel, in various wall-mounted shapes, sizes, and colors. The old-style vertical stall type urinals are rarely used.

When multiple urinals are used, it is advisable (sometimes required) to use urinal screens between them.

Urinals are usually required to have flush valves and it is highly advisable to use automatic flush valves for them.

Waterless urinals have been developed to conserve even more water. Various technologies are used, including special filtering devices and traps filled with special filtering oils, etc. Waterless urinals do have to be washed periodically (as do water-using urinals), but they can be highly effective and they are not necessarily "dirtier" or "smellier" than conventional urinals. It has been learned that waterless urinals can be used only where there is water flowing through the waste pipes from upstream fixtures, to dilute the urine and to flush out solid materials. (Even though many people think of urine as being all liquid, in reality there are small amounts of solids in it, especially minerals. In the absence of additional flushing water, those solids can build up in the waste pipes and cause major problems with clogging. The solution is to put the urinals downstream of lavatories to provide the necessary flushing water to keep the pipes from clogging.)

Here are some major manufacturers:

www.americanstandard-us.com
www.craneplumbing.com
www.eljer.com

Figure 10.2.2
Automatic flush-valve urinal

www.franke.com
www.gerberonline.com
www.kohler.com
www.totousa.com
www.zurn.com

10.2.3 Bidets

Bidets were invented in France in the 18th century, and are used commonly in various parts of the world. Although they are readily available in the US, they are uncommonly used. A bidet most closely resembles a low sink with an upward spraying faucet and it is used for cleaning the genitalia and lower body (it can also be used for washing the feet), although some people think of them as being more toilet-like. (They do not flush and they are not designed for removing waste, although they do have drains.) Bidets are most commonly made of vitreous china, usually to match water closets that could be nearby. Bidets do have integral faucets that are similar to lavatory faucets.

Some manufacturers make combination water closet-bidets, and some manufacturers offer special water closet seats that incorporate bidet functions (including water spraying, heating, and even air drying).

Here are some major manufacturers:

www.americanstandard-us.com
www.craneplumbing.com
www.eljer.com
www.gerberonline.com
www.kohler.com
www.totousa.com
www.zurn.com

10.2.4 Lavatories

A lavatory is a bowl for flowing and/or holding domestic water that usually has an overflow drain and which is usually used for hand washing. The term lavatory is also used to refer to a restroom, toilet room, or even bathroom. Like water closets, but even more so, lavatories are available in a bewildering array of sizes, types, materials, colors, and finishes—from wood, to metal, to glass, ceramics, enameled cast iron, stone, and plastics. And in many different mounting styles: freestanding pedestal, hung on the wall (stainless or vitreous china, usually), on top of a counter, on top of a cabinet, on top of a frame, dropped into a surface (every imaginable material from stone to metal to plastic laminate and everything else), suspended under a surface (again, every material imaginable), and even integrated into a surface (molded stone, cultured marble, and solid surface materials, to name a few). They can be round, oval, rectangular, or square; shallow or deep, and even with multiple bowls. (Multiple bowls would usually be referred to as multiple lavatories.)

Here are some major lavatory manufacturers:

Be very careful about costs for high-design plumbing fixtures of all types.

www.americanstandard-us.com
www.craneplumbing.com
www.eljer.com
www.franke.com
www.gerberonline.com
www.kohler.com
www.lacava.com
www.totousa.com
www.wetstyle.ca
www.zurn.com

Lavatory faucets are also available in a bewildering array of types and across a very wide range of costs. There are three "standard" arrangements: single-hole, two-hole, and three-hole. Two- and three-hole arrangements are available in narrow spacing (4″ o.c.) and wide spacing (8″ o.c.).

Single-hole simply means that whatever mechanism and piping there is goes through a single hole through the deck (or wall); two-hole means that there are two holes for the mechanism and piping, and three-hole means that there are three holes. Residential-style lavatory faucets most commonly use 4″ spacing and three holes, but that is hardly universal. Most lavatory faucets are manual (meaning a knob, lever, or handle has to be turned by hand) but automatic faucets are available for commercial applications. Standard automatic faucets have no temperature adjustment, so they usually deliver tempered water (not hot and not cold). Temperature-adjustable automatic faucets are available too.

Here are some major faucet manufacturers:

www.chicagofaucets.com
www.deltafaucet.com
www.dornbracht.com
www.franke.com
www.grohe.com
www.hansgrohe.com
www.kohler.com
www.kraususa.com
www.moen.com
www.sloanvalve.com
www.zurn.com

10.2.5 Sinks

While similar to lavatories—sinks are also bowls for flowing and/or holding water—sinks are far more flexible and used for more purposes: mopping, laundry, kitchen clean-up, kitchen prep, dog washing, hair washing, hand washing, etc. Sinks can be wall-mounted, floor-mounted, counter-mounted (above, on, or below), and even free-standing. Sinks can also be round, oval, rectangular, or square, all in a vast array of sizes: from 8″ square to 36″ square or even bigger. Commonly used materials for sinks include stainless steel (especially for traditional kitchen sinks), enameled cast iron, fiberglass (for laundry sinks mostly), and molded stone (mostly for mop sinks). But sinks are also made from wood, stone, terrazzo, and various metals.

When specifying stainless steel sinks, two major issues must be addressed: the alloy (blend) and the gauge of the metal. Stainless steel is manufactured in about 150 different grades, or types, only a few of which are commonly used to make sinks. All stainless steel includes chromium in the alloy, which is what makes it "stainless." Adding nickel to the alloy makes the metal non-magnetic (stainless steel that is magnetic is considered to be low quality and should not be used for architectural applications) and more flexible. The most commonly used alloys for sinks are 301 and 304, with 304 being the nickel-bearing option. The most common gauges used are 16, 18, and 20, with 16 being the heaviest and most costly. (It should be noted that 20 gauge sinks could be subject to denting from impacts.) Most stainless sinks have a brushed finish, sometimes with polished edges and trims; various degrees of polishing are available from a variety of manufacturers. The cost for stainless steel sinks varies from less than $75.00 to more than $1,500.00, depending on size, design, and quality.

Here are some major manufacturers:

www.elkayusa.com
www.kohler.com
www.kraususa.com
www.moen.com

Stainless steel sinks are also the most common type of sink in commercial kitchens, being used for hand washing (usually using wall-hung lavatories), prep, and cleaning. Prep and clean-up sinks in commercial kitchens often have built-in side drainboards on one or both ends, but not always.

Here are a few major manufacturers:

www.advancetabco.com
www.johnboos.com
www.sturdibilt.com

Faucets for sinks are even more variable than faucets for lavatories, with the main difference being that kitchen sink faucets often have a spray function, either by having a separate spray hose or a spray hose that is integrated into a faucet spout (a pull out or pull down type). Commercial sink faucets (which are sometimes wall-mounted) often have high capacity sprayers. Mop sink faucets are usually wallmounted and they have a special brace for the spout (running diagonally up to the wall) to support the weight of the mop bucket hanging on the faucet spout). All of the faucet manufacturers listed previously under lavatories also make sink faucets.

Here are a few major manufacturers for commercial sink faucets:

www.chicagofaucets.com
www.krowne.com
www.tsbrass.com

10.2.6 Bathtubs

Just as in sinks, many different bathtub options are available. Materials include vitreous china, enameled cast iron, enameled steel, fiberglass, acrylic, wood, and even decorative metals such as copper. Bathtubs are available as free-standing units, with or without legs, units intended to be installed with two enclosing walls (with one open end), units intended to be installed with three enclosing walls (in an alcove) and units that are integrated with shower walls, both with (usually acrylic) and without (usually fiberglass) ceilings. There are even units with doors in the sides (and lift sides) to make them more accessible. In recent years, there have been many innovations, including soaking tubs—double tubs where there is a tub within a tub. The inner tub is intended to be filled all the way to the rim and it spills over into the outer tub. Whirlpool features are available across a wide range, from standard 32″ × 60″ tubs for alcoves up through very large tubs that can accommodate a number of people at one time. It is quite common to see tubs large enough for at least two people in master bathrooms; large units (for 3–12 people) are more often seen on decks outdoors (often close to in-ground swimming pools) or in recreation areas indoors. The cost range for bathtubs is from less than $100.00 to more than $15,000.00.

Bathtub faucets as very similar to lavatory faucets and all of the same manufacturers make them. Some bathtub faucets include integral shower faucets as well. Shower faucets will be covered under 10.2.7.

Here are some major bathtub manufacturers:

www.americanstandard-us.com
www.craneplumbing.com
www.eljer.com
www.gerberonline.com
www.kohler.com
www.lacava.com
www.totousa.com
www.wetstyle.co
www.zurn.com

10.2.7 Showers

Showers are the most flexible of all fixtures, consisting of everything from a hose slung over a fence at the beach to elaborate sealed private steam rooms. Outdoor showers are commonly used in warm climates, especially in beach-front areas, both as a convenience and to reduce tracked-in sand and dirt. Indoor showers vary from basic heads mounted on the walls of tiled rooms to complex multi-headed massage showers that require computer control panels.

Shower rooms large enough for two people (or sometimes more) have become commonplace in high-end residential work, and very large shower rooms are commonly seen in commercial, school, recreational, and sports locker rooms. Prior to the 1990s, it was common to use "gang showers" for men in such facilities, using multiple individual shower heads around the walls of a large open room, or by using column mounted multiple shower heads in the middle of a large room—or even both. But by the 1990s, this had become unpopular for most men's showering facilities and today most such facilities are designed with private showers (as have been in women's shower rooms for decades). Gang showers are most likely to be found in sports locker rooms.

All of the usual faucet manufacturers make shower faucets too. In recent years, overhead (ceiling-mounted) rain type shower heads have become popular and they are available from numerous sources. The EPAct low-flow shower head rules are challenging and they would seem to make it both illegal and inappropriate to use multi-headed showers, especially those with very high flow rates. But such equipment is often classified as "water features" by the manufacturers, which avoids the rules (which are limited to "shower heads").

Here are some major shower manufacturers:

www.americanstandard-us.com
www.craneplumbing.com
www.eljer.com
www.gerberonline.com
www.kohler.com
www.lacava.com
www.totousa.com
www.zurn.com

10.2.8 Floor drains

Floor drains are usually code-required in restrooms that have more than two fixtures and they are functionally required anywhere that floors are likely to be wet on a regular basis (showers, kitchens, food prep areas, parking garages, repair garages, etc.). They range from 3" diameter to 24" (or even bigger), round or square, and in linear arrangements that are usually called trench drains. Drain bodies (the part below the floor) are available with and without traps and with and without clamping rings for water-proofing membranes. (Please note that there should always be a water-proofing area under the finish flooring in any wet area that is above occupied space; such a membrane can go a long way to reducing long-term problems from water damage. Sheet membranes are preferred, but poured, sprayed, or rolled on membranes are better than nothing.) Untrapped drains are used mostly in large walk-in freezers (there is condensate from the evaporative coils even in freezers) and in other situations where traps would cause problems.

Here are some major manufacturers for floor drains:

www.josam.com
www.jrsmith.com
www.siouxchief.com
www.wadedrains.com
www.watts.com
www.zurn.com

Drain bodies are available in PVC, CPVC, cast iron, and bronze, but PVC and CPVC should be avoided, especially under heavy use. (If there is a clog in the line, a plumber will often use a floor drain to gain access to the line; running a power snake through a PVC or CPVC drain body could break it.) Trench drains can be site-built as a part of a floor slab or by using pre-fabricated sectional bodies, usually made from composite plastic materials, with various types of grates, from perforated galvanized steel and stainless steel to cast iron and fiberglass. The grates (the top of the drain that keeps large objects out, including shoes and feet) on a trench drain are often bolted down for safety and to prevent theft.

Here are some major manufacturers for trench drains:

www.abtdrains.com
www.polycastdrain.com
www.watts.com

Non-trench drain gratings are available in chrome-plated steel, painted steel, painted cast iron, nickel bronze, plastic, and fiberglass, in both square and round configurations. Square gratings should be considered for drains in tiled floors because it is easier to cut the tile neatly around the grating. The preferred material for floor drain gratings is nickel bronze, due to its ability to withstand constant wetting and drying.

10.2.9 Floor sinks

Floor sinks are not really sinks at all—they do not have faucets and they cannot be used readily for cleaning or prepping objects. Instead, they are really floor drains with large bodies, usually square, and they are used for high-volume and indirect waste applications. In a commercial kitchen, it is usually required to connect sinks to the drain system by indirect means, which means that the discharge pipe from the sink drain hangs above a floor sink, with an air gap of at least one inch between the bottom of the pipe and the top of the floor sink. This prevents any possibility of back-siphoning dirty water back into the sink. Floor sinks are most often made of enameled cast iron (stainless steel is possible too) and they usually have heavy nickel bronze gratings. Partial gratings are commonly used ($\frac{3}{4}$ or $\frac{1}{2}$) to minimize splashing from the indirect waste.

Here are some major manufacturers for floor sinks:

www.josam.com
www.jrsmith.com
www.siouxchief.com
www.wadedrains.com
www.watts.com
www.zurn.com

10.2.10 Hub drains

Hub drains are used in commercial environments for indirect waste from walk-in coolers and freezers and from refrigerated cases. They are similar to floor drains except that they extend above the floor, so that other sources of water cannot enter the drains. They are manufactured by all of the companies who make floor drains and floor sinks.

10.2.11 Grease interceptors

If grease (of many types, but especially animal fats used in cooking) collects in piping, it restricts flow—eventually leading to blockages—and the portion of it that reaches the waste water treatment plant causes problems with the biology and chemistry of the treatment process. So there is a need to keep grease out of the waste water stream to the greatest feasible extent. As a result, grease interceptors have been required by codes for "grease laden waste" for decades. (Grease interceptors operate primarily by slowing down the flow rate to the point where the grease floats to the top of the water for skimming off. Some grease interceptors simply hold the floating grease in place but other types actively remove the grease, through some mechanical means, to a separate collection container.) The difficulty comes in determining what is "grease laden," and it should be noted that this does not include pouring liquid grease (or semi-liquid grease) down a drain—floor sink, floor drain, or sink—because such quantities of grease should *never* enter the plumbing system at all. Large quantities of grease are disposed of in grease dumpsters (usually for recycling) or by specialized systems that pump the grease into trucks that arrive on site specifically to carry away the grease (also for recycling). Prior to 2005 (or so), it was widely understood in the plumbing engineering community that grease laden waste includes the discharge from a three-compartment sink in a commercial kitchen (for pot washing) and occasionally floor drains in unusually greasy environments (such as meat preparation rooms in supermarkets or butcher shops). In fact, some codes prohibited connecting commercial dishwashers and garbage disposers; the high volume of very hot water leaving a dishwasher prevents grease from accumulating in an interceptor (thereby making it non-functional) and the solids that come through a garbage disposer turn the interceptor into a small septic tank—not a good thing at all. But in recent years, code officials have been requiring the connection of more and more drains in commercial kitchens to grease interceptors, and today we commonly see requirements for all floor drains, mop sinks, dishwashers, three-compartment sinks, and even garbage disposers. There are three major effects of these increasing requirements:

1. Larger grease interceptors. This is not necessarily a big problem, except for increased costs. The preferred design for a kitchen with any significant amount of grease laden waste is to install a large (usually about 1,000 gallons) pre-cast in-ground grease interceptor outside the building, mostly to make servicing easier and to reduce odors inside the building. This is not necessarily a big problem—unless there is no room for a large inground unit. (Grease interceptors have to be cleaned out periodically to make sure that too much grease does not build up. If the unit is inside, on the floor, in the floor, or under the floor, it will usually be relatively small and it will probably be cleaned out by hand—a dirty and smelly job. If it is outside, it will probably be serviced by a large vacuum truck.)

2 Making a grease interceptor into a septic tank requires special maintenance arrangements and different piping. This can be difficult to deal with and should be avoided if at all possible.
3 It will no longer be feasible to install a grease interceptor above the floor of the kitchen. In the past, in small kitchens, standard practice was to install a small unit on the floor under the three-compartment sink, but that will not work if floor drains have to be connected to it. In that case, it must be recessed into the floor (easy if it is a slab-on-grade floor) or installed below the floor (if there is a level below). This can be very challenging, depending on what is under the floor. (This has become a problem in new multi-use multi-story buildings that have below-grade parking because it means that the grease interceptor has to be located in the parking garage, where there is usually low headroom, limited floor space, and difficult access for servicing.)

10.2.12 Sand interceptors

It is also bad to have large amounts of sand (more generally known as "grit") in waste water piping too. It can wear out pipes (by scouring the pipe walls from the inside out) and it can interfere with the waste water treatment process. So floor drains or trench drains in vehicle service areas are required to discharge through a sand interceptor.

Sand interceptors also work by slowing down the flow rate of the water, but in this case the reason is so that the heavy sand will fall to the bottom—sort of the opposite of a grease interceptor. So sand interceptors are essentially just large pits (with solid lids or grates, depending on the application) that have to be cleaned out by hand periodically.

Here are some sand interceptor manufacturers:

www.highlandtank.com
www.jensenprecast.com
www.jrsmith.com
www.tandcplastics.thomasnet.com
www.zurn.com

10.2.13 Oil interceptors

Oil (motor oil from vehicles, mostly) is very similar to grease, so floor drains and trench drains in vehicle service areas inside buildings are required to discharge to oil interceptors, which are very similar to grease interceptors. The floating oil is trapped for later removal when the device is cleaned. The only challenge is that the rules for sizing oil interceptors are rather vague and it is not uncommon to see disagreements with local code enforcement officials.

Here are some oil interceptor manufacturers:

www.highlandtank.com
www.jensenprecast.com

www.jrsmith.com
www.tandcplastics.thomasnet.com
www.zurn.com

10.2.14 Sump pumps

Sump pumps fall into two broad categories: standard and grinder. Standard sump pump simply means that the water that is being pumped is either clean (as in ground water or storm water) or gray water (sanitary waste with no solids—wash water, etc.). Grinder pumps are required if solid matter is involved; a grinder pump is similar to a garbage disposer—a pump with sharp grinding teeth to break down solids.

Sump pumps are used for various different purposes:

- To remove water leaking into a building, usually as basement level. This can be ground water or storm water.
- To lift water from a drain that is below the sewer or septic system connection, as in a basement shower, lavatory, sink, or floor drain.
- To lift sanitary water from a urinal or water closet whose drain is lower than the sewer or septic system connection.

Given these widely varying applications, sumps and sump pumps vary widely too. The simplest, and smallest, units usually consist of fiberglass tubs (24″ diameter or smaller, usually 36″ deep or so) that are dropped down into the floor slab, with piping connections into the sides of the tub; this technique works for footing drains and for floor drains. The pump is installed in the tub, and it is usually a type of pump that is called "submersible," which means that the pump is under water at the bottom of the sump, with the pump motor above the top of the sump (theoretically above the water level). The pump and the motor are connected together with a long shaft. These pumps are available in plastic (very low cost and very unreliable) and bronze (much more costly and much more reliable). If a sump is expected to see direct collection of storm water (also known as rainfall), an engineer should be consulted because it is extremely challenging to size a pump correctly to handle rain. A lid is not required for a sump like this, but it is a good idea to keep pets, children, and adults from accidentally falling in.

Sumps can also be formed in concrete, which is what is seen in most houses built in the 1970s or earlier. In this case, a double tube form is placed and concrete is poured in between the tubes. When the inner form is removed, there is a round sump. If the purpose of the sump is to collect ground water, the bottom might simply be the earth under the foundation and water will come into the sump through the open bottom. If the purpose is to collect water from floor drains, laundry equipment, etc., the sump needs to have a bottom to keep that water from leaking into the ground. The pumps used in such a sump would be the same as the pumps used in a fiberglass tub. This type of sump also does not require a lid.

If a sump is required to have a grinder pump, first, it is strongly advised (although not necessarily required in one- and two-family dwellings) to put in two pumps—one to run normally and one for backup purposes. Second, the sump will probably have to be large because the pumps are a little larger and there should be two of them. Third, this sump is required to have a sealed lid and a plumbing vent to prevent sewer gas (remember that there are solids in there) from entering the building. If it is necessary to open a sump like this for servicing or inspection, it will be very smelly.

Here are some sump pump and grinder pump manufacturers:

www.flotecpump.com
www.hydromatic.com
www.libertypumps.com
www.lilgiantpumpproducts.com
www.saniflo.com
www.zoeller.com

10.2.15 Domestic hot water recirculation pumps

When a faucet is turned off, there is no water flowing, so the water in the pipes leading to the faucet is not moving. If that is hot water, it gradually loses heat to the surroundings until it is not hot anymore. For uninsulated piping (as is found in many buildings built prior to the 1970s) the water can cool in just a few minutes; insulation slows down, but does not stop, this process. This is why it usually takes some period of time to get "hot" water at any given faucet. If there is 60 feet of $\frac{1}{2}$"-diameter copper piping between the water heater (where the water is always hot, no matter what kind of water heater it is) and the faucet, and the flow rate is 1.5 gpm, here is how to calculate how long it will take for the hot water to "arrive":

1 Find the cross-sectional area of the pipe, in square feet.

$$\frac{1}{2}\text{"-diameter pipe} = \pi \times r^2$$

$$= \pi \times (0.25")(0.25")$$

$$= 0.20 \text{ square inches} = 0.0014 \text{ square feet}$$

2 Find the volume of the pipe in cubic feet.

$$60' \times 0.0014\,\text{sf} = 0.082 \text{ cubic feet}$$

3 Find the volume of the pipe in gallons.

$$1\text{ cf} = 8.33 \text{ gallons, so } 0.082 \times 8.33 = 0.68 \text{ gallons}$$

4 Divide the volume by the flow rate to get the time in minutes, then convert to seconds.

$$0.68 \text{ gallons} \div 1.5\,\text{gpm} = 0.45 \text{ minutes} \times 60 \text{ seconds/minute}$$

$$= 27 \text{ seconds}$$

Twenty-seven seconds should sound like a long time, because it is a long time. Generally, one would hope to have hot water within 10 seconds or so of turning on the faucet. If it takes 27 seconds to get water 60′ away, how much $\frac{1}{2}$″-diameter piping could there be for it to take 10 seconds (at 1.5 gpm)? Approximately $10 \div 27 \times 60 = 22$ feet. If the flow rate were 3 gpm, then this distance would halve to 11 feet. In reality, it is best to have faucets within 20′ or so.

If longer distances cannot be avoided, there are two approaches to solving the problem:

1 Use multiple water heaters and shorter runs of piping. This can be a cost-effective solution, depending on the details (including the type of water heater being used, space availability, maintenance issues, etc.).
2 Add hot water recirculation.

Both techniques are used in large facilities.

Hot water recirculation is achieved by running a small additional hot water pipe (usually 1″ but $\frac{1}{2}$″ might suffice for smaller systems) with a small in-line pump to keep water moving through the hot water piping at all times (actually, not quite all the time, but whenever a building is occupied and hot water might be needed). This pipe is run out to the most remote corner of the floor, or building, connected to the hot water line at that point, and connected to the cold water source of the water at the other end. If the branch lines from the hot water main to any given fixture are kept to 20′ or less, then there will be little delay in getting hot water at any point in the system.

10.2.16 Domestic water pressure booster pumps

The water pressure delivered by the utility (from water mains under the streets usually) or from a well, usually runs in the general range of 50–80 psi (pounds per square inch). Occasionally, isolated areas are seen with lower pressure, but higher pressure is rarely seen because it can cause problems with valves, faucets, and piping. When water moves vertically up through a multistory building, it loses pressure due to gravity (the weight of the water resists the pressure). One pound per square inch is equal to about 17.1 inches of vertical height, so almost 1.0 psi is lost for every 18 inches or so of vertical rise, or about 0.67 psi per one foot vertical. This is rarely a problem in low-rise buildings, but it gets to be a problem in most buildings (with typical water distribution) above six stories or so. Also, most water utilities (and codes) require a device called a reduced pressure zone backflow preventer (RPZBP) on a domestic water service because such a device makes it impossible for contaminated water in a building to move backward along the pipes and into the public system. For many years, RPZBPs were required only for healthcare and food service occupancies, but then all occupancies with mop sinks were added. These days, there are very few buildings that are not required to have RPZBPs. This matters because an RPZBP uses up 10–12 psi of the incoming pressure, equivalent to the first 15 to 18 feet of vertical rise.

To illustrate this, consider the following example of an existing eight-story building in Indianapolis (public housing for senior citizens, about 80 feet vertical to the highest hot water piping).

When the building was built in the 1970s, it had no RPZBP. The incoming pressure from the utility was about 70 psi. The pressure drop due to 80 feet of vertical rise is 80 feet × 0.67 psi/foot ≈ 53 psi, which left about 17 psi on the top floor of the building. 17 psi is low, but workable, especially when using flush-tank style water closets (which they had in the apartments in this building).

During a project to add fire suppression sprinklers to a 20-story building across the street, the owner was forced to add an RPZBP to the domestic water service in order to get a fire water service for the sprinkler system. The owner thought that it would be a good idea to add an RPZBP to the eight-story building too, and hired a contractor to put it in. The complaints about water pressure on the 8th floor started immediately (it was now down to 5–7 psi), and the owner had no choice other than to re-pipe around the RPZBP temporarily until a domestic water pressure booster system could be designed and installed.

Why was there not a problem at the 20-story building? Because it already had a domestic water pressure booster system.

Domestic water pressure booster pumps are special pumps that are designed to boost pressure in low-rise buildings where the incoming pressure is unusually low and in mid-rise and high-rise buildings where the incoming pressure simply cannot work. These systems usually use two or even three pumps, called duplex and triplex, respectively, in an integrated package with controls. They are not used when they are not needed for two reasons: (1) because they are expensive and (2) because if pressure is boosted too high, there will be problems with the plumbing.

In the case of the eight-story building in the example, the new booster system added about 50 psi to the incoming pressure. This meant that the pressure on the ground floor was 120 psi, which is too high. When boosters are used, it is necessary to add pressure-reducing valves on the main branch lines on the lower floors to prevent problems with valves, faucets, and piping.

Here are some manufacturers of domestic water pressure booster systems:

www.amtrol.com
www.mypumppro.com
www.syncroflo.com

10.2.17 Water heaters

Heaters for domestic water are available in two basic types: storage and non-storage (sometimes erroneously called "instantaneous"; the correct term is "continuous flow") and can be fueled by electricity, natural gas, propane, and even fuel oil.

Storage-type water heaters keep a tank of hot water ready at all times, and these systems are done in two different ways: integral tank and separate, or auxiliary, tank.

An integral tank storage-type water heater is what most people are most familiar with; they are usually round, anywhere from 18″ to 30″ diameter and from 20″ to 80″ tall. A separate tank simply means that there is a boiler (it is not usually called a water heater) that sends hot water into the tank; such tanks are usually very large to accommodate high but inconsistent hot water demand (as in hotels).

Before delving into the details of the equipment, an important question must be answered: are electric storage-type water heaters "slower" than natural gas storage-type water heaters? Answer: no, but they appear to be. Here is why most electric water heaters appear to be "slow."

If one were to go to a store to buy an inexpensive water heater, one would find 30-gallon natural gas-fired units with heat input of 30,000 BTUH (or sometimes more); the comparable electric water heater has power input of 1.5 kW. How does 30,000 BTUH relate to 1.5 kW? A watt is equal to 3.42 BTUH, so the 30,000 BTUH unit is equivalent to 30,000 BTUH ÷ 3.42 BTUH/watt = 8,772 watts, which is equal to 8.8 kW. 8.8 kW is 5.9 times 1.5 kW, which is why the electric water heater seems to be slow. It is slow, but it is only slow because it has low heat input. If the natural gas water heater had input of 5,000 BTUH, it would be equally slow. So why do electric water heaters not have higher input? Some of them do. The next family up from the very basic unit already described (30 gallons and 1.5 kW) would be a 40-gallon, 4.5 kW unit. That is almost three times the power input, but that is still barely more than half of the gas input. Would you want a 4.5 kW water heater at home? Maybe—that is less power than an electric range or an electric dryer—but such a water heater is substantially more costly. Higher power inputs are possible, but the costs climb rapidly. The author once designed a water heater for a restaurant that had to be electric and it had to fit into a small space, so it ended up with 54.0 kW of power input (that is 150 amps at 208V, 3-phase!); it worked extremely well, but it was expensive to buy and it was even more expensive to operate.

Does increasing the size of the tank make a water heater more effective (that is what plumbers usually tell people with a weak water heater—get a bigger one)? If the heat input goes up in proportion to the size, there is hardly any difference; if the heat input is smaller relative to the larger tank, it might be worse; if the heat input is larger relative to the larger tank, it might be a little better. Adding capacity in the tank helps very little, and here is why:

When water starts flowing out of a water heater (when a faucet is opened, for example), cold water (usually at 60 °F or even less) starts flowing in, which rapidly decreases the average temperate in the tank.

Consider this commercial example: 80-gallon, 4.5 kW input electric water heater; heat recovery capacity is 21 gallons per hour (0.35 gallons per minute); standing water temperature is 120 °F. Hot water on for five minutes at 2.0 gpm (for two showers); tank is now about 71.5 gallons at 120 °F and 8.5 gallons at 60 °F; average temperature is now about 114 °F.

Hot water on for five more minutes (10 minutes total) at 2.0 gpm—the tank is now about 63 gallons at 120 °F and 17 gallons at 60 °F; average temperature is now about 107 °F.

Hot water on for five minutes longer (15 minutes total) at 2.0 gpm—the tank is now about 55 gallons at 120 °F and 25 gallons at 60 °F; average temperature is now about 101 °F—at this point, the water would barely feel warm.

How would this look for a 40-gallon tank with the same 4.5 kW input?

After five minutes, average temperature would be about 107 °F (as compared to 114 °F). After 10 minutes, average temperature would be about 95 °F (as compared to 107 °F). After 15 minutes, average temperature would be about 83 °F (as compared to 101 °F).

The larger tank does help, but the bottom line is that the hot water would "run out" in less than 10 minutes with the 40-gallon tank and in about 15 minutes with an 80-gallon tank.

The answer to improving performance in storage-type water heaters is to increase heat input, and not necessarily to increase the tank size.

What if the heat input were doubled for the 40-gallon unit in the previous example?

After five minutes, average temperature would be about 116 °F. After 10 minutes, average temperature would be about 111 °F. After 15 minutes, average temperature would be about 107 °F.

In all cases, the higher heat input 40-gallon unit would out-perform the lower heat input 80-gallon unit. In fact, the hot water would last nearly twice as long, which makes sense if the heat input is doubled. Doubling the tank capacity, while helpful, would not double output.

Of course, one of the answers to this is not to use storage-type water heaters. But the details of storage water heaters must be covered first.

Tanks are usually glass-lined to minimize problems with corrosion and mineral build-up. Electric units have an electric heating element, or elements, immersed directly in the water, and many units are "dual element, non-simultaneous." This means that there are two elements, but only one works at a time. Each time the unit turns on, it automatically changes which element turns on to even out the wear. While this is typical in residential units, it is less common in commercial units, where both (or more) elements are likely to operate at the same time to increase capacity. These water heaters are generally inexpensive and maintenance is very simple.

Remember that the key rating for a water heater is recovery rate and not gallons of capacity; do not let someone sell you a large water heater with a low recovery rate. No one will be impressed.

Here are some large manufacturers for both electric and gas-fired hot water heaters:

www.aosmith.com (brand names: AO Smith, American, and State)
www.bradfordwhite.com
www.rheem.com
www.ruud.com/products/tank_water_heaters/ (the water heater division of Ruud)

Natural gas-fired storage-type water heaters are similar (also with glass-lined tanks), but they have burners instead of heating elements. The conventional style of burner is roughly 80% efficient and it simply heats the bottom of the tank; such units draw combustion air from the surrounding room, so they cannot be located in closets or very small unventilated rooms, and they have vents (metal flues or chimney flues) for the combustion products. The flue can be combined with a conventional furnace flue. High-efficiency units are more like fire-tube boilers, where the fire and its heat actually extend upward through the water tank; such units have direct combustion air intake for the burner (called sealed combustion, via PVC piping) and they have low-temperature combustion vents that are also run in PVC piping. The venting arrangements are very similar to the venting of high-efficiency furnaces (see 7.5.3). Propane-fired water heaters are virtually identical to natural gas-fired water heaters (the burners are slightly different) and oil-fired water heaters are very similar.

In the 20th century, it was believed that it is safe to store hot water, meaning water that is 120°F or hotter, especially if it is hotter than 140°F. In the 19th century, it was common to store cold water too. This is why hotels used to use domestic hot water boilers with large (usually thousands of gallons) storage tanks; no one believed there to be a health risk. But by the beginning of the 21st century, it was understood that organisms can live in hot water, which means that as little water—cold, hot, or otherwise—should be stored as possible. A small hotel that might have stored 2,000 gallons of hot water in 1960 might have stored only 400–500 gallons in multiple high-efficiency natural gas-fired storage-type water heaters since 2000 or so; in the future, such a building might use many tankless continuous-flow natural gas-fired water heaters instead. (Old habits die hard though, and the author has seen boiler and storage tank systems in buildings that were built in 2011.)

What is a tankless continuous-flow natural gas-fired water heater? A small cabinet that contains a large variable fire burner with a heat exchanger (to transfer the heat to the water passing through), controls, and combustion connections. Such units are available without sealed combustion, but it is not advisable to use those models, mostly due to potential back-drafting problems which can cause burner failure and freezing if the problem happens when the outdoor temperature is below freezing. Remember that the burner for a low-cost residential water heater has input of 30,000 BTUH; the heat input of a 100-gallon high-efficiency commercial water heater is 199,000 BTUH, and larger inputs are possible with larger commercial water heaters. The typical input for a continuous-flow unit is 199,000 BTUH that can turn down (reduce) to about 19,000 BTUH. When water flows through the unit, the burner turns on and burns at whatever level is required for the flow rate. For typical units, 100% burner corresponds to about 5.0 gpm of water flow, so 10% of burner would correspond to about 0.50 gpm of water flow—one slow lavatory faucet maybe. The greatest limitation for these units is the 5.0 gpm, because that can support only four showers at the same time. If a hotel has 200 showers, 150 of which might be on at peak times, it might be necessary to use dozens of these water heaters! That would be very awkward, although they can be grouped together and they can even use recirculation. If a 200-room hotel were to be designed this way, there would probably be a smaller grouping on each floor, or even two groupings on each floor, to simplify the piping and reduce the need for recirculation. Finally, these water heaters are more sensitive to water quality than conventional units (either regular or high-efficiency) because they have smaller water passages. Primarily, they have a tendency to accumulate calcium (also called lime), which causes scaling and eventually heavy build-up. To address this, these units should only be used with softened incoming water, which is much lower in minerals (including calcium). These units are also available for propane fuel.

Here are some major manufacturers for continuous-flow water heaters:

www.bradfordwhite.com
www.navienamerica.com
www.noritz.com
www.rinnai.us
www.takagi.com

10.2.18 Drinking fountains and electric water coolers

A drinking fountain is a fixture that includes a bowl with a drain and a cold domestic water source, usually in the form of a "bubbler" spout with a faucet-like handle. There is no refrigeration to chill the water in a drinking fountain, so we actually see very few true drinking fountains today. They are available as individual wall-mounted, pedestal-mounted (indoor and outdoor) units, and wall- and pedestal-mounted groups.

What are more commonly seen (at least in the US), are electric water coolers. These fixtures consist of drinking fountains with integral water chillers. The most common form is the dual-height wallmounted unit, with the refrigeration units in cabinets under the bowl of the drinking fountains. But these can be found in a number of other forms: free-standing integral floor-mounted units (usually considered old-fashioned these days—and not accessible); wall-mounted units with recessed refrigeration (recessed into a thick wall behind the drinking fountains); or even completely remote water chillers (that can be mounted above the ceiling, below the floor, in a nearby closet, etc.). When a remote chiller is used the units that are seen are truly drinking fountains again, with the only difference being that they receive chilled water from the refrigeration system instead of domestic cold water.

Most codes require access to public drinking water, which can be met with drinking fountains, electric water coolers, or even bottled water in some cases. (Restaurants are exempt in most cases if they serve water at the tables.)

10.3 Piping materials

Plumbing would be nothing were it not for piping, and piping is where technology has changed the most over the years. In buildings, we use piping for several different purposes: cold domestic water, hot domestic water, sanitary drainage, sanitary venting, steam condensate water removal, cooling condensate water removal, ground water removal, storm water removal, heating hot water, chilled water, fire protection (sprinkler and/or standpipe) water, water-source heat pump loop water, natural gas, propane, pressurized air, and medical gases (including nitrogen, oxygen, nitrous oxide, air, and vacuum). Each of these will be covered in turn, but first it is necessary to go through the various piping materials, because each material can be used for multiple different piping systems.

The first piping, built by the Romans, was made from lead (now understood to be extremely hazardous for domestic water piping); piping in the US from the 19th to the middle of the 20th century was usually galvanized iron (also known as galvanized steel) with threaded joints or cast iron, in bell & spigot form where the joints were sealed with a backer rod of oakum (tarred hemp or jute fibers in a rope-like form) and molten (liquid) lead (there was no health hazard from the lead in the joints which was not exposed to water under normal use), or glass for acid waste in laboratories; in the middle of the 20th century, copper tubing became available, followed by polyvinyl chloride (PVC), then chlorinated polyvinyl chloride (CPVC), acrylonitrile butadiene styrene (ABS), polyvinyldine fluoride (PVDF), black iron, flexible stainless steel, and cross-linked polyethylene (PEX). By 2000 or so, cellular (foam) core PVC became available as a lower cost substitute for standard PVC.

10.3.1 Galvanized iron (galvanized steel) piping

This type of piping was used almost exclusively for indoor domestic water piping from the latter part of the 19th century up through the 1960s, and it is still used for large sizes today. The pipe material comes in lengths up to 20 feet and the joints are threaded (screwed). Using this type of piping in the field means that it is necessary to cut pipe lengths to fit, and thread the ends of the cut pipe, which is labor intensive and very costly. Today, this is usually only seen in hydronic (heating hot water and chilled water) and fire protection sprinkler piping systems. The wall thickness of the pipe is determined by its "schedule"; the common schedules include thinwall (the thinnest), 5, 10, 20, 40, and 80. The most commonly used schedule for hydronic piping is schedule 40, although the lighter schedules are commonly seen in sprinkler systems (see Chapter 11 for more information). Schedule 80 is rarely used.

All pressurized piping (including domestic water piping) is sized according to flow velocity, which is based on flow rate, pipe diameter, and pipe internal roughness. The calculations for this are too complex for this text—so much so that even very sophisticated plumbing designers usually use tables for pipe sizing instead of doing the detailed calculations.

10.3.2 Cast iron

Cast iron piping is used primarily for sanitary waste and vent and storm piping, and it comes in two basic styles: bell & spigot (also called hubbed) and hubless.

Bell & spigot is the older system, and it is commonly seen in buildings built prior to the mid-1960s. The pipe is black and the bell & spigot joints are easily identifiable: the bell (or hub) is an enlarged end of a pipe section that is designed to receive the

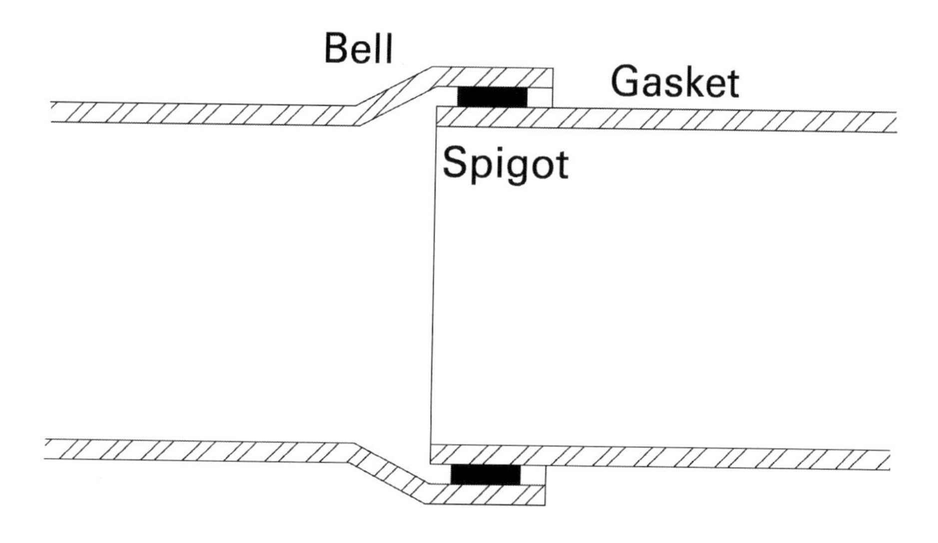

Figure 10.3.2.A
Bell & spigot cast iron piping joint

regular end of another section of pipe. When the pipes are put together, a rope-like material called oakum is packed into the joint to seal it (partially) and to hold the pipes in position while molten (liquid) lead is poured into the joint to seal it; the lead fills the whole depth of the bell (less the thickness of the oakum) and it is finished off flush with the end of the bell. Obviously, this is a dangerous and labor-intensive process and it has been used only rarely since the 1960s, except in the City of Chicago (where it was code-required until very recently).

Hubless cast iron simply means that the pipe sections have plain, flat, and smooth ends, and the joints are made by inserting the pipe ends into a rubber sleeve (about 4″ to 6″ long, depending on pipe size). The rubber sleeve is then clamped onto the pipe sections using stainless band clamps, usually four to six clamps per joint. This system is easier and much faster to build, and therefore less costly.

This system is commonly used in many buildings today, largely because the cost premium can be small and because it is inherently suitable for use in a plenum. Cast

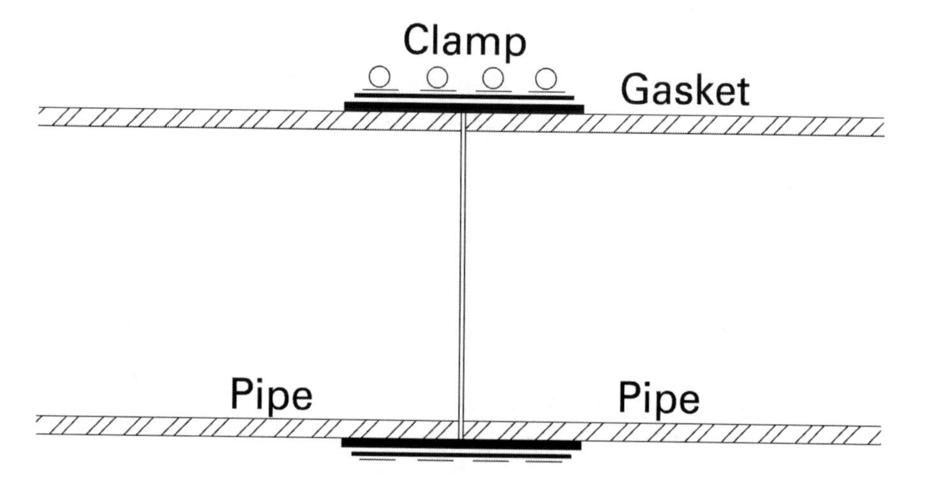

Figure 10.3.2.B
Hubless cast iron piping joint

iron does not have the smoothest interior walls (of the types of pipe that are commonly used for waste piping) but it does have excellent durability, usually lasting at least 60 and sometimes up to 100 years.

The International Building Code says that materials that do not meet the 24/50 flame spread/smoke developed requirements may not be exposed within a plenum. PVC piping cannot meet these requirements, so it is necessary to wrap it with material that does meet the requirements if it is located in a plenum. The wrapping material that is most commonly used is closed-cell elastomeric foam insulation. Bare cast iron piping is fine in a plenum, and some CPVC piping might be acceptable as well.

Non-pressurized piping for sanitary waste is sized by "fixture units," using tables in the plumbing code. It should be kept in mind that minimum sizes are important: $1\frac{1}{4}''$ for a lavatory drain, $1\frac{1}{2}''$ for a sink drain, 3″ for a water closet drain, etc. Even small buildings require 4″ piping but 6″ piping (and larger) is only required in relatively large facilities. (Outdoors, most municipal sewer utilities require 6″ minimum, even for a single-family house.) The most important issue is that gravity flow piping (including sanitary and storm) must slope to drain, usually at a rate of $\frac{1}{4}''$ per foot or $\frac{1}{8}''$ per foot, depending on the application and the pipe size. But it must be noted that 100 feet of pipe slopes downward 12.5″ even at $\frac{1}{8}''$ per foot; overhead waste and storm piping tends to get in the way as a result.

10.3.3 Glass piping

Glass was the only material available until the late 20th century to carry highly acidic waste water, which is commonly seen in laboratories (high school, college and university, and industrial). It consisted of pyrex (heat resistant and very strong glass) glass tubing, jointed using rubber sleeves and clamps similar to those used for hubless cast iron. But the glass is so smooth that it is difficult to keep the joints tightly connected, and the system was abandoned as soon as another alternative was developed. See 10.3.8 for more information.

10.3.4 Copper

Copper comes in two types—drawn (usually called "hard") and annealed (usually called "soft") and in six sub-types:

- Type K, thickest walls, in both drawn and annealed
- Type L, medium wall thickness, in both drawn and annealed
- Type M, thinnest walls, in drawn only
- Type ACR (for refrigeration only), in drawn and annealed
- Type DWV (for drain, waste, and vent), in drawn only
- Medical gas, in drawn only

Drawn tubing comes in lengths from 18 to 20 feet and annealed tubing comes in continuous spools. Both types can be joined using soldered (sealed with metal that melts at low temperature) fittings or brazed (sealed with metal at high temperature) fittings; annealed tubing can be joined using flare joints and compression fittings, and both types can be joined using various mechanical systems.

Annealed tubing is used when it is highly desirable to have a minimum number of joints (no joints at all if the run is no longer than the length of the spool of tubing), but it is soft and flexible and difficult to install using straight lines. In other words, it is going to be ugly, so it does not work well in exposed environments. In exposed environments and for large sizes (annealed tubing does not work very well larger than $\frac{1}{2}$"), drawn tubing is nearly always used. Annealed tubing is usually used underground due to the limited number of joints.

Both annealed and drawn tubing can be insulated, but it is more difficult to insulate annealed tubing due to its flexibility.

It is best to use type K drawn tubing due to its increased wall thickness, but it is also substantially more costly. There is little reason to use type M for anything. Type DWV is rarely used because it is far more costly than cast iron, PVC, or CPVC for DWV applications. Drawn medical gas tubing is the only piping that can be used, with brazed joints only, for medical gas applications.

10.3.5 Polyvinyl chloride (PVC)

PVC piping has the smoothest interior walls (along with CPVC), so it is the least likely to clog in waste applications. It is manufactured with solid walls and with bell & hub solvent-weld joints. Solvent welding looks like gluing, but it is not really gluing at all; instead, solvent welding means that the glue-like chemical that is applied to the joint actually melts both pieces of PVC and fuses them together. Such joints truly are stronger than the pipe itself. PVC piping is usually white in color, and it is available in all normal sizes that are needed for waste and vent applications, from $1\frac{1}{4}$" to 24". Like galvanized steel pipe, PVC is available in schedules, the most common of which are 20, 40, and 80. The vast majority of PVC piping used in buildings is schedule 40. The wall thickness of a schedule 40 4" PVC is nearly $\frac{1}{4}$".

Near the start of the 21st century, the manufacturers designed a less costly version of PVC, which is "cellular core PVC." This material consists of very thin inner and outer shells of solid PVC with PVC foam filling the space in between. The foam actually gives the pipe better bending strength (it is less likely to sag) but the thin shells also make the pipe more susceptible to damage, both on the inside (from grit and other objects in the water stream) and on the outside (from sharp rocks in an excavation if buried, or from someone bumping into it with a sharp tool). From the author's perspective, this material—which can save only a small amount of money—should be avoided at all costs, especially underground, although there is little doubt that someone will disagree with that position.

Figure 10.3.5
PVC piping

10.3.6 Chlorinated polyvinyl chloride (CPVC)

CPVC is very similar to PVC, except that it is stronger and more flexible (so that it can be used for domestic water piping) and it meets the 25/50 flame spread/smoke developed requirements for use in a plenum. It is installed in the same way as PVC, with bell & spigot solvent-weld joints. It is offwhite to distinguish it from PVC. When used for water distribution piping, a full range of valves is available too. This is the preferred material to be used for waste and vent piping if cast iron is not going to be used. It is very acceptable for domestic water use as well.

10.3.7 Acrylonitrile butadiene styrene (ABS)

ABS is a viable alternative for PVC but not for CPVC, so it is of limited usefulness. It is usually black and it is installed in the same way as PVC and CPVC.

10.3.8 Polyvinyldine fluoride (PVDF)

PVDF is the replacement for glass acid-waste piping. It is plastic, but it is a plastic that has been engineered to be acid resistant. It is blue and it is installed in the same way as PVC, CPVC, and ABS.

10.3.9 Black iron (plain steel)

Black iron piping is a very old system that is used primarily for natural gas, using welded (required for 3″ pipe and larger) or threaded joints. Most natural gas piping in residences and commercial buildings is black iron. But natural gas can be dangerous, especially in enclosed spaces, so it is illegal to have joints in natural gas piping in concealed locations (inside walls, above inaccessible ceilings, etc.) This is why the natural gas piping at your house is probably more exposed than you expect it to be. Final connections to equipment (the range or dryer at home) are allowed to use short lengths of stainless steel flexible hose.

10.3.10 Corrugated flexible stainless steel

This product was developed as a replacement for black iron natural gas piping. It is delivered on spools (like annealed copper) and it is installed very quickly (low labor) without joints (or with very few joints). Because it is flexible, it can be installed in concealed locations without joints, but it is bright yellow (it has a vinyl jacket over the flexible stainless steel core for identification purposes) and not very attractive for exposed installations. It is limited in size, so it can be difficult to use for very large facilities. Because it is not amenable to branch fittings—tees, especially—this piping is installed in a manifold arrangement. This means that the main gas line ends at a manifold, which is a large pipe with connections (and valves) for many small pipes. A separate length of piping runs from each connection to a single appliance at the other end, which eliminates most of the joints from the system.

Here are some manufacturers:

www.gastite.com
www.omegaflex.com
www.tru-flex.com
www.wardflex.com

10.3.11 *Cross-linked Polyethylene (PEX)*

PEX piping is very flexible and it is shipped on very long spools, which makes it excellent for use in buried water-source heat pump loops and for concealed domestic water piping. PEX is installed as domestic water piping by using a manifold system similar to that described for flexible stainless steel natural gas piping, but PEX manifolds are usually double: one for cold and one for hot. In a PEX installation in a single-family dwelling, there is usually one manifold that sends two tubes to each sink, lavatory, shower, or washing machine and one tube to each water closet and dishwasher. If someone needs to work on a lavatory faucet, that fixture can be turned off at the manifold without affecting the rest of the system. It is very low in cost and very effective, but not attractive. Just like annealed copper tubing, it is flexible and difficult to train into neat patterns. In residential applications, it is usually color-coded too: blue for cold and red for hot. That is convenient, but it makes it even less attractive. It is not available in large pipe sizes, so it is usually necessary to marry it to some other more conventional piping material (usually copper or CPVC) in large facilities.

10.3.12 *Piping insulation*

There are four basic materials for piping insulation:

1 Cellular glass
2 Elastomeric
3 Fiberglass
4 Mineral wool

Each of these is used for various applications, as follows.

Cellular glass is used most commonly for hydronic (heating hot water and chilled water) applications, especially underground, from 1″ to 3″ thicknesses, usually with jackets. Jackets are membranes (from paper through foil) that cover the insulation to protect it and to give a finished look (many jackets are paintable). Jackets can be factory-installed or field-installed. This material can also be used for insulating domestic water. The material amounts to a glass foam, having millions of tiny bubble cells that are made of glass. It is more costly than competing products for many applications.

Elastomeric insulation is the dominant material that is used to insulate domestic water piping and refrigeration piping, but it can be used on storm piping (and the underside of roof drain bodies) and on other systems as well. This material is a soft closed-cell foam (similar to neoprene) that is usually black and most often seen in $\frac{1}{2}$″ thickness. If thicker insulation is required, thicker products can be purchased or multiple layers of $\frac{1}{2}$″ thickness can be used. The most common tradename for this material is "Armaflex." This material should be protected by a special coating (indoors) or by a tough jacket (outdoors), but it is frequently installed without such protection. If exposed to the elements outdoors (especially direct sunlight), it will dry out, crack, and often fall off the piping. This is a very cost-effective product.

Fiberglass is similar to cellular glass, except that it is composed of fibers (like typical fiberglass insulation) instead of bubbles. It is used for similar purposes; it must be jacketed, and it is very cost effective.

Mineral fiber is similar to fiberglass, except that the fibers are made from minerals and not from glass. It is used for similar purposes; it must be jacketed, and it is very cost effective.

10.4 Piping systems

Having reviewed all of the various piping materials that are, or can be, used in buildings, the various systems that we use for piping can be covered, as follows.

10.4.1 Non-pressurized piping systems

This sloping is the most problematic plumbing issue for many design projects. If someone is trying to add a water closet 100′ from the nearest underground waste piping that is only 8″ below the floor, it simply will not work because there is not enough vertical space for the necessary slope. For overhead piping, long runs fall considerable distances and can cause headroom problems.

Non-pressurized means that the piping does not run full and that it flows by gravity—downhill, in other words. This applies to storm water, sanitary waste water, condensate water, and vents. If these pipes run full, something is wrong with the plumbing.

The first requirement is that the piping (storm, sanitary waste, and condensate) has to slope downward toward the discharge point, usually at a rate of $\frac{1}{8}$″ vertical per foot as noted previously, roughly 1%. (Vents slope upward by 1% or more.)

Second, given the high potential for clogs and stoppages (due to solids in the waste stream), all joints have to be "wyes" and not "tees" so that there is a gradual transition (see Figures 10.4.1.A and 10.4.1.B).

Third, there must be a cleanout at every 90° turn and at no more than 100 feet on center for long runs of piping.

Figure 10.4.1.A PVC wye fitting

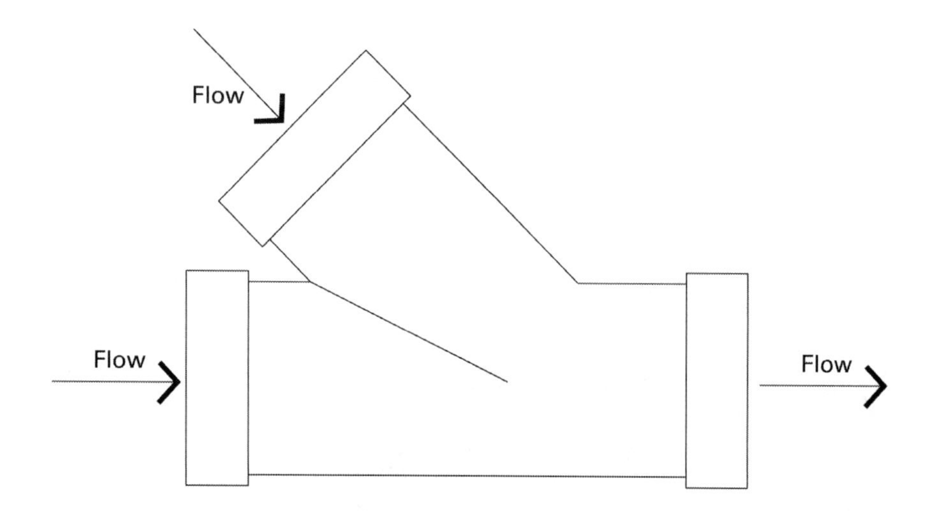

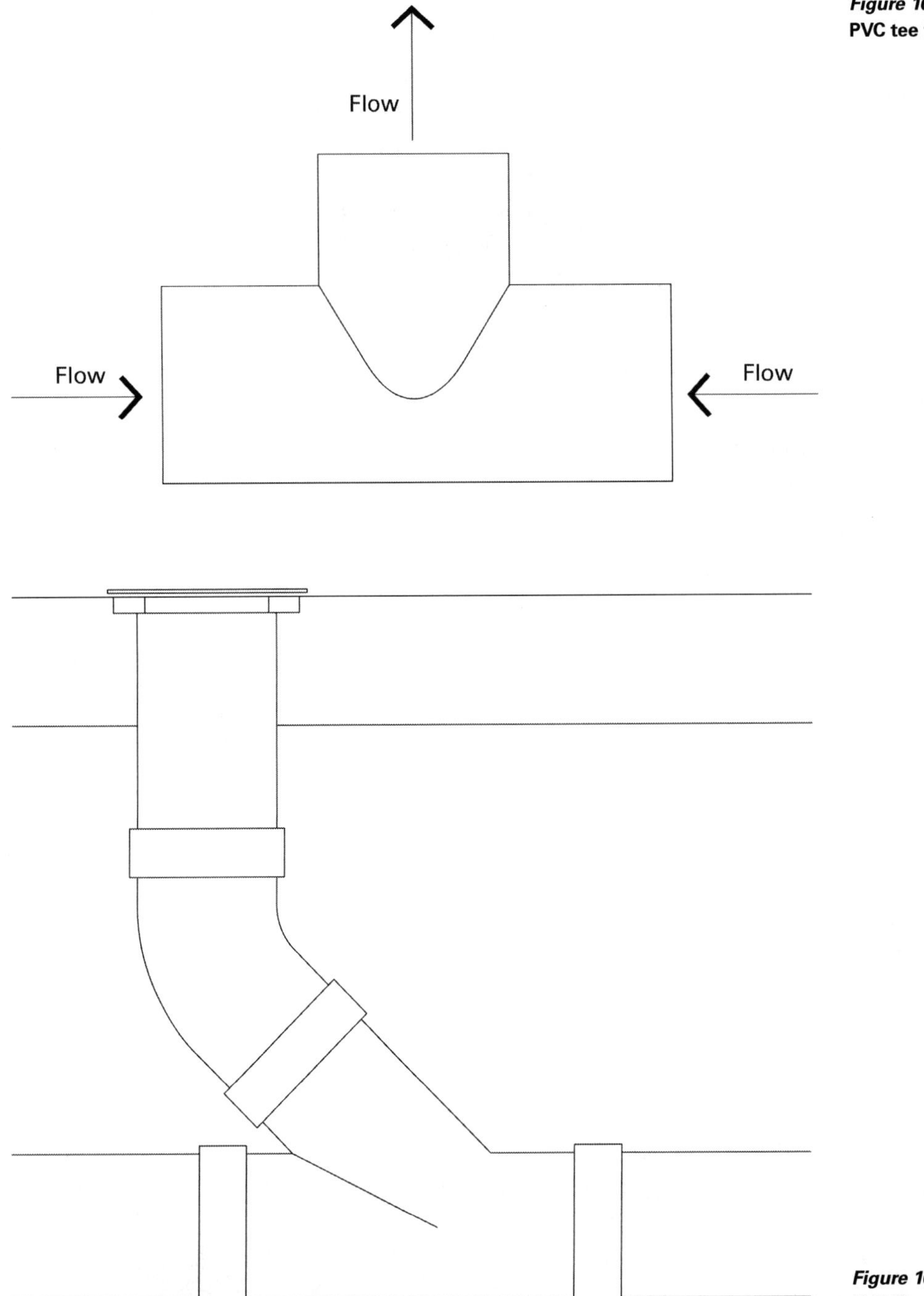

Figure 10.4.1.B
PVC tee fitting

Figure 10.4.1.C
PVC floor cleanout

Vents are used to expel sewer gas from the building and to equalize pressure in the system. Every plumbing system is required to have at least one full-sized (meaning equal to the largest piping in the system) vent to the atmosphere, which is usually referred to as "vent through roof, or VTR." Every fixture, except floor drains, is required to be vented. The vent piping must be separate, meaning that a first-floor fixture cannot vent into a wet drain pipe coming down from the second floor. In some instances, it is acceptable to use air admittance valves (trade name for the company that invented it: Studor) for individual fixtures where it is excessively difficult to put in a conventional vent. Air admittance valves are one-way air valves that let air into the system, without letting air and sewer gas out.

Acceptable materials include cast iron (bell & spigot and no-hub), copper with soldered joints, PVC with solvent-weld joints (if not in an air plenum), CPVC with solvent-weld joints, ABS with solventweld joints (if not in an air plenum), and PVDF with solvent-weld joints (for acid waste only). Copper is commonly used for condensate drainage systems; if it is, it is necessary to insulate the first five feet at the connection point to a drain pan to prevent condensation. For steam condensate removal, it is necessary to use at least 30 feet of cast iron from the connection to the drain due to the high temperature of steam condensate.

10.4.2 Pressurized piping systems

Pressurized systems are not required to slope to drain (except for certain buildings that are designed to be "winterized" where the water piping is drained before a hard freeze and for sprinkler systems) because the pressurization can overcome vertical ups and downs. This allows for far more flexibility in installation.

10.4.2.1 Domestic water

Most buildings have domestic water piping for restrooms, bathrooms, kitchens, break rooms, some coffee makers, some vending machines, ice makers, laundries, etc. Conventional domestic water piping is designed in a branch arrangement where the largest line comes in at the service with branches to each fixture or grouping of fixtures, moving away from the location of the service. This is done by using eccentric (i.e. not equal; an example would include a $\frac{3}{4}''$ tee in a 2″ line) tees and reducers (to reduce the size of the piping as fixture loads are taken off). Direction changes are accomplished with 45 and 90 degree elbows. Well-designed systems have zone valves to shut-off areas, or wings, of the facility to make servicing easier, and all systems have valves at each fixture for the same purpose. Valves at individual lavatories, sinks, and tank-style water closets are called "stops" and they are available with permanent handles or with removable handles (usually called "keys"). The most commonly used materials for domestic water piping are copper (drawn above grade and annealed below grade) and CPVC, although galvanized iron is still used for large sizes (large meaning larger than 3″ or so).

PEX is also commonly used for domestic water piping, especially in residential work (virtually all one- and two-family dwellings use PEX now). These systems use manifolds and direct runs from the manifold to each fixture, with no joints in between. In large facilities, especially multi-family housing, a conventional copper or CPVC system might feed manifolds in each dwelling unit.

Most systems use both hot and cold water, and some systems have water softening (for cold, hot, or both). Water softening is generally desirable where water hardness exceeds 11 grains or so.

Pipe sizing is done on the basis of water velocity, and the goal is to keep the water moving at less than 8 feet per second (faster means excessive noise and pressure drop due to friction). As noted previously, the calculations are quite complicated for this, so even sophisticated designers usually use special tables for this pipe sizing.

Generally speaking, all copper domestic cold water piping should be insulated to reduce the possibility of condensation (except possibly in very dry climates) and all domestic hot water is usually required to be insulated by the energy code.

10.4.2.2 Hydronic piping (heating hot water and chilled water)

These systems are very similar to domestic water systems except that they recirculate (they are closed loops), so that they have pumps. The most commonly used materials are copper (for 3″ lines and smaller) and galvanized iron for large lines. All hydronic piping is heavily insulated (usually 1–2″) to prevent heat loss from the hot water and heat gain to the cold water. It is feasible to use CPVC and PEX for chilled water applications, but not for heating hot water, unless the heating hot water temperature is relatively low—120 °F or so.

10.4.2.3 Heat pump loop water

Heat pump loop water piping is very similar to hydronic piping—closed loop with pumps—except that the temperatures are more limited. Non-earth-coupled loop water is generally limited to 80–105 °F or so, and earth-coupled water will usually run between 27 and 105 °F. The most commonly used materials are copper and CPVC for non-buried piping and PEX for buried piping. Very large systems could use galvanized iron piping too for the largest sizes.

10.4.2.4 Natural gas and propane

Natural gas is usually delivered to a building (by the utility company) at medium to high pressure, which usually means 15 psi or greater. At the meter, the utility installs a regulator to reduce the pressure to the range that is needed in the building. The most common pressures that are used in buildings are 6″ water column (w.c.) (roughly 0.35 psi) and 2.0 psi. Occasionally, 5.0 psi is used too, which is the highest pressure allowed inside buildings by code. When 6″ w.c. is used, it is piped directly to each appliance with no additional regulators. When 2.0 psi (or 5.0 psi) is used, it is necessary to regulate the pressure at each appliance. For local regulators like this (if small), "ventless" types can be used, but most such regulators require a vent that is piped to the open atmosphere to dump small amounts of gas that leak out in the regulating process. The reason to use 2.0 psi (or 5.0 psi) natural gas inside a building is to reduce pipe sizes, especially for long runs.

Large natural gas lines (greater than 2.5″ and at any pressure) are required to be built using black steel piping with welded joints; smaller lines can be built using black steel piping with threaded joints, copper piping (with various types of joints), and flexible stainless steel tubing. Flexible stainless steel tubing is available only up to 2″, so another type of piping must be used if larger sizes are required.

Propane is most often piped using annealed copper tubing, but other methods—drawn copper piping, black steel piping, or corrugated flexible stainless steel piping—are acceptable also. When propane is used, there is a local storage tank on site, which is the fuel source. These tanks are usually installed on the ground at a distance from buildings, but they can be installed underground at additional cost.

10.4.2.5 Air (compressed air)

Compressed air is used for various different purposes in buildings: to drive pneumatic tools, to provide air for machines, to blow out pipes or pieces of equipment, to blow off dirt, etc. The most commonly used material for compressed air piping is steel, although copper and even CPVC can be used. (In the old days, pneumatic HVAC control systems used mostly annealed copper tubing but newer such systems use flexible plastic tubing.) Compressed air piping is found in many laboratories, most auto repair garages, fire stations, and most machine shops and factories. In fire stations, conventional compressed air is used but specialty compressed breathing air (to fill portable canisters to be taken to fires in the field) is used as well. The latter requires a separate compressor (and other equipment) and it is most commonly called a "Cascade" system (after the most common equipment manufacturer).

10.4.2.6 Medical gases

The most common piped medical gases are oxygen and vacuum (seen in all hospitals and virtually all operating rooms), but medical compressed air, nitrogen, and nitrous oxide are also very common. Medical gases must be piped using only drawn copper (medical gas grade) with brazed joints. There are extensive regulations and only qualified professionals should design such systems. Liquid nitrogen is also used occasionally in medical (or laboratory) environments, but the piping is very similar to the conventional medical gas piping, except that it is insulated. The most important points about medical gases for interior designers is that the rooms that contain them must be rated and carefully ventilated, so most such rooms are located along an exterior wall to make ventilation simpler and to allow for cylinder delivery without going into the building.

10.5 The plenum

In addition to the requirements noted under 10.3.2, the plenum rules apply to all other piping systems too. So all piping located in a plenum must meet the 25/50 flame spread/ smoke developed rules with no exceptions.

This can be awkward when doing minor renovations in existing buildings where wrapped PVC waste and vent piping already exists in a plenum. The applicable rules in any given jurisdiction might, or might not, be clear about what to do in such a situation. At a minimum, all new piping should be compliant with the letter of the code; preferably, any disturbed or modified piping would be compliant as well, but few owners will be willing to pay to replace all existing piping. It really has to be handled on a case-by-case basis.

10.6 Accessibility

The second most challenging aspect of accessible design is probably in plumbing. Once all accessible routes, doors, ramps, stairs, etc. have been worked out, it is necessary to make sure that kitchen sinks are accessible (especially in break rooms, but in any accessible living unit) and that proper fixtures and installation standards are followed in accessible restrooms (most of them these days) and bathrooms. Higher water closet seats are required (at least 17″ versus the standard 15″); lower urinal rims are required (no more than 15″ versus the standard 24″); sinks and lavatory rims cannot exceed 34″ in height; knee spaces must be provided at accessible sinks and lavatories; showers must be hand-held and adjustable; drinking fountains and water coolers must be dual-height; accessible water closets must be between 17″ and 19″ (to the center) from the sidewall; there cannot be more than a $\frac{1}{2}$″-high threshold at the entrance to an accessible shower; piping exposed under a lavatory or sink must be insulated to prevent burns, etc. And "clear spaces" at all fixtures must be included. None of this is optional; all rules simply must be followed at all times.

10.7 Plumbing design documentation

Plumbing drawings are actually some of the most difficult because many pipes are small and located very close together—too close together to show up on readable drawings. So plumbing drawings are more abstract than HVAC drawings, by necessity. But, as is always the case, what matters most is clarity. Waste and vent piping usually shows up on separate drawings from the domestic water piping; natural gas piping could show up on the domestic water piping drawings or on separate drawings (even on HVAC drawings in some cases); and medical gas piping always appears on its own drawings. Condensate piping could appear on the waste and vent drawings or on the HVAC drawings.

Plumbing drawings are always required to have a "waste and vent riser" or "schematic," or "isometric" which is drawn as a 3-dimensional isometric projection. This is done to show the overall arrangement of the system and to verify that all components are included (especially critical for vents).

Fixture information is shown on a fixture schedule and details are used for water heaters, water softeners, hot water recirculation, certain joints, mounting methods, etc.

Summary

Interior designers do commonly select plumbing fixtures, largely due to aesthetic concerns. That is perfectly OK, but flushing performance should not be ignored for water closets—no matter how cool the design might be. And it is not OK to cheat on accessibility. As long as fixtures function and facilities are accessible, few problems are likely to be encountered with plumbing. Of course, well-designed plumbing fixtures can be very costly, so the budget should be kept in mind as well.

Outcomes

10.1 Understanding the whys and hows of water usage in buildings, especially water conservation, by knowing how to reduce usage without compromising functionality and health.

10.2 Understanding plumbing fixture types, including, but not limited to, water closets, urinals, lavatories, sinks, showers, bathtubs, etc.

10.3 Understanding faucets, especially different types (sink, lavatory, bathtub or shower; one, two, three or more handles, finishes, etc.) and their application.

10.4 Understanding basic piping types and systems, especially which material is used for which system and that gravity flow piping must slope to drain.

10.5 Understanding code issues related to the plenum, especially limitations on allowable materials.

10.6 Understanding code issues related to accessibility, especially fixture mounting limitations and clear floor spaces.

Note

1 US Public Law 102-486-0CT. 24, 1992, Energy Policy Act of 1992, pages 106 STAT.2826–106 STAT.2827.

Chapter 11

What are fire protection systems?

Objectives

11.2 To understand standpipes
11.3 To understand wet-pipe and dry-pipe fire suppression sprinkler systems
11.4 To understand fire pumps
11.5 To understand fire alarm systems
11.6 To understand smoke removal systems
11.7 To understand fire protection design documentation

11.1 Introduction

Fire protection systems fall into two main categories:

1 Systems to assist the fire department
2 Systems to assist building occupants

Both types of systems may have manual and/or automatic features, according to the specific situation. In general fire alarm requirements are spelled out in the International Building Code although some provisions are found in the International Fire Code. Detailed requirements are usually spelled out in various NFPA standards. All of these can have an impact on the look of a project, so it is important for interior designers to understand the basic requirements and system arrangements.

11.2 Standpipes

Standpipes are category A systems that consist of vertical pipes that are located in stairwells (most commonly on the intermediate landings) to be used primarily by firemen to fight fires in tall or very large buildings. According to the International Building Code, there are three classes for standpipes:

- Class I: $2\frac{1}{2}$" hose connections for use by fire-fighters only
- Class II: $1\frac{1}{2}$" hose connections for use by building occupants
- Class III: $1\frac{1}{2}$" hose connections for use by building occupants and $2\frac{1}{2}$" hose connections for use by fire-fighters.[1]

Class II and Class III standpipes are rare, mostly because fire departments prefer not to have building occupants attempting to fight fires. When standpipes are installed, they are usually Class I systems.

In general, standpipes are required in buildings having floors more than 30 feet above the level of fire department access, and Class III standpipes are required unless the building is sprinklered. Given that most buildings are sprinklered today, the most common arrangement is to see Class I automatic wet standpipes in all stair towers in most buildings that are three stories or taller. (Dry standpipes are allowable in some situations, especially parking garages in freezing climates; and manual standpipes—where there is no permanent water connection—are also allowed in some situations.) Inside a building, there is little difference between a wet standpipe and a dry standpipe.

11.3 Fire suppression sprinklers

Fire suppression sprinklers are used primarily to keep fires from spreading long enough for the fire department to arrive and put out the fire. (There are special sprinkler systems that are designed to put out fires—mostly early suppression fast response (ESFR) systems and dry agent systems—but such systems are only used in special situations, such as high pile storage, data centers, cooking equipment in commercial kitchens, etc.) These systems do not work the way they are often shown in movies.

First, the individual sprinklers (often erroneously called "sprinkler heads") go off one at a time when a small device called a fusible link melts due to excessive heat (usually more than 140 °F or so). So even though more than one sprinkler can go off if a fire spreads, there is no way to set off all of them (whether that is two or 100) from any single location. (Holding a lighter under a single sprinkler could set off that particular sprinkler but no others.)

Second, the main valve at the source water line cannot be turned on to activate the sprinklers; that valve must be open at all times so that water can flow if, and when, one or more sprinklers activate. That valve is used to turn off the water flow after the fire has been extinguished.

Also, even if all of the sprinklers did go off, they could not all flow water because the water supply system is not designed to be able to flow all of the sprinklers in the system. This is a practical issue related to water availability and cost. The most common design calls for designing the water supply to flow the sprinklers in the "most remote 1,500 sf." That means that the sprinkler designer figures out the 1,500 sf area in the building that is farthest from the water supply and calculates for that flow. Given that sprinklers are usually installed to cover about 225 sf each (in offices and similar spaces), 1,500 sf is usually covered by just seven sprinklers—which would flow about 8 gpm per sprinkler or 56 gpm (gallons per minute) for all seven. Flowing 50 sprinklers, would flow 400 gpm, and there simply is not a good reason to design a system to be able to do that. (That said, fire-fighters usually use about 250 gpm for a single hose, so they tend to use large amounts of water.)

Sprinklers come in three basic types:

1 Wet-pipe
2 Dry-pipe
3 Pre-action

Wet-pipe means that the main water valve is open (as noted above) and that all of the piping is full of water; if a sprinkler fusible link breaks, the water immediately flows through the opening that the breakage creates.

Dry-pipe means that the pipes are filled with air, back to a point where a compressor is located to maintain air pressure that is higher than the water pressure, thereby holding back the water. There is no valve; it is just air pressure holding the water. If a fusible link melts, the air leakage through the opening allows the water to flow and fill the entire system. (It is usually required that the system fill within 60 seconds of such an event, so these systems are limited in size.) Dry-pipe systems are usually used in environments that are subject to freezing: outdoors, cold attics, parking garages, etc., although it is acceptable to use a dry-pipe system to cover a normal area too. In other words, a dry-pipe system can cover a heated building and a cold attic above it, but a wet-pipe system cannot.

Pre-action is a combination of wet-pipe and dry-pipe where a small area is set up with a solenoid valve (electrically operated valve) to keep water out of some piping. This is done to minimize the likelihood of an accidental water flow that could cause major damage—for example, in rooms containing large and very valuable electronic equipment, such as medical imaging machines. In this case, if a fusible link melts (or if the fire alarm system detects heat or fire), the valve would open to fill the pipes. But if the fusible link had not melted and there was simply a false reading in the fire alarm system, no water would enter the room and no damage would be done. This type of system is required for sprinklers in elevator machine rooms (the water cannot flow until the elevator machine turns off), so such sprinklers are commonly omitted by building a two-hour rated machine room (which is allowed under the International Building Code).

Large pipes are required for sprinkler systems. The incoming main line is most commonly a 6″ pipe, although small buildings often use 4″ pipes, and very large buildings use 8″ (or larger) pipes. Where the pipe enters the building, it has a main shut-off valve (the valve mentioned previously) which must be monitored by the fire alarm system to make sure that it stays open. After the main valve, a backflow device is required to prevent water from leaving the sprinkler system in the building and going back into the public water supply (this is usually a double-check detector valve but other types of valves are also acceptable, depending upon the local jurisdiction) and then riser valves are used to divide the system into zones. A sprinkler riser is not allowed to serve more than 52,000 sf (on a single floor), so systems for larger buildings have multiple risers. (If a system has four risers, that means that the building is at least (52,000 sf × 3) + 1 sf = 156,001 sf.) The individual risers can be wet or dry as needed, and it is not uncommon to see combination wet/dry systems. Some jurisdictions require only horizontal backflow valves but others allow for vertical backflow valves too; vertical backflow valves take up much less floor space. Typical horizontal fire services are 8–10′ long and about 1.5′ deep, but vertical fire services are usually only about 2′ wide by 1.5′ deep.

The branch piping varies from 4″ (sometimes larger in large systems) to 1″ (a branch line feeding a single sprinkler) and the piping must slope for drainage. (It does not have to slope very much, but it does have to be possible to drain the system.)

The most common type of piping that is used is plain steel with grooved and gasketed joints (the most common trade name is Victaulic), but welded, threaded, and twistlocked joints can be used too. Dry-type systems require galvanized piping. Special

Figure 11.3.A
Typical horizontal fire riser

Figure 11.3.B
Typical vertical fire riser

CPVC piping is allowable too as long as it is rated for sprinkler use (most common tradename: Fireguard). CPVC piping for sprinklers is orange.

The sprinklers themselves come in five types:

1 Upright (facing upwards, used in exposed structure applications to bounce flowing water off the underside of the structure above and then down to the fire)
2 Pendent (facing downwards, with partially exposed piping that is covered by a large escutcheon)
3 Recessed (facing downwards and mostly recessed into the ceiling with only the lower deflector below the ceiling)
4 Concealed (similar to recessed but completely above the ceiling with a flat plate in a hole in the ceiling below the sprinkler; flowing water will knock the plate out of the hole to allow the sprinkler to do its job)
5 Sidewall (horizontal, recessed horizontal, or concealed horizontal)

If no one asks what kind of sprinklers to use for a given project, the interior designer should intervene to find out and make corrections if necessary. A beautiful project should not be ruined by an ugly sprinkler system.

The most common finish for upright sprinklers is brass, and for other types the most common finishes are chrome-plated and white-painted. The cover plates for concealed sprinklers are painted—and there are options for the finish—but they cannot be painted in the field. If a special color is required, it must be done by the manufacturer in their factory.

Sprinkler systems are designed according to "hazard" level, running from "light," to "ordinary I," to "ordinary II," to "extra." Light hazard includes most spaces, except electrical, mechanical, and storage spaces, which are usually ordinary I; ordinary II and extra cover special situations that are unusually hazardous (paint rooms, chemical storage, etc.). As noted previously, the most common spacing for sprinklers in light hazard areas is 15′ o.c. in both directions (or 225 sf per sprinkler); for ordinary hazard I, it is 130 sf per sprinkler (about 11.4′ o.c. in both directions). Other classifications are special cases that should be reviewed with the sprinkler designer.

In most jurisdictions, only registered Fire Protection Professional Engineers are allowed to design sprinkler systems, but it is common for other engineers, architects, and even interior designers to provide guidance for how to install a system in a given situation—symmetrical sprinkler placement, hidden piping in specific locations, etc.

11.3.1 Dry agent systems

Water-based sprinkler systems will not work at all to put out grease fires in kitchens (the water can actually spread the fire and make things worse) and they can do a lot of damage in other situations (valuable electronic equipment as noted previously). So dry agent (i.e. chemical) systems are used for hoods in commercial kitchens and for some highly specialized applications. The common trade name for hood systems is "Ansul." See fire alarms under 11.5 for details about hood fire suppression operation.

Dry agent systems are sometimes used for data centers (large collections of computer servers and/or networking equipment) and for some medical equipment

SIN	Type	Response	K Factor	Element	Thread	Temperature °F
GL2801	Pendent	Quick	2.8	3mm Bulb	1/2"NPT	135, 155, 175, 200, 286
GL4201	Pendent	Quick	4.2	3mm Bulb	1/2"NPT	135, 155, 175, 200, 286
GL5601	Pendent	Quick	5.6	3mm Bulb	1/2"NPT	135, 155, 175, 200, 286
GL8106	Pendent	Quick	8.1	3mm Bulb	3/4"NPT	135, 155, 175, 200, 286
GL2815	Upright	Quick	2.8	3mm Bulb	1/2"NPT	135, 155, 175, 200, 286
GL4215	Upright	Quick	4.2	3mm Bulb	1/2"NPT	135, 155, 175, 200, 286
GL5615	Upright	Quick	5.6	3mm Bulb	1/2"NPT	135, 155, 175, 200, 286
GL8118	Upright	Quick	8.1	3mm Bulb	3/4"NPT	135, 155, 175, 200, 286
GL2826	Hz. Sidewall	Quick	2.8	3mm Bulb	1/2"NPT	135, 155, 175, 200, 286
GL4226	Hz. Sidewall	Quick	4.2	3mm Bulb	1/2"NPT	135, 155, 175, 200, 286
GL5626	Hz. Sidewall	Quick	5.6	3mm Bulb	1/2"NPT	135, 155, 175, 200, 286
GL8127	Hz. Sidewall	Quick	8.1	3mm Bulb	3/4"NPT	135, 155, 175, 200, 286
GL5632	Vert. Sidewall	Quick	5.6	3mm Bulb	1/2"NPT	135, 155, 175, 200, 286
GL8133	Vert. Sidewall	Quick	8.1	3mm Bulb	3/4"NPT	135, 155, 175, 200, 286
GL5606	Conc. Pendent	Quick	5.6	3mm Bulb	1/2"NPT	135, 155, 175, 200
GL5650	Flush Pendent	Quick	5.6	2.5mm Bulb	1/2"NPT	135, 155, 175, 200
GL2851	Pendent	Standard	2.8	5mm Bulb	1/2"NPT	135, 155, 175, 200, 286, 360
GL4251	Pendent	Standard	4.2	5mm Bulb	1/2"NPT	135, 155, 175, 200, 286, 360
GL5651	Pendent	Standard	5.6	5mm Bulb	1/2"NPT	135, 155, 175, 200, 286, 360
GL8156	Pendent	Standard	8.1	5mm Bulb	3/4"NPT	135, 155, 175, 200, 286, 360
GL2861	Upright	Standard	2.8	5mm Bulb	1/2"NPT	135, 155, 175, 200, 286, 360
GL4261	Upright	Standard	4.2	5mm Bulb	1/2"NPT	135, 155, 175, 200, 286, 360
GL5661	Upright	Standard	5.6	5mm Bulb	1/2"NPT	135, 155, 175, 200, 286, 360
GL8164	Upright	Standard	8.1	5mm Bulb	3/4"NPT	135, 155, 175, 200, 286, 360
GL2870	Hz. Sidewall	Standard	2.8	5mm Bulb	1/2"NPT	135, 155, 175, 200, 286, 360
GL4270	Hz. Sidewall	Standard	4.2	5mm Bulb	1/2"NPT	135, 155, 175, 200, 286, 360
GL5670	Hz. Sidewall	Standard	5.6	5mm Bulb	1/2"NPT	135, 155, 175, 200, 286, 360
GL8171	Hz. Sidewall	Standard	8.1	5mm Bulb	3/4"NPT	135, 155, 175, 200, 286, 360
GL5675	Vert. Sidewall	Standard	5.6	5mm Bulb	1/2"NPT	135, 155, 175, 200, 286, 360
GL8176	Vert. Sidewall	Standard	8.1	5mm Bulb	3/4"NPT	135, 155, 175, 200, 286, 360
GL5653	Conc. Pendent	Standard	5.6	5mm Bulb	1/2"NPT	135, 155, 175, 200

Figure 11.3.C
Typical sprinklers
Source: Globe Fire Sprinkler Corporation

installations, but such systems are very costly compared to sprinkler systems (even pre-action systems), so they remain relatively rare. The most common system of this type is called "FM 200."

11.4 Fire pumps

Fire pumps are used for sprinkler and/or standpipe systems when there is inadequate water volume or pressure coming to the facility (or for ESFR systems where they are simply required). There are two basic types of fire pumps: electric and engine-driven.

Electric fire pumps are simple (although they take up lots of space) but the wiring is challenging and very costly.

Engine-driven (usually diesel engine) fire pumps do not have complicated wiring but they tend take up even more space (to the point where they are often located in their own mini-buildings).

For interior design purposes, pumps like this are designed by the fire protection engineer and the electrical engineer working together; the only concern for the interior designer is if space is required. The IBC does require a separate fire-rated room for a fire pump, but it is convenient (and sensible) to put the domestic water service (see Chapter 10) in the same room. At least some jurisdictions allow for the domestic water service to be located in the fire pump room, although a modification might be required in order to do so.

Figure 11.4
Typical electric fire pump

11.5 Fire alarm systems

Modern fire alarm systems can be very sophisticated and can be linked to security, HVAC, AV, and even lighting control systems. But there are only two basic ways to wire fire alarm systems:

1 Hard-wired (zoned)
2 Networked (addressable)

11.5.1 Hard-wired and networked fire alarm systems

Even though hard-wired (zoned) systems are still allowable under most codes, such systems are rarely used. But many existing systems are still installed and working, so it is useful to have a basic understanding of how they operate. Hard-wired systems use "dumb" devices (even smoke detectors) that simply send an electrical signal to the fire alarm control panel (FACP), which only knows that *something* has happened in the zone. When an alarm goes off due to a smoke detector (or other detection device), it is necessary for someone (usually a fire-fighter) to find the device that is in alarm in order to know exactly what is going on. These systems were used because they were the only available option and were better than no system at all.

Networked systems, by contrast, use smart devices on network loops so that the FACP knows which device sent a signal, which it describes in text on the fire alarm annunciator that is built into the FACP and on a remote annunciator (if there is one). In a networked addressable system, the fire department knows exactly what is going on as soon as they see the annunciator. (In large facilities, it can still take time to figure out where the device is located, but it is faster than searching for a hard-wired device somewhere in a zone.) Networked systems also provide automatic self-testing (which meets requirements for monthly testing—rarely done for hard-wired systems because they required manual testing of each device in the system).

The devices that are used in the two systems look very similar, so it is not feasible to tell which type of system is in use just by looking at smoke detectors. It is necessary to go to the FACP, which will have either (1) a zone annunciator or (2) an addressable text window; the former indicates a hard-wired zoned system and the latter indicates an addressable system.

Whichever way the system is wired, there are three types of systems:

1 Manual
2 Automatic
3 Automatic with voice-evacuation notification

11.5.2 Manual fire alarm systems

Manual systems do not detect fires, unless they are monitoring water flow in sprinkler systems or smoke detectors for elevator recall. (The latter two functions provide indirect detection only.) Manual systems include manual pull stations so that occupants can trigger the fire alarm and they include full audible (sound) and visual (light) notification. (Manual pull stations can be omitted in most sprinklered buildings.)

11.5.3 Automatic fire alarm systems

Automatic systems detect heat and fire directly by having heat and smoke detectors. Smoke detectors are required in common use spaces (usually corridors and lobbies), usually at no less than 30′ o.c., and areas having increased risk of fire; if those areas of increased risk could involve steam or other harmless smoke, heat detectors are used to detect fire while preventing false alarms (usually in storage, electrical, and mechanical spaces). Large air-handling units (> 2,000 cfm) are required to have return-air smoke detection, which will turn off the unit if smoke is sensed if there is no fire alarm system, or turn off the unit and report to the FACP if there is a fire alarm system. Special types of smoke detectors (i.e. beam detectors, etc.) are available for special applications. Automatic systems also include manual pull stations (except in most sprinklered buildings) and full audible and visual notification.

Ansul systems in Type I hoods in kitchens are required to self-activate if there is a fire or they can be activated by a manual switch located next to a nearby egress door. Activation means that the dry agent chemicals are released to control the fire; the make-up air to the hood is turned, but the exhaust keeps running (mostly to remove smoke). These systems must be tied into a fire alarm system so that the fire alarm system can monitor, but not control, these systems.

Fire alarm systems also tie into smoke removal systems; see 11.6 to follow.

Networked (addressable) fire alarm systems do self-testing and monitor device performance. Most smoke detectors are prone to problems if they get too dirty, so the FACP monitors "dirtiness" of the smoke detectors. If a detector is a little dirty, the system sends a "trouble" (non-alarm) signal saying that the detector is dirty; if that signal is ignored long enough, the system will send a trouble signal saying "very dirty"; if that signal is ignored long enough, the system could go into alarm to assure that the detector gets cleaned. (False fire alarms are a big problem for fire departments and they often bill property owners for false alarms. They should be avoided as much as possible.)

11.5.3.1 Automatic fire alarm systems with voice-evacuation notification

For high-rise and very large buildings (occupant load >1,000 in many cases), the audible notification side is required to be capable of sending out voice messages via a speaker system, instead of the typical loud "alarm" sounds in normal systems. This greatly increases the cost of the system and is therefore undesirable for many projects. The messages include both pre-recorded instructions (e.g. "evacuate floors 10–30," "evacuate all floors," etc.) and live instructions from the on-scene fire-fighters.

Here are several well-known fire alarm vendors:

www.boschsecurity.us
www.firelite.com
www.gamewell-fci.com
www.honeywelllifesafety.com (Honeywell owns Fire-Lite, Gamewell, Notifier, and SilentKnight)
www.notifier.com
www.silentknight.com
www.simplexgrinnell.com

11.5.4 Detection

11.5.4.1 Heat detection

Heat detectors come in two basic types: fixed temperature and rate-of-rise. Fixed temperature devices go into alarm when the temperature at the detector reaches a pre-set level (usually around 135°F). Rate-of-rise detectors go into alarm when the temperature at the detector increases faster than a pre-set rate beyond a pre-set time (40°F over 5 minutes, for example). Both types can be highly effective. The devices themselves come in two configurations: hard-wired (usable in any system) and with snap-in bases that are interchangeable with similar smoke detectors. The latter makes it very simple to change a smoke detector to a heat detector, or vice versa. Their cases are usually, but not always, white plastic.

11.5.4.2 Smoke detection

Smoke detectors are available in several types: ionization, photo, beam, and laser, and for use as ceiling-mounted (usually white plastic) individual devices, to wall-mounted (beam or laser in large spaces, sometimes with mirrors) devices, to devices inside air-handling equipment (devices like this use sampling tubes inside the ducts to "see" smoke).

They are also available in "system" and "non-system" devices. System devices are devices that talk to a FACP (either hard-wired in zones or networked with addresses) and do not include audible or visual notification. Non-system devices are the kind of device that is used in a single-family residence: 120 volt primary power with 9 volt back-up power from an integral 9 volt battery, and they include both audible and visual notification. Non-system detectors must be linked together within a given dwelling unit so that they all go into alarm/notification mode if any one of them senses smoke.

In the past, it was common to see non-system detectors in multi-family projects, coupled with a manual or automatic system in common areas; this was done to reduce costs. But typical codes require audible and visual notification throughout a facility if a common alarm is triggered, so it is no longer helpful to use non-system devices within the dwelling units (because they cannot go into notification mode by a signal from the FACP). Instead, it makes sense to use system detectors and notification devices throughout the project, just as in any other commercial project.

11.5.5 Notification

11.5.5.1 Visual notification

Notification always requires visual devices in all "common use" areas. For many years, common use was understood to include only lobbies, corridors, meeting spaces, etc., but the definition started to become more inclusive after enactment of the Americans with Disabilities Act (ADA) in the early 1990s. Today, the IBC defines common use as essentially every space except those in which two or more persons work. So an office for two people, or an office for one person with a single guest chair, are both considered to be common use areas. Within dwelling units, this is a little different, but visual notification for common alarms is required within each dwelling unit.

It is not necessarily required that every occupant of a space be able to see a visual device from every possible location within a space, but it is necessary to be able to tell that a device is flashing from every possible location within a space.

Visual notification devices are wall-mounted (at +80″ aft) devices that can be in red or white plastic cases (they are common with speakers in voice-evac systems so they are ceiling-mounted). It is very common to use combination audible/visual devices.

11.5.5.2 Audible notification

The general rule for audible notification is that the signal be heard at a level of at least 15 dBA above the ambient noise level everywhere throughout a facility. There is an additional rule for facilities having sleeping or dwelling units that requires loudness of at least 75 dBA on the pillow at each bed. In many situations, the pillow requirement is more stringent, but both requirements must be met. There are also special rules for "loud" facilities, such as manufacturing plants. Even though it is not written in as a code requirement, it is becoming common to see AV systems linked to fire alarm systems so that the sound system will be automatically turned off (or turned way down quickly) so that the fire alarm can be heard. This sort of issue may be encountered anywhere there is a likelihood of loud amplified sound: concert venues, auditoriums, gymnasiums, even some churches.

Standard audible notification devices (usually called "horns") are wall-mounted in red or white plastic cases. It is very common to use combination audible/visual devices.

Voice-evacuation notification does not use horns; it uses speakers instead, which greatly complicates the system and increases its costs. A speaker, or speakers, is required in each occupied space and the FACP must be able to power all speakers simultaneously. Speakers might resemble conventional audible devices (or combination audiblevisual devices) or they might be completely separate devices. Speakers are most likely to be in white plastic or metal enclosures, but many other options are available.

11.5.6 Remote annunciation

It is least costly to locate the FACP itself (which usually has a lockable door on the cabinet) at the fire-fighters' entrance. But such panels can be large and it could be troublesome if someone accessed the interior and disturbed the wiring or the programming. So it is best if the FACP is located in a secure area and not in a public area. If the FACP is not at the fire-fighters' entrance, it is necessary to add remote annunciation to the system. This means duplicating the annunciation at the FACP (whether zoned or addressable and whether standard or voice-evac) at another device at the fire-fighters' entrance. In an addressable system, such devices are usually about 4″ high by 10″ long, so they are easier to incorporate into the design. Remote annunciators tell the fire-fighters everything they need to know about an alarm, but they sometimes have to access the FACP to re-set the system. Remote annunciators do add a little cost to a system (less than $2,000.00 in most cases), so the owner needs to approve the expense before the feature is added to the system. In voice-evac systems, the remote annunciator is required to be a "Remote Fire Command Center" which adds radio communication to the standard remote annunciator; unfortunately, doing so means using a cabinet that is about 20″ wide and about 40″ tall—about two-thirds of the FACP itself in most cases.

11.6 Smoke control

Smoke control systems are required in two situations: (1) atriums and (2) high-rise buildings.

Atriums require smoke removal. Atriums are openings between three or more floors and a special system is required to bring in large amounts of outside air at the bottom of the space, while exhausting more air from the top of the space to remove smoke. These systems can be difficult to plan into spaces due to louver sizes (and duct sizes, sometimes) and they require a generator to assure emergency power even if nothing else in the building requires a generator.

In high-rise buildings, stair towers require either large smoke control vestibules on each floor or pressurization of the stair towers to exclude smoke (the opposite of smoke removal). The vestibules are rarely done because they use up too much floor space. Pressurization is effected by putting in large fans (the air volume is usually around 1,000 cfm per floor, so a 20-story stair tower is going to move lots of air around) that push air into the stair tower, with a relief system to maintain positive pressure between 0.15 psi and 0.35 psi. (If the pressure were higher than 0.35 psi, it might be difficult to open doors.)

Elevator hoistways also require either lobbies (similar to stair vestibules) or hoistway pressurization but, in this case, the lobbies are far more common.

Any of these systems—atrium smoke removal, stair pressurization, elevator hoistway pressurization—would need to trip automatically via direct smoke detection or upon a general alarm signal.

11.7 Fire protection design documentation

In many jurisdictions, only licensed fire protection engineers are allowed to design sprinkler (and standpipe and fire pump) systems; nevertheless, the MEP engineer usually writes the basic specifications for these systems to identify the scope and limitations (pipe and joint types, sprinkler types, etc.). When special conditions are encountered—such as installing sprinkler piping in an historic structure—the MEP engineer's directives often become more detailed, often taking the form of guiding the designer as to where piping can, and cannot, be located. Architects and interior designers sometimes get involved in piping placement or even sprinkler placement in sensitive locations.

Fire alarm systems are usually designed by electrical engineers (and sometimes by fire alarm vendors), so they usually appear on the electrical power drawings.

Summary

Even though interior designers will not design standpipes, sprinkler systems, fire pumps, smoke control, or fire alarm systems, it is useful to understand the basics of each of these, especially the selection and location of sprinklers and the location of fire alarm devices (especially notification devices). This is true because interior designers do have input about sprinkler types and finishes and they could well be responsible for locating fire alarm devices.

If a particular design does not lend itself to brass pendent sprinklers, the interior designer can, and should, provide input to whoever is writing the sprinkler specifications (most often the MEP consulting engineer) in order to get the right sprinklers in the right places in the project.

Similarly, the interior designer should either locate fire alarm devices herself, or review the locations selected by the MEP consulting engineer (usually) to make sure that the locations do not interfere in a negative way with various design features—wood trim, decorative elements, art work, furnishings, etc.

All of these elements are part of the overall design, so it is perfectly reasonable for an interior designer to be involved in the design presentation of these systems. A fire pump is unlikely to have a direct aesthetic effect on a design, but its presence must be accounted for in the space plan.

Outcomes

11.2 Understanding dry and wet standpipes, their locations, and their uses.

11.3 Understanding wet-pipe and dry-pipe sprinkler systems, sprinkler types and finishes, and the impact of sprinkler systems on typical interior design projects.

11.4 Understanding basic electric and engine-driven fire pumps and their impact on typical interior design projects.

11.5 Understanding hard-wired and networked manual and automatic fire alarm systems and their impact on typical interior design projects, especially in the location of notification devices.

11.6 Understanding basic smoke control systems.

11.7 Understanding basic fire protection documentation.

Note

1 2015 International Building Code, page 219

Glossary

Access panel: a special panel, used in a wall or ceiling, to gain access to something without using tools or without cutting. Such panels are usually steel and hinged; sometimes, they are key-lockable, and sometimes they are fire-rated.

Acoustical consultant: a specialist in architectural acoustics and sound systems. There are no legally binding credentials for this title.

Acoustically absorptive: materials that absorb more sound than they reflect.

Acoustically diffusive: materials that are constructed and shaped to reflect spread—diffuse—sound.

Air conditioning: filtering cooled air to remove dirt, dust, etc.

All-thread: a steel rod for hanging piping, wiring, ductwork, walls, bulkheads, or equipment where the whole length of the rod is threaded so that bolts and nuts can be attached at any point along the length.

Ampere: (the flow rate) of electricity. The unit is named after the French scientist Andre-Marie Ampere.

Astragal: a special plate to cover gaps between doors.

Base: the lamp (or source) end of the electrical connection, or connections, for an electric light source.

Batten: a strip of wood, usually applied to the surface to conceal a joint.

Bearing wall: a wall that supports structure above, either as a solid wall (concrete or masonry) or a hollow wall (wood or metal studs, structural tile, etc.).

Bio-fuel: a naturally grown material, including, but not limited to, wood, plant stalks, grasses, etc.

Blocking: additional reinforcement or attachment devices, most commonly wood, for attaching wood trim, objects hanging on walls, etc.

Bright: no technical definition, but a high light level.

Bus: an electrically conductive bar, or plate, of metal, usually copper or aluminum. Used in load centers, panelboards, distribution boards, busways, etc.

Busway: a means to move large amounts of power around in a facility without wires, by using bus bars, or plates, in a special enclosure.

Cast-in-place concrete: concrete that is placed on-site in its wet form into site-built forms.

Centralized: a single system for an entire building or for a large section of a larger building.

Change order: a formal change to a construction contract.

Chilled and hot water: water that is cooled by a chiller to be used for cooling and water that is heated by a boiler for heating.

Chiller: a machine to produce cold water for space cooling.

Chiller barrel: the part of a chiller where the refrigerant removes heat from the water.

Circuit: a wiring loop that can serve a single device or an entire facility.

Clerestory: a window between two levels of roof.

Code consultant: a design professional who provides code advice to other design professionals and building owners. There are no specific established credentials for this title.

Cold-rolled: cold sheet metal that is rolled into a shape, as opposed to structural steel which is rolled into shapes when it is still hot from the foundry.

Color rendering: the ability of an artificial light source to render color accurately, expressed as a number between 0 and 100, with 100 being the maximum.

Color temperature: the equivalent temperature at which a piece of tungsten wire closely resembles the color of the light, using the unit of Kelvin.

Compressor: a machine that compresses a refrigerant in gas form into liquid form so that it can absorb heat from the space when it evaporates into gas in the evaporator coil.

Concrete masonry units: often called "concrete blocks," these are modular units that are factory-formed and poured using various different types and weights of concrete, cured, and shipped to a site to be assembled on-site, usually using Portland cement mortar (as is used for brick).

Condensate: water condensed out of air by a cooling system.

Cones: photoreceptors for color vision.

Constant air volume: a system that runs at a constant speed (although multiple speeds are possible).

Construction manager: in its broadest sense, simply someone who manages construction, although some industry organizations use far more formal definitions. This is a similar term to contractor or builder.

Contractor: in its broadest sense, someone who builds, from individual workmen to organized trade contractors (e.g. brick masons, carpenters, electricians, etc.) to large-scale general contractors who package the work of individual trade contractors and coordinate their activities.

Convenience receptacle: the standard outlet for a standard plug, universal throughout the US, but different in other countries. Almost universally in the US, this provides 120 volts at up to either 15 amps (old system) or 20 amps (almost universal now); the power capacity is in volt-amps (see kVA), which is the same as watts when the power factor is 1.0.

Cooling: reducing indoor air temperature.

Cooling tower: an outdoor machine used to cool the compressors in an indoor chiller.

Damper: a mechanical device, usually having multiple blades, that is used to open or close a duct or air opening.

Dark: no technical definition, but a low level of light.

De-centralized: multiple systems serving small areas around a building.

Deflection: the normal bending of a structure under load, which occurs with all structural systems: concrete, steel, and wood. Typical limits are $\frac{1}{360}$ for floors and $\frac{1}{240}$ for roofs. Over a 40′ 0″ span, $\frac{1}{360}$ is 1.33″ and $\frac{1}{240}$ is 2″, so this is not at all trivial.

Direct expansion: cooling achieved by using an electric compressor with condenser and evaporator coils in a refrigerant circuit.

Distribution board: an assembly to provide over-current protection for **feeders**.

Domestic water, also called potable water: clean water used for drinking, bathing, cleaning, etc.

Double studs: literally two walls back-to-back and very close together, usually without facing on the interiors.

Downlighting: light that is direct downward only from a ceiling or suspended luminaire in space.

Ductwork: air delivery "piping," for lack of a better description.

Dutch door: a door where the upper half (roughly) is separately hinged from the bottom half.

Efficacy: the term for light output in lumens/watt for lamps and other sources.

Elastic materials: materials that return to their normal state after an applied force has been removed. Elastic materials include air, water, and most building materials.

Embodied energy: a term from sustainable design that means the amount of energy that it takes to produce a raw material for use in a building or building product. The highest embodied energy materials are metals, especially aluminum, and some plastics.

Energy recovery: the ability to transfer heat, coolness, or humidity, from one air stream, or water stream, to another.

Energy Recovery Ventilator (ERV): a mechanical device that moves heat from one air stream to another. In heating mode, the warm exhaust air is routed through the ERV to pre-heat the incoming outside air; in cooling mode, the cool (and relatively dry) exhaust air pre-cools and de-humidifies (partially) the incoming outside air.

Environmental: engineering of storm water drainage and sanitary water treatment (sewage).

Ex-filtration: air leakage out of a building through the walls or roof, or around the windows and doors.

Faucet: a special valve, or valves, for delivering domestic water—hot, cold, or mixed—for handwashing, bathing, showering, dishwashing, etc.

Feeder: a major wiring run for a large piece of equipment or from a distribution board to a panelboard or load center.

Fidelity: Accuracy of reproduction.

Field-finished: finishing is done in the field by painters, usually using hand tools (e.g. brushes and rollers), but sometimes by spraying.

Fire pump: an electric or diesel engine-driven pump to increase pressure for a fire sprinkler system.

Fire-rated: a building assembly that has been tested by an independent lab (e.g. UL, ASTM, etc.) and shown to meet particular criteria, usually in terms of hours of fire exposure.

Fire riser: usually the main vertical pipe in a fire sprinkler system.

Fire standpipe: vertical piping to be used by the fire department (usually) in the case of a fire in a tall building.

Flashing: a membrane to exclude water.

Flitch: the collected slices of veneer (leaves) from a single quarter log.

Flush valve: a special high-capacity fast-acting valve for flushing water closets, urinals, and hopper sinks.

Footcandle: a unit of lighting density, best described as the amount of light cast by a single candle flame onto the inside of a sphere that is 1′0″ in radius.

Fovea: the area of the retina where the cones are concentrated.

Full-height wall: a wall from the floor to the underside of the structure above.

Galvanizing: an electrochemical process to bond zinc to steel for protection against corrosion.

Gauge: the thickness of steel.

Geotechnical: soils engineering, primarily for foundations.

Glare: unwanted light.

Grout: essentially very loose mortar that will flow into all cracks and crevices; commonly used in exterior concrete masonry walls for structural reinforcement purposes but rarely used in interior walls.

Gypsum board: a building product made by sandwiching compressed gypsum (gypsum is a naturally occurring mineral) powder between sheets of heavy paper.

Heat exchanger: a device that can pass heat from one air stream, water stream, or refrigerant stream to another similar stream. Many different types exist for various purposes.

Heat pump: a reversible compressor that provides cooling when running in one direction and heating when running in reverse.

Hot gas: refrigerant in its gaseous state, after it has absorbed heat in the evaporator coil (the evaporator coil evaporates the liquid refrigerant from the compressor into gas).

Incandescent: glowing from high heat.

Infiltration: air leakage into a building through the walls or roof, or around the windows and doors.

Interior designer: a design professional who is responsible for interior components: walls, doors, windows, millwork etc., space planning, and colors and finishes, all specifically defined as non-structural.

Iso-footcandle: a map of lighting levels similar to a topographic drawing that shows contour lines connecting all points at the same level; sometimes shown in conjunction with point-by-point plots.

IT consultant: a specialist in computers, networks, and/or communications. While there are no legally binding credentials for this title, there are a number of industry standards.

Jamb: the side of an opening.

kVA: kilovolt ampere, or 1,000 volt ampere, a quantity of electrical power. A volt ampere is commonly understood to be a watt, which is correct only part of the time.

Landscape architect: a design professional focused on site planning and planting design.

Lath: support for plaster, either wood or metal.

Lay-in ceiling: a gridded ceiling system, usually 2′ × 2′ or 2′ × 4′, with removable panels.

Light: an architectural term for window.

Lighting controls: everything from basic wall switches of various types to wall-box dimmers, wired and wireless sensors, and elaborate systems.

Lighting designer: a design professional focused on the design of interior and exterior lighting. While there are no uniformly accepted credentials for this title, the Lighting Certified (LC) designation from the National Council on Qualifications in the Lighting Professions is very widely accepted.

Liquid refrigerant: refrigerant in its liquid state, after passing through the compressor (the compressor compresses the refrigerant into liquid).

Load center: a lightweight assembly to provide over-current protection for branch circuits.

Long-lead: referring to fabrication and delivery time. Long-lead usually refers to items that take more than 30 days to get to a job-site. In the case of door frames, long-lead can mean 60 days or even longer in busy times.

Luminaire: an assembly of lamp(s), socket(s), transformer(s), ballast(s), driver(s), and/or housing to produce light; often called a "fixture."

Mark-up: a fee charged by one entity for work done, or a product provided by, another entity. A good example is that a flooring contractor buys the flooring from a flooring supplier and then marks up the cost to charge the general contractor, who then marks it up again for the owner.

Mesopic: mid-range light, between scotopic and photopic, or from 0.2 fc to 2.0 fc.

Mixing valve: a device to limit the maximum temperature for domestic hot water used for hand washing and bathing.

Modification: deviation from written and adopted code agreed to by the authority having jurisdiction.

Modular: equipment, usually boilers or chillers, made in small sizes that are joined together to make up larger units. Such equipment often has cases that are designed for edge-to-edge installation.

Mortar: essentially, mortar is concrete without the coarse aggregate; in masonry work, mortar is the material that bonds together the stones, bricks, CMUs, etc.

Noise reduction: the application of materials to reduce sound within a space.

Nosing: the projection of a tread beyond the face of a riser below.

Occupancy sensor: a device that senses when movement or heat is present. Movement is sensed by ultrasonic (very high frequency sound) detection and heat is sensed by infrared heat detection; both are available in dual-technology devices, which can be wall- or ceiling-mounted.

Optics: the science of shaping and controlling light.

Over-current protection: used to prevent over-heating of wiring because over-heated wiring tends to cause fires. Over-current protection can be done with fuses (non-resettable, meaning that they have to be replaced if they fail) or with circuit breakers (re-settable electro-mechanical or electronic devices that can be re-set if they trip). Over the years, fuses have become less and less popular and we see mostly circuit breakers today.

Panelboard: a piece of equipment that subdivides a large circuit into multiple smaller circuits.

Parabolic: in the shape of a parabolic curve, used to focus light.

Phosphor: a chemical compound that glows when subjected to light.

Photopic: the range of light above 2.0 fc.

Plain (or flat) sliced: a log sliced into very thin layers perpendicular to the center of the log.

Plaster: available in gypsum, lime, and Portland cement versions. Gypsum plaster is used for interior applications only where it is desirable to do fine molding work (often found in decorative crown molds and other details), but it has no water resistance and little strength, so it is a poor material for wall surface base coats. Lime plaster was used up through the mid-to-late 19th century primarily for interior wall surfacing; though tougher than gypsum plaster, it is extremely slow to cure and is therefore impractical for most projects today (with the exception of some historic restoration work). Portland cement plaster is used for interior wall surfacing and exterior wall surfacing; in the latter case, it is called "stucco." When Portland cement plaster is used for interior wall surfacing, the finish coat is usually gypsum plaster to provide a smooth surface. All conventional plastering is done in three coats on some form of **lath**. The first coat is very rough, about $\frac{1}{2}$" thick, and usually called the "brown coat." The second coat is rough, about $\frac{1}{4}$" thick, and is called the "scratch coat." And the third coat is either smooth, sand finished, or rough, about $\frac{1}{8}$" thick, and called the finish coat. A finished thickness of $\frac{7}{8}$" is typical for plaster.

Point-by-point: lighting calculation output showing the footcandle level at every point of a 10′ × 10′ (or other size) grid.

Power house: a general term that usually refers to a free-standing building that supplies electric power, heating, and often cooling to multiple buildings on the same campus, often using tunnels to distribute services.

Pre-cast concrete: concrete that is formed and placed in a factory, cured, and brought to the job-site as a finished product.

Prismatic lens: a flat (on one side) clear lens with small prisms on the other side to disperse light and to conceal lamps.

Professional Engineer: a design professional focused on a particular discipline of engineering: civil (site grading and drainage, soils, foundations, structural, waste water treatment, domestic water treatment, etc.), structural, mechanical, electrical, plumbing.

Quarter-sawn: similar to flat-sliced except that the slicing is nearly parallel to the radius line, which results in consistent and beautifully patterned graining, well suited to cherry and similar species.

Rail & cable: a system to hang artwork using vertical cables suspended from a high rail; eliminates the need for holes in the walls.

Recirculation pump: usually a small pump to return unused domestic hot water to a water heater.

Refrigerant: a chemical substance that can "phase-change" from liquid to gas (and back again) at relatively low temperature and pressure.

Refrigeration cycle: using chemicals that change phase from liquid to gas, or vice versa, under relatively low temperature and pressure to cool air or water.

Registered Architect: the broadest of all design and construction professionals, encompassing all facets: from site planning, landscaping and engineering to structural design, building layout, walls, roofs, windows, doors, and all interior components, overlapping with interior designer.

Retina: the photoreceptor area at the back of the eye.

Rift-sawn: a log sliced into very thin layers at a particular angle to the grain, which results in very consistent grain patterns; most suitable to white oak and other straight-grained woods.

Riser: the vertical portion of a stair step.

Rod: a photoreceptor for low-level light perception without color.

Rotary-sliced: veneer cut by rotating a log and peeling it in a very thin layer. This is the lowest quality manner of veneer slicing because it results in the least consistent grain patterns.

RPZBP (Reduced Pressure Zone Backflow Preventer): a device to prevent backward contamination of upstream water.

Running bond: building masonry (brick, tile, concrete masonry, or even cut stone) by overlapping units at $\frac{1}{2}$ (concrete masonry, tile, and some brick) or $\frac{1}{3}$ points (usually brick) along their length to interlock the units, horizontally. (In old solid masonry walls built using mostly brick, this applies between **wythes** as well, with bricks turned perpendicular to the wall face to lock the wythes together; such patterns are called bonds, and several different bonds have been developed over the centuries.)

Scotopic: the range of light below 0.2 fc.

Service disconnect: a special switch for disconnecting power to a multi-phase load.

Set-point: the temperature setting at a thermostat or other control device.

Shop-finished: also called factory-finished, meaning that the finishing is done in a controlled environment, usually a spray booth.

Socket: the luminaire end of the electrical connection, or connections, for an electric light source.

Sound isolation: the resistance of construction to the passage of sound from one side to the other.

Speech intelligibility: the ability to be understood when speaking.

Spline: a flat narrow piece of wood used to strengthen a joint.

Staggered studs: if the sole plate is a 2 × 6 and the studs are 2 × 4, two sets of studs are used, offset by half a space (usually 8") and alternating between faces of the plate.

Steel tube column: a steel shape (round, square, or rectangular) rolled from hot steel plates with a welded seam along one side to close the shape. They are available in many different sizes, shapes, and wall thicknesses.

Stile and rail: horizontal and vertical framing members, respectively; most commonly used in doors.

Stoking: adding fuel to a fire, usually by loading coal or wood into a burner.

Stringer: the sloping supports for the treads and risers of a stairway.

Sub-contractor: a contractor who works for another contractor (e.g. a painter working for a general contractor).

Toe-nailing: diagonal nailing of a stud to a plate or of a wood floor board to a wood subfloor or wood framing.

Tool: finishing of mortar joints in concrete masonry, brick, and glazed structural tile work.

Transformer: a piece of equipment that changes voltage.

Transom: a panel or light above a door.

Tread: the horizontal portion of a stair step, or the walking surface.

Troffer: a large-scale recessed luminaire, usually designed for use in a suspended lay-in (grid) ceiling.

Uplighting: bouncing light off a ceiling or exposed structure, usually for a diffuse effect.

Urinal: a plumbing fixture for removing human urine only, usually used only for males. Female urinals exist but are rarely used.

Vapor pressure: the pressure exerted by water vapor contained in air on the surface of the liquid.

Variable air volume: changing air volume on the basis of dynamic load information.

Ventilation: the process of moving "fresh," usually outside air through an occupied space to reduce the concentration of carbon dioxide (exhaled as we breathe) and contaminants. Ventilation can be done naturally by using open windows and natural air currents, or mechanically by using fans, other equipment, and ductwork.

Ventilator: a device, usually a fan, that moves air into and out of a building.

Vertical control: the necessity for masons to keep horizontal joints constant thickness so as to hit vertical dimensions.

Vestibule: a small space between sets of doors, mostly used to minimize weather intrusion. Sometimes called an air-lock.

VOC: volatile organic compounds; chemicals found in many different products, especially products that are applied wet, that cause odors, breathing difficulties, and sometimes even cancer.

Volt: a difference in electrical potential, best understood as the amplitude—or height—of the sine wave for alternating current. The unit is named after the famed Italian scientist Alessandro Volta.

Water closet: a plumbing fixture for removing both liquid and solid wastes; often called a toilet.

Water heater: a device to warm domestic water for bathing, cooking, etc. Available in electric, oil-fired, and natural gas-fired tank types and electric and natural gas-fired tankless types (also call instantaneous).

Watt: a unit of power that is named after the famed Scottish inventor: James Watt.

Wet-curing: conventional coatings, such as shellac, lacquer, varnish, and paints, all cure by releasing moisture into the atmosphere: they "dry." Wet-curing urethanes cure by absorbing moisture from the atmosphere.

Wrap-around: a frame that extends beyond the face of the wall on both sides, with returns (similar to a conventional wood frame with applied trim).

Wythe: a single thickness of brick, concrete masonry, or other masonry; usually $3\frac{5}{8}''$ for brick; $3\frac{5}{8}''$, $5\frac{5}{8}''$, $7\frac{5}{8}''$, $9\frac{5}{8}''$ or $11\frac{5}{8}''$ for concrete masonry units, and between 3″ and 6″ for solid stone.

Resources

AIA	American Institute of Architects, www.aia.org
ALA	Association of Licensed Architects, www.alatoday.org
ANSI	American National Standards Institute, www.ansi.org
ASHRAE	American Society of Heating, Refrigeration, and Air-Conditioning Engineers, www.ashrae.org
ASID	American Society of Interior Designers, www.asid.org
AWI	Architectural Woodwork Institute, www.awinet.org
BHMA	Builders Hardware Manufacturers Association, www.buildershardware.com
CMS	Centers for Medicare and Medicaid Services, www.cms.gov
ETL	ETL Testing Laboratories, Inc., www.intertek.com/marks/etl
FSC	Forestry Stewardship Council, http://us.fsc.org/en-us
Hurt	Samuel L. Hurt, PE, RA, RID, LC, Sam@tec-mep.com
IALD	International Association of Lighting Designers, www.iald.org
ICC	International Codes Council, www.iccsafe.org
IES	Illuminating Engineering Society, www.ies.org
IIDA	International Interior Design Association, www.iida.org
ISO	International Organization for Standardization, www.iso.org
NFPA	National Fire Protection Association, www.nfpa.org
NSPE	National Society of Professional Engineers, www.nspe.org
SDI	Steel Door Institute, www.steeldoor.org
UL	Underwriter' Laboratories, www.UL.com

Index

Page numbers in italics refer to figures. Page numbers in bold refer to tables.